Tree Species Effects on Soils: Implications for Global Change

NATO Science Series

A Series presenting the results of scientific meetings supported under the NATO Science Programme.

The Series is published by IOS Press, Amsterdam, and Springer (formerly Kluwer Academic Publishers) in conjunction with the NATO Public Diplomacy Division.

Sub-Series

I. Life and Behavioural Sciences	IOS Press
II. Mathematics, Physics and Chemistry	Springer (formerly Kluwer Academic Publishers)
III. Computer and Systems Science	IOS Press
IV. Earth and Environmental Sciences	Springer (formerly Kluwer Academic Publishers)

The NATO Science Series continues the series of books published formerly as the NATO ASI Series.

The NATO Science Programme offers support for collaboration in civil science between scientists of countries of the Euro-Atlantic Partnership Council. The types of scientific meeting generally supported are "Advanced Study Institutes" and "Advanced Research Workshops", and the NATO Science Series collects together the results of these meetings. The meetings are co-organized by scientists from NATO countries and scientists from NATO's Partner countries — countries of the CIS and Central and Eastern Europe.

Advanced Study Institutes are high-level tutorial courses offering in-depth study of latest advances in a field.
Advanced Research Workshops are expert meetings aimed at critical assessment of a field, and identification of directions for future action.

As a consequence of the restructuring of the NATO Science Programme in 1999, the NATO Science Series was re-organized to the four sub-series noted above. Please consult the following web sites for information on previous volumes published in the Series.

http://www.nato.int/science
http://www.springeronline.com
http://www.iospress.nl

Series IV: Earth and Environmental Sciences – Vol. 55

Tree Species Effects on Soils: Implications for Global Change

edited by

Dan Binkley
Colorado State University,
Fort Collins, CO, USA

and

Oleg Menyailo
Institute of Forest SB RAS,
Krasnoyarsk, Russia

Springer

Published in cooperation with NATO Public Diplomacy Division

Proceedings of the NATO Advanced Research Workshop on
Trees and Soil Interactions, Implications to Global Climate Change
August 2004
Krasnoyarsk, Russia

A C.I.P. Catalogue record for this book is available from the Library of Congress.

ISBN-10 1-4020-3445-8 (HB) Springer Dordrecht, Berlin, Heidelberg, New York
ISBN-10 1-4020-3447-4 (e-book) Springer Dordrecht, Berlin, Heidelberg, New York
ISBN-13 978-1-4020-3445-9 (HB) Springer Dordrecht, Berlin, Heidelberg, New York
ISBN-13 978-1-4020-3447-3 (e-book) Springer Dordrecht, Berlin, Heidelberg, New York

---

Published by Springer,
P.O. Box 17, 3300 AA Dordrecht, The Netherlands.

Printed on acid-free paper

Printed in the Netherlands.

Contents

List of Contributors

Gert Bachmann, Institut für Ökologie und Naturschutz, Althanstr. 14, A-1090 Wien, Austria

I.N. Bezkorovaynaya, V.N. Sukachev Institute of Forest, Russian Academy of Sciences, Siberian Branch, Academgorodok, Krasnoyarsk, Russia, 660036

Dan Binkley, Department of Forest, Rangeland and Watershed Stewardship; Graduate Degree in Program in Ecology; and Natural Resource Ecology Laboratory, Colorado State University, Ft Collins, CO 80523 USA

N. Brüggemann, Forschungszentrum Karlsruhe GmbH, Institute for Meteorology and Climate Research, Atmospheric Environmental Research (IMK-IFU), Kreuzeckbahnstr. 19, D-82467 Garmisch-Partenkirchen, Germany

Klaus Butterbach-Bahl, Forschungszentrum Karlsruhe GmbH, Institute for Meteorology and Climate Research, Atmospheric Environmental Research (IMK-IFU), Kreuzeckbahnstr. 19, D-82467 Garmisch-Partenkirchen, Germany

A. I. Buzykin, V.N. Sukachev Institute of Forest, Russian Academy of Sciences, Siberian Branch, Academgorodok, Krasnoyarsk, Russia, 660036

Mark D. Coleman, USDA Forest Service, Southern Research Station, Savannah River, PO Box 700, New Ellenton, SC 29809, USA

Jordi Cortina, Departament d'Ecologia, Universitat d'Alacant Ap. 99 03080 Alacant, Spain

Anne-Marie Domenach, UMR Ecofog, BP 709, 97387 Kourou Cedex, Guyane Française

Michael Englisch, Institut für Waldökologie und Boden, Bundesamt und Forschungszentrum für Wald, Seckendorff-Gudent Weg 8, A-1131 Wien, Austria

Zdenek Filip, Marie Curie Chair, Dept. of Biochemistry and Microbiology, Institute of Chemical Technology, Technická 5, CZ-16628 Prague, Czech Republic

R. Gasche, Forschungszentrum Karlsruhe GmbH, Institute for Meteorology and Climate Research, Atmospheric Environmental Research (IMK-IFU), Kreuzeckbahnstr. 19, D-82467 Garmisch-Partenkirchen, Germany

Christian P. Giardina, USDA Forest Service, North Central Research Service, 410 MacInnes Drive, and The Ecosystem Science Center, School of Forest Resources and Environmental Science, Michigan Technological University, Houghton, MI 49931, USA

Stephan Hättenschwiler, Centre d'Ecologie Fonctionnelle et Evolutive, Centre National de la Recherche Scientifique, F-34293 Montpellier, France

Evelyn Hackl, Institut für Waldökologie und Boden, Bundesamt und Forschungszentrum für Wald, Seckendorff-Gudent Weg 8, A-1131 Wien, Austria

Jessica E. Hancock, USDA Forest Service, North Central Research Service, 410 MacInnes Drive, and The Ecosystem Science Center, School of Forest Resources and Environmental Science, Michigan Technological University, Houghton, MI 49931, USA,

Bruce A. Hungate, Department of Biological Sciences and Merriam Powell Center for Environmental Research, Northern Arizona University, Flagstaff AZ 86001, USA

Ralf Kiese, Forschungszentrum Karlsruhe GmbH, Institute for Meteorology and Climate Research, Atmospheric Environmental Research (IMK-IFU), Kreuzeckbahnstr. 19, D-82467 Garmisch-Partenkirchen, Germany

John S. King, The Ecosystem Science Center, School of Forest Resources and Environmental Science, Michigan Technological University, Houghton, MI 49931, USA

V.V. Kuzmichev, V.N. Sukachev Institute of Forest, Russian Academy of Sciences, Siberian Branch, Academgorodok, Krasnoyarsk, Russia, 660036

Juliya A. Kurbatova, A.N. Severtzov Institute of Ecology and Evolution RAS, Leninski Prospect 33, 117 071 Moscow, Russia

Erik A. Lilleskov, USDA Forest Service, North Central Research Service, 410 MacInnes Drive, and The Ecosystem Science Center, School of Forest Resources and Environmental Science, Michigan Technological University Houghton, MI 49931, USA

Wendy M. Loya, The Ecosystem Science Center, School of Forest Resources and Environmental Science, Michigan Technological University, Houghton, MI 49931, USA

Fernando T. Maestre, Department of Biology, Duke University, Phytotron Building, Box 90348, Durham, NC 27708 USA

Mikhail I. Makarov, Department of Soil Science, Moscow State University, 119992 Moscow, Russia

T.I. Malysheva, Department of Soil Science, Moscow State University, 119992 Moscow, Russia

Oleg V. Menyailo, V.N. Sukachev Institute of Forest, Russian Academy of Sciences, Siberian Branch, Academgorodok, Krasnoyarsk, Russia, 660036

L.V.Mukhortova, V.N. Sukachev Institute of Forest, Russian Academy of Sciences, Siberian Branch, Academgorodok, Krasnoyarsk, Russia, 660036

Hans Papen, Forschungszentrum Karlsruhe GmbH, Institute for Meteorology and Climate Research, Atmospheric Environmental Research (IMK-IFU), Kreuzeckbahnstr. 19, D-82467 Garmisch-Partenkirchen, Germany

Michael Pfeffer, Institut für Waldökologie und Boden, Bundesamt und Forschungszentrum für Wald, Seckendorff-Gudent Weg 8, A-1131 Wien, Austria

Kurt S. Pregitzer, The Ecosystem Science Center, School of Forest Resources and Environmental Science, Michigan Technological University, Houghton, MI 49931, USA

Cindy E. Prescott, Faculty of Forestry, University of British Columbia, Vancouver, British Columbia, Canada

L.S. Pshenichnikova, V.N. Sukachev Institute of Forest, Russian Academy of Sciences, Siberian Branch, Academgorodok, Krasnoyarsk, Russia, 660036

Jacques Roy, Centre d'Ecologie Fonctionnelle et Evolutive, Centre National de la Recherche Scientifique, F-34293 Montpellier, France

P. Rosenkranz, Forschungszentrum Karlsruhe GmbH, Institute for Meteorology and Climate Research, Atmospheric Environmental Research (IMK-IFU), Kreuzeckbahnstr. 19, D-82467 Garmisch-Partenkirchen, Germany

Andreas Rothe, Bayerisches Staatsministerium für Landwirtschaft und Forsten Ludwigstr. 2 80539 München, Germany

Michael G. Ryan, USDA Forest Service, Rocky Mountain Research Service, 120 West Prospect, and Graduate Degree Program in Ecology and Department of Forest, Rangeland and Watershed Stewardship, Colorado State University,Fort Collins, CO 80523, USA

Hana Santruckova, University of South Bohemia, Faculty of Biological Sciences and Institute of Soil Biology Branisovska 31, 370 05, Ceske Budejovice, Czech Republic

Olga B. Shibistova, V.N. Sukachev Forest Institute, Akademgorodok 660 036, Kraskoyarsk, Russia

L.S. Shugalei, V.N. Sukachev Institute of Forest, Russian Academy of Sciences, Siberian Branch, Academgorodok, Krasnoyarsk, Russia, 660036

Miluse Smejkalova, University of South Bohemia, Faculty of Biological Sciences, Branisovska 31, 370 05, Ceske Budejovice, Czech Republic

D. S. Sobachkin, V.N. Sukachev Institute of Forest, Russian Academy of Sciences, Siberian Branch, Academgorodok, Krasnoyarsk, Russia, 660036

Roman S. Sobachkin, V.N. Sukachev Institute of Forest, Russian Academy of Sciences, Siberian Branch, Academgorodok, Krasnoyarsk, Russia, 660036

Pamela H. Templer, Department of Environmental Science, Policy and Management, University of California, Berkeley, CA 94720, USA

M. Tesařová, Department of Soil Science and Microbiology, Mendel University of Agriculture and Forestry, Zemědělská 1, 613 00 Brno, Czech Republic

V.A. Tretyakova,V.N. Sukachev Institute of Forest, Russian Academy of Sciences, Siberian Branch, Academgorodok, Krasnoyarsk, Russia, 660036

Carl C. Trettin, USDA Forest Service, Southern Research Station, Charleston, SC, USA

Eva Uhlirova, University of South Bohemia, Faculty of Biological Sciences and Institute of Soil Biology Branisovska 31, 370 05, Ceske Budejovice, Czech Republic

E.F.Vedrova, V.N. Sukachev Institute of Forest, Russian Academy of Sciences, Siberian Branch, Academgorodok, Krasnoyarsk, Russia, 660036

Lars Vesterdal, Forest and Landscape, Royal Veterinary and Agricultural University, Horsholm, Denmark

G. Willibald, Forschungszentrum Karlsruhe GmbH, Institute for Meteorology and Climate Research, Atmospheric Environmental Research (IMK-IFU), Kreuzeckbahnstr. 19, D-82467 Garmisch-Partenkirchen, Germany

Sophie Zechmeister-Boltenstern, Institut für Waldökologie und Boden, Bundesamt und Forschungszentrum für Wald, Seckendorff-Gudent Weg 8, A-1131 Wien, Austria

Chapter 1

GAINING INSIGHTS ON THE EFFECTS OF TREE SPECIES ON SOILS

Dan Binkley[1] and Oleg Menyailo[2]

[1]Department of Forest, Rangeland and Watershed Stewardship; Graduate Degree in Program in Ecology; and Natural Resource Ecology Laboratory, Colorado State University, Ft Collins, CO 80523 USA, [2]Institute of Forest SB RAS, Krasnoyarsk 660036, Russia

INTRODUCTION

The interactions of trees and soils have fascinated scientists, farmers, and foresters for centuries. The success of trees depends strongly on the properties of the soils beneath them, as recognized by A.A. Nartov in the 17th Century (cited in Remezov and Pogrebnyak 1969):

> "The properties of spruce and pine are such that 100 years are required for pine on dry soil, and for spruce on humid soil… However, pine growing on humid soil will hardly reach a height of 6 feet in this span of time… spruce cannot succeed on hard and dry hillocks or similar locations unless its roots, which spread far underground, will reach the moisture they require."

By the early 20th Century the influence of trees on soil development was well recognized. G.F. Morozov noted (cited in Remezov and Pogrebnyak 1969):

> "The idea that forest is an agency of soil formation was never really alien to forestry; the idea became more and more definite in the course of its development…foresters began using such expressions as "beech soil," "oak soil," etc., not merely in the sense of a soil suitable for the given species, but with emphasis on the idea that the soils are actually being influenced by the tree stand."

D. Binkley and O. Menyailo (eds.),
Tree Species Effects on Soils: Implications for Global Change, 1–16.

By the late 20th Century, interests in the effects of tree species on soils had broadened to include ideas of sustainability of long-term soil fertility, responses to air pollution, and influences on understory vegetation diversity. Concerns about concentrations of trace gases that absorb infrared radiation and warm the atmosphere intensified interest in forest biogeochemistry, and the differences in trace gas fluxes from forests dominated by different species. The effects of changing climate on forest biogeochemistry may be moderated directly by the effects on species composition of forests (and the differences in biogeochemistry among species) rather than by the simple effect of temperature or moisture. For example, the rate of nitrous oxide production and methane consumption differed by 3-fold under the influence of different species in the Siberian afforestation experiment (Oleg and Hungate, this volume).

This volume presents a summary of ideas, data, and perspectives from a NATO Advanced Research Workshop held in Krasnoyarsk, Siberia (August 26-29, 2004) on the effects of tree species on soils, including implications for global changes. The chapters cover a broad range of projects; some deal with landscape-scale patterns across forested landscapes, and others deal with species planted in common garden settings. The experimental designs are equally diverse, with some scientists confident that the effect of species is far stronger than any pre-existing differences among soils in their plots. Other scientists used replicated designs that did not require this assumption. Our introductory chapter provides a context to help readers evaluate the strength of evidence in later chapters, and also highlights some of the key findings from our workshop.

UNRAVELING THE EFFECTS OF TREE SPECIES FROM OTHER SOIL-FORMING FACTORS

The scientific investigation of soils blossomed in the late 19th Century, with the leadership of Vasily V. Dokuchaev in Russia and Eugene W. Hilgard in the United States (Jenny 1961a). These scientists began to see soils as something more than geology or chemistry, emphasizing interactions among climate, geology, and biology:

> "The still young discipline of these relations is of an exceptional inspiring scientific interest and meaning. Each year it makes greater and greater strides and conquests; gains daily more and more of active and energetic followers, eager to devote themselves to its study with the passionate love and enthusiasm of adepts." (Dokuchaev 1898, quoted by Jenny 1961a)

At the end of the 19th Century, Dokuchaev (1951) summarized his view of soil formation in an equation:

$$S = f(cl, o, p)t^o$$

where S = soil, cl = climate, o = organisms, p = geologic substrate, and t^o is a measure of relative age (this version of the equation was related by Jenny 1961b). A similar equation was proposed by an American ecologist, Charles Shaw (1930), who was unaware of Dokuchaev's work:

$$S = M(C+V)^T+D$$

where S = soil, M = parent material, C = climate, V = vegetation, T= time and D = deposition or erosion. The most familiar form of a soil-forming equation is probably Hans Jenny's (1941):

$$S = f(cl, o, r, p, t, \ldots)$$

where S =s oil, cl = climate, o = organisms, r = topography, p = parent material, t = time, and "…" is a place holder for other factors that might later be determined to be important. All of these expressions provide a key variable for the effect of biological factors such as tree species.

Recognition of the potential effect of vegetation on soils was an important step, but a century of forest soil investigations leaves a great deal to be investigated in the next century. Earl Stone's (1975) classic summary of the state of knowledge on tree species effects on soils concluded that many beliefs were no better than myth; aside from the N-fixing species, he thought the evidence supporting generalizations was too weak to support confident interpretations. A body of evidence has accumulated since Stone's review to document that species dramatically affect soils (see reviews by Binkley 1995, Binkley and Giardina 1998, Augusto *et al.* 2002). The mythological themes remain strong; Sverdrup *et al.* (2002) claimed that the idea that tree species differ in their effects on mineral weathering is only myth, yet their rationale included no empirical evidence to support or refute their own claims. We remain far from our goal of a generalizable understanding about the magnitude of species effects, and how consistent these effects might be across soils types and along environmental gradients. We hope this volume is a notable step in spurring progress on these themes.

DESIGN OF STUDIES

The chapter by Zechmeister-Boltenstern and coauthors (this volume) examined rates of turnover of microbial N and C, as well as gas flux rates, in twelve types of natural forests. These forests span the range of forest types in Central Europe, and the differences among these forests relate to both the dominant tree species, and the environmental conditions at each site. These confounding differences among sites were avoided in one study reported in the chapter by Cortina and Maestre (this volume), where the survival of *Pistacia* shrubs was mapped at a single site. However, the apparent within-site uniformity of the slope was disproven by the spatial pattern of shrub mortality; clusters of high mortality indicated that the slope was very non-uniform. The Siberian afforestation experiment described in several chapters tried to overcome within-site variations in soil properties by removing the 0-20 cm mineral soil, mixing it, and redistributing the soil across the site. Unfortunately, the yield of a wheat crop planted on the site revealed that within-site variation was still sizable (Shugalei, this volume). These limitations can be addressed by experimental designs that include replication of treatment; however, even designs with replicate plots may not meet the definition of experimental replication. Common-garden experiments that test only monoculture effects may not extrapolate well to diverse forests; Roy et al. (this volume) use a creative approach of examining soil influences of species near the boundaries of monoculture plots, relating soil features to the chemistry of litter inputs rather than simply to species names.

DEFINING REPLICATION

We may believe many things about the effects of trees on soils. For example, we might believe that decomposition of deciduous hardwood litter is more rapid than that of conifers. Beliefs may be true, but they need to be tested with challenging experiments before strong confidence is warranted. The decomposition belief could be tested in a beech forest using litterbags filled with beech and spruce litter. If the beech litter decomposed more quickly than the spruce litter, the confidence warranted in the belief would be increased – but not by very much because of a long list of unexamined assumptions. Would the same results have developed if the bags were placed on a spruce soil rather than a beech soil? Did the results depend on the chosen mesh size for the litterbags? Not all spruce needles are alike; would the use of leaves from another site (with higher nutrient supply) have shown different results? Beech and spruce may not be representative of the full range of hardwood and conifer species, so the pattern in this experiment provides no degrees of freedom for a statistical inference about the classes of species. And in any case, could a difference in decomposition rates of fresh litter really tell us very much about the longer-term differences in soils that would develop as humified material accumulated?

Classical experimental design in a chemistry laboratory would test ideas about chemical reactions by holding all variables constant (including temperature, air pressure, volume, and procedures) except for the variables involved in the hypothesis test. The experiment may involve testing 4 concentrations of a chemical, and the scientist might do 4 replications of each to be sure the observed results are consistent and repeatable. This general design has been adopted in many forestry studies (and strongly advocated, e.g. Mead *et al.*, 1990), often without realization of the weakness that comes from holding all other variables constant. In many cases, we need answers that apply across diverse sites, and testing the effects of treatments in a replicated study within a single site cannot provide these answers.

Confidence in ideas about the effects of species needs to be developed from a clear context that defines the population of interest (in a statistical sense), and experimental designs that draw representatively from the population. In the case of decomposition of hardwood and conifer litters, the populations of all hardwood species and all conifer species could not be represented by only beech and spruce, and the population of soil and forest types could not be represented by a single beech forest.

The definition of "replication" needs to be considered carefully in designing an experiment, and in the interpretation of results. Statistical analyses rely on carefully designed replication to account for the influence of factors (such as prior differences in soils) other than the factor of interest (such as tree species effects).

We illustrate some of these key points with a case study that contrasted the N content of soils under *Eucalyptus saligna* and N-fixing *Falcataria moluccana* (Garcia-Montiel and Binkley 1998). A comparison of two adjacent plots indicated that the N-fixer increased soil N by 36 g m^{-2} yr^{-1} over a 12-year period (Figure 1). However, the random assignment of species to each of these plots could have placed the N-fixing species in a plot that already had higher soil N initially, so any prior difference would confound the estimate of N accretion. These plots were relatively small (30 x 30 m), and perhaps larger plots would reduce the likelihood of the species overlying prior differences in the soil. Alternatively, the soils could have been removed, mixed, and reapplied to the landscape as in the Siberian afforestation experiment.

To account for the possible prior variation in initial soil conditions, or in the influence of initial conditions on N fixation and accretion, this study in Hawaii was replicated in 4 blocks. With 4 replicate plots of each species, it would be unlikely that all 4 replicates of one species would fall on higher N soils than the 4 replicates of the other species. Across all 4 replicates, the average rate of N accretion was 18 g m^{-2} yr^{-1}, just half of the rate indicated by the single pair of plots.

Even this replicated design has limitations for making inferences about N accretion under *Falcataria*; no amount of replication (or prior soil mixing to increase uniformity) within a single site can provide degrees of freedom for testing a hypothesis about a population of sites. Even with high confidence in the species effects at this site, we don't know if the species effect would be consistent on similar soils at other sites. Fortunately this experiment was replicated at a total of 3 sites (with 4 replicate blocks at each site), and across all these plots the average rate of N accretion was just 12 g m^{-2} yr^{-1}. The strength of this experimental design is rare in studies that have examined the

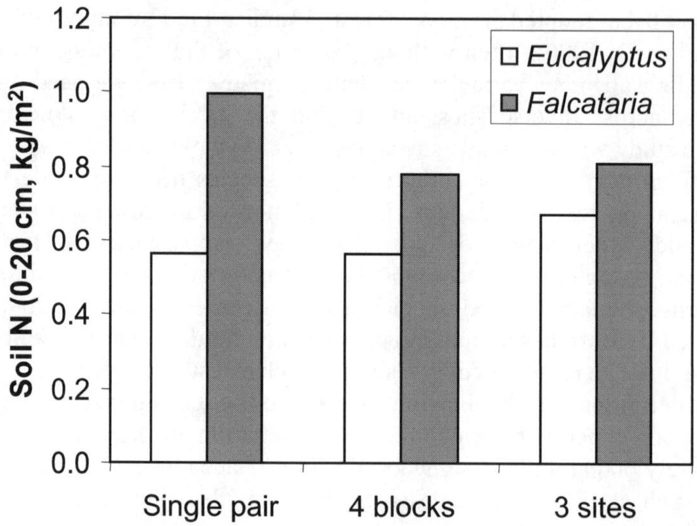

Figure 1. The difference in soil N beneath *Eucalyptus saligna* and *Falcataria moluccana* indicated an annual rate of N accretion (over 12 years) of 36 g N ha^{-1} yr^{-1} when comparing a single pair of plots; 18 g N ha^{-1} yr^{-1} based on 4 blocks (with a pair of plots in each block) within one site; and 12 g N ha^{-1} yr^{-1} across 3 sites (with 4 blocks/site; from data of Garcia-Montiel and Binkley 1998).

effects of tree species on soils, but even this design was limited by the fact that the three replicate sites were all on the same soil series; we have no statistical basis for inferring the likely effect of *Falcataria* on any other type of soil (Figure 2).

The bottom line is that replication in statistics is based on the definition of the population of interest, and what comprises a representative sample of that population. If the population in the N-fixation study were defined as "these two plots," then the first estimate of N accretion would be valid, because the entire population was assessed. If the population were defined as a 10-ha landscape, then the single pair of samples would represent a single sample, with no degrees of freedom to assess the variability across the 10 ha that would be unrelated to N fixation. The 4 replicate blocks provide degrees of freedom relative to the population of 10 ha, but nesting all 4 replicates within this single location gives no estimate of the variation that would be encountered on similar soils at other sites. Repeating the study on 3 sites provides an estimate of the variation among sites across the 10,000 ha of this Kaiwiki soil series, but if the population of interest included other types of soils, even this design would be insufficient. If this study could afford a total of 12 pairs of plots (as it had 4 blocks at 3 sites), the most powerful design would have involved placing single pairs of plots (one *Eucalyptus*, one *Falcataria*) at 12 separate locations across the entire population of interest (see Stape *et al.* 2004 for a fertilization trial using this approach).

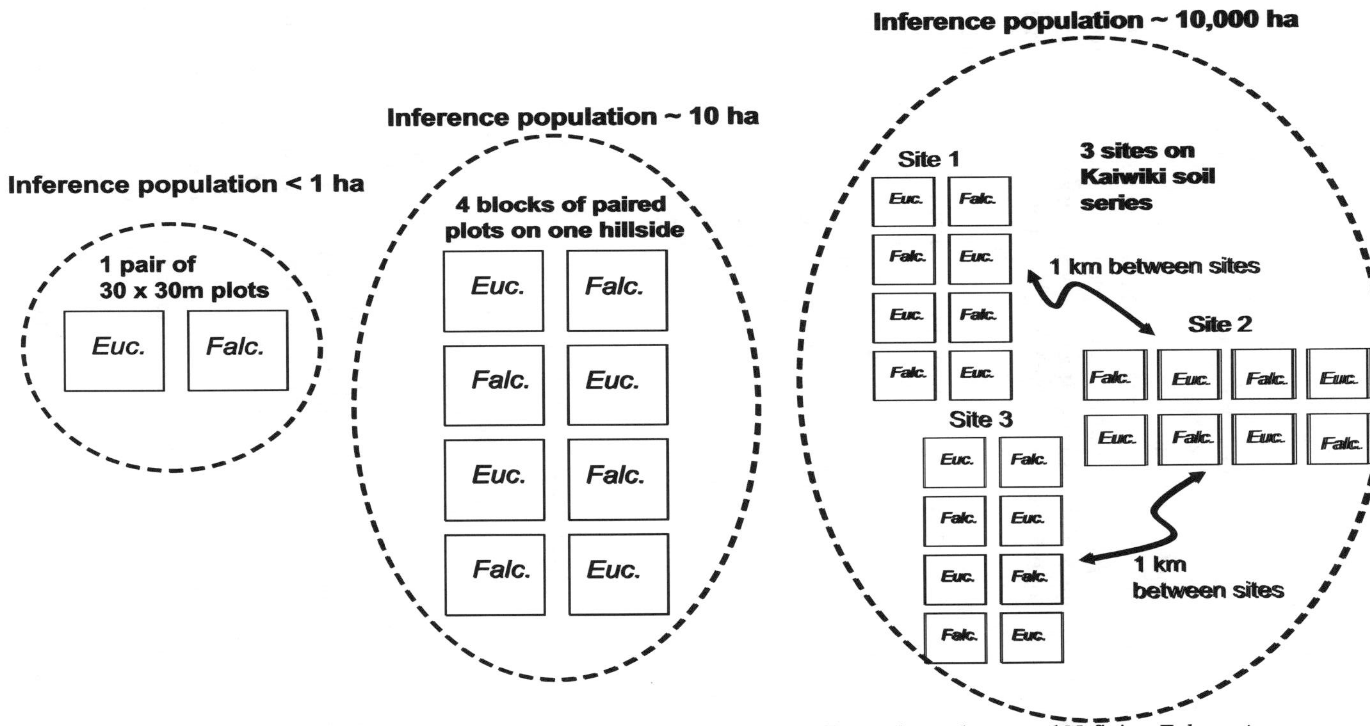

Figure 2. Diagram of experimental design for study in Figure 1, examining effects of *Eucalyptus* and N-fixing *Falcataria* on soils. A stronger design would have spread the 12 pairs of plots across a random set of locations on this single soil type (or even across soil types), giving a population of inference of 100 km2 or more

Many forestry experiments use a classic replicated, randomized (sometimes blocked) design, with 3 or 4 replicates plots for each treatment (e.g., Mead *et al.* 1990). This design is powerful if the replicates (or blocks) are spread throughout the population of interest, but if they are clustered at a single location (and the population not even defined!), then any inference to the population is limited to non-statistical sorts. Non-statistical inferences may still be drawn, based on professional judgment about whether interactions across sites will be zero, linear, or non-linear, but these inferences are typically too weak to support important decisions.

In some cases, a broad question about the effects of a tree species can be examined by a "meta-analysis" that considers the consistency of effects across many studies. If all studies with N-fixing trees show strong increases in soil N, then we have high confidence that this effect is a general one that is not limited to a particular type of site. We may have enough case studies of N fixing species for this sort of analysis (see chapter in this volume), but meta-analyses have not yet been developed for other species. For example, Binkley and Giardina (1998) noted that 5 common-garden experiments found that Norway spruce acidified soils more strongly than other species, and that larch seemed to reduce soil acidity. Larch also significantly reduced soil acidity compared to most other coniferous and deciduous species at the Siberian afforestation experiment (Menyailo *et al.*, 2002). The need for further meta-analysis of common garden experiments is evident, especially if factors could be identified that accounted for some of the variation in results among studies. For more insight on limitations of designs commonly used in forest research, see Bennett and Adams (2004).

DETERMINING THE EFFECT OF TIME

We also note that the points raised above on spatial variation and replication also apply to testing hypotheses about changes over time. An experiment might compare the effects of beech and spruce on soil invertebrates, and how these effects change over a 20 year period. If sampling were limited only to the beginning and the end of the 20 year period, one could test whether the two periods differed, but 0 degrees of freedom for the effect of time would preclude a statistical inference about the effects of time. If the initial sampling occurred when the O horizon was moist, and the second sampling (20 years later) happened during a dry period, then the significant difference between samplings could result from moisture rather than time. If the sampling in time had been repeated at years 0, 5, 10, 15, and 20, then one could explicitly test for the effect of time, expecting that any variation in moisture content would represent "noise" (unexplained variance) that would not mask a strong effect of time.

In some cases, an evaluation of change over decades might focus on a soil property that shows very little variation among seasons or soil moisture conditions. An evaluation of changes in the C content of a soil horizon over a 20 year period may be unreplicated in time, but scientists (and readers) may still be confident that the significant difference between the time periods was a result of the passage of time (and all the processes that happened over that period).

Statistical inferences provide us with "blind" evaluations of the probability that an observation may have happened at random; we are free to use other criteria (such as our confidence that soil C contents do not vary substantially within seasons) to gauge the confidence warranted in experimental results, as long as we are very clear about when we are using "blind" statistics, and when we are using other information to make an inference. We also note that the issues raised above apply to other factors in the development of forests and soils. For example, the use of a single genotype (or provenance) to represent a species provides no statistical basis for inferring the effects of other genotypes (or provenances) of the same species. We may be willing to infer that a single genotype is representative of the population of genotypes of the same species, but this inference is based on judgment rather than statistics.

EXPERIMENTAL LIMITATIONS

A dozen studies around the world have examined the effects of tree species on soils by planting species in common gardens. This is an elegant design that removes a broad range of factors that could confound the test of species' effects. However, common garden designs typically include several important limitations. Many of them have been established on former agricultural soils, so the apparent effect of tree species develops from a very unnatural starting point. Former agricultural soils would typically have a legacy of prior fertilization, a dearth of seeds of typical understory species, and the absence of an O horizon. The species included in some common gardens may be unlikely to be found on the same soil type across forested landscapes, so the impacts on an unusual soil may not represent broad-scale impacts of each species.

The changes that develop over the time span of most individual research projects may not represent important, long-term changes in soils. We have too little information on the effects of tree species to chart the time course over which soils change. For example, soils may change more rapidly under white pine than under Norway spruce, but the "endpoint" conditions could be the same after enough time had passed.

We might also want to infer the mechanism behind the effects that tree species exert on soils, but even well-designed common garden experiments cannot test alternative ideas. For example, Son and Gower (1991) examined the effects of 5 tree species on annual net N mineralization in Wisconsin, USA. The species differed by more than 2.5 fold in N mineralization, and

76% of the variation in net N mineralization was explained by the ratio of lignin:N in aboveground litterfall (Figure 3). The evidence failed to refute the idea that lignin:N influences N supply, but it could not prove lignin:N was the driver. Indeed, the differences among species in N mineralization related even better with the total fungal biomass, with 89% of the variance accounted for (Scott 1998). The fungal biomass may or may not have been the key to the variation among species in N mineralization; the total biomass of bacteria showed a strong negative trend with fungal biomass (Figure 3), highlighting the risk of assigning special importance to any individual covariate. Indeed, the direct drivers of change beneath the tree species might be the soil community of microbes and animals; we know that soil animal communities differ strongly under the influence of tree species (see Bezkorovaynaya this volume, Elmer et al. 2004), but we know almost nothing about the implications of these changes for rates of biogeochemical cycling.

Some features of soils can be very dynamic, complicating any questions about the effect of tree species. In the Wisconsin common garden, which species showed the greatest N mineralization? The answer to this question depended on the time period of incubation; the ranking of species shifted dramatically across time periods (Table 1).

Issues of scale are also important in tree and soil studies. Long bridges span wide rivers, but lengthening bridges does not widen rivers. In soils, higher pH and base saturation may be associated across landscapes with greater growth rates of trees, but increasing pH and base saturation within a single site may not increase growth. Forest productivity increases across Scandinavia with increasing soil pH and base saturation and pH (Dahl *et al.* 1967, Lahti and Vaisanen 1987), but dozens of liming experiments have shown that raising pH within a single site does not increase growth in this region (Nihlgård and Popovic 1984, Popovic and Andersson 1984, Derome *et al.* 1986, Andr9ason 1988, Derome and Pätilä 1989). This apparent contradiction between within-site and across-landscape perspectives may relate to a positive correlation across landscapes in soil N supply and soil pH (Giesler *et al.* 1998), and changing pH within sites may or may not increase the supply of growth-limiting nitrogen.

ADVANCES IN METHODOLOGY, AND SURPRISES

Over the past 20 years, our ability to delve into soil processes has expanded dramatically. In the past, experiments on the influence of tree species were often limited to characterizations of one or more soil horizons, or in some cases to a nutrient budget of a whole stand (Bergkvist and Folkeson 1995). A lack of overall mass balance typically limited the confidence warranted by any particular component of these budgets. The Krasnoyarsk Workshop (and this book) highlighted several promising approaches for new investigations, as well as surprises. Dr. Phil Ineson presented information on the stable isotopes of carbon in vegetation and worms in microcosms; the rapid appearance of the labeled C indicated that the worms must be feeding in part on plant roots and not just on plant detritus (which remained unlabelled in this short-term experiment). Santruckova (this volume) and Mukhortova (this volume) demonstrated the insights that could be developed by analyzing the kinetics of reaction rates in the laboratory and the field. Butterbach-Bahl and Kiese (this volume) demonstrated that variation in time may be as important as variation in space; the annual total flux of trace N gases depended on short-lived "spikes" in gas efflux. Intermittent sampling that missed these rare spikes would underestimate the efflux in a plot by several-fold, and catching these spikes would be fundamental to determining the influence of different species. Dr. Cindy Prescott discussed ideas about differences among species in preference for forms of nitrogen; the ericaceous shrub salal (*Gaultheria shallon*) was expected to prefer organic forms of N, but its utilization of ammonium and nitrate rivaled that of other species. In fact, molecular techniques demonstrated that salal shrubs formed mycorrhizal associations not only with ericoid mycorrhizae, but also with arbuscular and ectomycorrhizae.

THE CASE FOR EVIDENCE-BASED FOREST SCIENCE

As scientists, we benefit from a legacy of decades and centuries of development of science in dozens of fields, and this includes valuable approaches to conducting science. We suggest that forest science should embrace standards of evidence established for other scientific fields, and consider explicitly the level of confidence warranted in our ideas. One example of this approach is "evidence-based medicine," which developed as a means for health practitioners to evaluate the confidence warranted in the value of medical treatments. Cochrane (1972) launched the idea that randomized controlled experiments are vital for assessing the effectiveness of medical treatments. Prior to this initiative, the efficacy of treatments was often judged with anecdotal evidence, or evidence from trials with poorly

Table 1. The ranking of species in relation to cumulative net N mineralization depended strongly on the time period of comparison (from Son and Gower 1991, Scott 1996).

Time period	Species ranking
Annual, in situ incubations	Norway spruce < red pine = red oak < white pine < European larch
0-20 days laboratory	Norway spruce < red oak < red pine < white pine < European larch
0-60 days laboratory	White pine < red pine < red oak < Norway spruce < European larch
0-387 days laboratory	White pine = red pine < European larch < Norway spruce < red oak

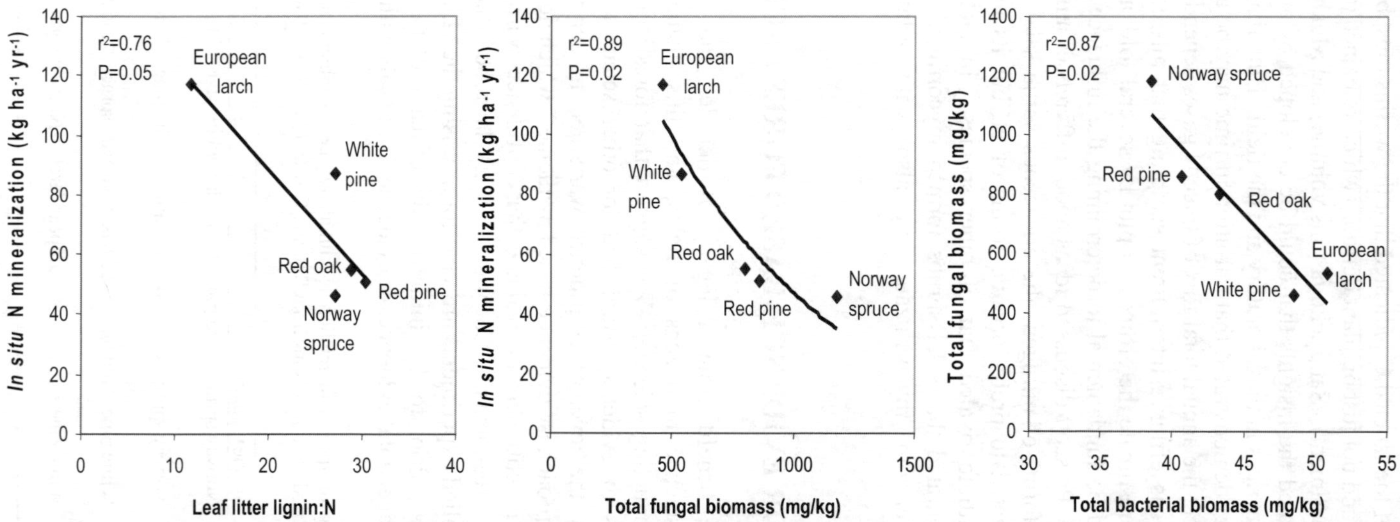

Figure 3. Annual net N mineralization in the field related well to the lignin:N of aboveground litterfall in a common garden experiment (left; Son and Gower 1991), but lignin:N may not have a causal effect on net N mineralization as evidenced by other patterns among these stands. Net N mineralization correlated strongly with fungal biomass (middle), and fungal biomass related well with bacterial biomass (Scott 1998). Simple correlations cannot be relied upon to test mechanistic hypotheses.

Table 2. A preliminary scheme for rating the confidence warranted in forestry experiments. Many forest studies (including those of species effects on soils) are type 3 or 4; few study designs support Level-1 or Level-2 confidence in applying results to forest landscapes.

Level	Type of evidence
1	Meta-analysis of several similar experiments, showing consistent effects
2a	Replicated experiment at several sites, with explicit extrapolation to the population of interest
2b	Replicated experiment at several sites, but no formal *a priori* plan for extrapolating to the population
3a	Experiment at a single site with replication of treatments
3b	Case studies across environmental gradients with potentially confounding spatial factors
4	Case studies, no replication of treatments within a single site
5	Expert opinion or inferences from expected first principles

controlled factors. Fletcher and Sackett (1979) took the next step and developed "levels of evidence" to rate the value of medical treatments. Formal rating systems are now standardized in medicine. (See: Oxford Centre for Evidence-Based Medicine http://www.cebm.net/index.asp, The Cochrane Collaboration and Library http://www.cochrane.org /docs/descrip.htm, and the University of Toronto's Centre for Evidence-Based Medicine, http://www.cebm.utoronto.ca/). Professional medical societies have developed similar schemes to evaluate the strength of evidence, and the power of this approach is so persuasive that over 100 grading scales are used among medical journals to gauge the confidence warranted in experiments and recommendations for patient care (Ebell *et al.* 2004).

We should endeavor to be as straightforward and effective in relating science and practice in forestry, and provide explicit statements about the power of our experimental evidence. This power goes beyond the simple P value from a statistical test to include critical details about the design of individual experiments, and groups of experiments. We offer a preliminary example of this approach in Table 2, and readers can refer to this table when evaluating confidence warranted by the experimental designs in this volume and in other publications.

LOOKING TO THE FUTURE

Our workshop identified major advances in our understanding of the effects of species on soils, and how these might shape the responses of forests to changing climate. The advances in methodology and the surprises that came from recent studies, combine to illustrate the potential for future gains in knowledge. We would highlight several key areas for development.

1. Future research projects should define the populations of interest, and design experiments to address these populations. If the population of interest happens to be beech and Norway spruce forests on limestone soils in Bavaria, then experimental plots need to be established to provide a fair, representative sampling of this population.
2. The effects of tree species should be examined by testing challenging hypotheses that go beyond simple descriptions of the magnitudes of effects. For example, if a species happens to increase the supply of N in soils, does this increase lead to an increase in growth? Only an experimental manipulation of N supply could test this hypothesis. Hypothesis testing will be especially important for determining the mechanisms that underlie observed effects of species.
3. Few forests are dominated by single tree species, yet almost all of our information on the effects of tree species on soils comes from monoculture experiments. Are the effects linear functions of the proportion of species, or do interactions reduce or enhance these effects? We may gain preliminary insights on the effects of species in mixtures by examining the borders between monoculture plots, but explicitly designed experiments will be needed to test many important hypotheses.
4. The effects of tree species on soils may be moderated (or enhanced) by the influence of overstory trees on understory vegetation. We have seen many cases where understory vegetation differs dramatically beneath the influence of overstory species, but no studies have manipulated understory vegetation to isolate this effect. Similarly, tree species have major influences on soil biota, and experiments must address these biotic differences that may determine the overall effect of tree species.
5. We recommend that studies on tree-species effects take advantage of the constraints offered by mass balance in nutrient cycling studies. Confidence in experimental results will be high if the various pools of nutrients among treatments add up to the same total contents for the entire tree+soil system. Substantial deviations from mass balance would indicate caution is needed in interpreting the findings (Fisher and Binkley 2000). A classic estimate of fine root production in a forest was so high that canopy photosynthesis could not provide enough C to grow the roots (as pointed out by Ryan 1991). Another study reported greater rates of fine root growth on a poor site than on a fertile site, but calculations of the N requirement for both stands (J. Aber, personal communication) indicated the poor site would have to have a greater supply of N than the fertile site (which seems unlikely). An unreplicated comparison of burned and unburned spruce forests claimed that burning increased soil calcium (Ca) by 5000 kg/ha, yet the total Ca content of the trees was just 1000 kg/ha; the 4000 kg/ha gap in the budget suggests that little confidence is warranted in this estimate of fire impact. Multiple lines of evidence can provide insights about the "reasonableness" of production and nutrient budgets.

ACKNOWLEDGEMENTS

We thank the participants in the NATO Advanced Research Workshop for their stimulating ideas, data, and conversations. The broad range of perspectives and experiences made for a delightful workshop, and a fertile ground for future experiments on the effects of tree species on soils in relation to global changes.

REFERENCES

Andr9ason O 1988 Suiting forest management to a changed environment. In Forest Health and Productivity. pp 67-75. The Marcus Wallenberg Foundation Symposia Proceedings #5, Falun, Sweden.

Augusto L, Ranger J, Binkley D and Rothe A 2002 Impact of several common tree species of European temperate forests on soil fertility. Ann. For. Sci. 59, 233-254.

Bennett LT, and Adams MA 2004 Assessment of ecological effects due to forest harvesting: approaches and statistical issues. J. Appl. Ecol. 41, 585-598.

Bergkvist B and Folkeson L 1995 The influence of tree species on acid deposition, proton budgets and element fluxes in south Swedish forest ecosystems. Ecol. Bull. 44, 90-99.

Binkley D 1995 The influence of tree species on forest soils -- processes and patterns. *In* Proceedings of the Trees and Soil Workshop. Eds. D J Mead and I S Cornforth. pp 1-33. Agronomy Society of New Zealand Special Publication #10. Lincoln University Press, Canterbury.

Binkley D and Giardina C 1998 Why trees affect soils in temperate and tropical forests: the warp and woof of tree/soil interactions. Biogeochemistry 42, 89-106.

Cochrane A L 1972 Effectiveness and efficiency. Royal Society of Medicine Press, London.

Derome J and Pätilä A 1989 The liming of forest soils in Finland. Meddelser fran der Norsk Institut for Skogforskning. 42, 147-155.

Derome J, Kukkola M and Mälkönen E 1986 Forest liming on mineral soils. Results of Finnish experiments. National Swedish Environmental Protection Board Report 3084, Solna, Sweden.

Dokuchaev V V 1951 Writings. Akademia Nauk, Moscow. 6, 381.

Ebell MH, Siwek J, Weiss BD, Woolf SH, Susman J, Ewigman B, and Bowman M. 2004 Strength of recommendation taxonomy (SORT): a patient-centered approach to grading evidence in the medical literature. Amer. Fam. Physician 69, 548-556.

Elmer M, La France M, Förster G, and Roth M 2004 Changes in the decomposer community when converting spruce monocultures to mixed spruce/beech stands. Plant Soil 264, 97-109.

Fisher RF, and Binkley D. 2000. Ecology and management of forest soils. Wiley, New York.

Fletcher S and Sackett D 1979 Canadian Task Force on the Periodic Health Examination: The periodic health examination. Can. Med. Assoc. J. 121:1193-1254.

Garcia-Montiel D C and Binkley D 1998 Effect of *Eucalyptus saligna* and *Albizia falcataria* on soil processes and nitrogen supply in Hawaii. Oecologia 113, 547-556.]

Giesler R Högberg M and Högberg P 1998 Soil chemistry and plants in Fennoscandian boreal forest as exemplified by a local gradient. Ecology 79, 119-137.

Jenny H 1941 Factors of soil formation: a system of quantitative pedology. McGraw-Hill, New York.

Jenny H 1961a E.W. Hilgard and the birth of modern soil science. Collana Della Revista "Agrochemica," Pisa.

Jenny H 1961b Derivation of state factor equations of soils and ecosystems. Soil Sci. Soc. Am. Proc. 25, 385-388.

Mead DJ, Whyte AGD, Woollons RC, and Beets PN 1991. Designing long-term experiments to study harvesting impacts. In Long-term Field Trials to Assess Environmental Impacts of Harvesting. Eds. WJ Dyck and CA Mees. Pp 107-124 Proceedings IEA/BE T6/A6 Workshop, Forest Research Institute, FRI Bull 161, Rotorua, New Zealand.

Lahti T and Vaisanen R A 1987 Ecological gradients of boreal forests in South Finland: an ordination test of Cajander's forest type theory. Vegetatio 68, 145-156.

Menyailo O V, Hungate B A and Zech W 2002 Tree species mediated soil chemical changes in a Siberian artificial afforestation experiment. Plant Soil 242, 171-182.

Nihlgård B and Popovic B 1984 Effekter av olika kalkningsmedel I skogsmark -- en litteraturversikt. Statens Naturvardsverk PM 1851, Solna.

Popovic B and Andersson F 1984 Markkalkning och skogsproduktion -- litteratur□versikt och revision av svenska kalkningsförsk. Statens Naturvrdsverk PM 1792, Solna.

Remezov N P and Pogrebnyak P S 1969 Forest Soil Science. Translated by A. Gourevitch, U.S. Department of Commerce, Springfield.

Ryan MG 1991 A simple method for estimating gross carbon budgets for vegetation in forest ecosystems. Tree Phys. 9, 255-266.

Scott N A 1996 Plant species effects on soil organic matter turnover and nutrient release in forests and grasslands. PhD dissertation. Colorado State University, Fort Collins.

Scott N A 1998 Soil aggregation and organic matter mineralization in forests and grasslands: plant species effects. Soil Sci. Soc. Am. J. 62, 1081-1089.

Shaw C F 1930 Potent factors in soil formation. Ecology 11, 239-245.

Son Y and Gower S T 1991 Aboveground nitrogen and phosphorus use by five plantation-grown trees with different leaf longevities. Biogeochemistry 14, 167-191.

Stape J L, Ryan M G and Binkley D 2004 Testing the 3-PG process-based model to simulate *Eucalyptus* growth with an objective approach to the soil fertility rating parameter. For. Ecol. Manag. 193, 219-234.

Stone E 1975 Effects of species on nutrient cycles and soil change. Philos. T. Roy. Soc. B. 271, 149-162.

Svedrup H, Hagen-Thorn A, Holmqvist J, Wallman P, Warfinge P, Walse C and Alveteg M 2002 Biogeochemical processes and mechanisms. *In* Developing Principles and Models for Sustainable Forestry in Sweden. Eds. H Sverdrup and I Stjernquist. pp 91-196. Kluwer, Dordrecht.

Chapter 2

EFFECTS OF BRITISH COLUMBIA TREE SPECIES ON FOREST FLOOR CHEMISTRY

Cindy E. Prescott[1] and Lars Vesterdal[2]
[1]*Faculty of Forestry, University of British Columbia, Vancouver, British Columbia, Canada;*
[2]*Forest and Landscape, Royal Veterinary and Agricultural University, Horsholm, Denmark*

INTRODUCTION

Although the question of tree species effects on soils has been of scientific interest for decades, two big questions remain largely unanswered. First, are there consistent effects of tree species on soil and forest floor properties? And second, how large are species effects compared with influences of other site factors? Here we address these questions by reviewing studies of the effects of tree species native to British Columbia, Canada, on forest floor chemistry, to determine if there are consistent patterns. Then we examine studies that provide some indication of the relative influences of tree species and other site factors, to assess the importance of species effects.

The province of British Columbia (BC), by virtue of its size and variety of climatic zones, has a large number of tree species, which also occur in neighboring provinces and U.S. states. In addition to comprising the natural and planted forests of BC, several of these species have been introduced into reforestation and afforestation projects in many European countries. The effects of BC tree species on forest floor properties have been investigated in several studies in North America and in Europe. In this paper we review what is known about the forest floors created by tree species native to British Columbia, particularly with respect to nutrient concentrations, rates of nitrogen mineralization, proportions of ammonium and nitrate, and microbial communities. Given the substantial influence of foliar litter on nutrient cycling in the forest floor (Prescott, 2002), we also discuss nutrient concentrations and rates of decay of foliar litter of BC tree species. Much of this information is drawn from common garden experiments both in BC

D. Binkley and O. Menyailo (eds.),
Tree Species Effects on Soils: Implications for Global Change, 17–29.

Figure 1. Location of the province of British Columbia

(EP571 - Prescott et al.,2000a; Skimikin - Thomas and Prescott, 2000);UBC Research Forest - Prescott and Preston, 1994), in Ireland (Clonsast Bog - Prescott et al., 1995), Denmark (Vesterdal and Rauland-Rasmussen 1998), France (Gloaguen and Touffet 1982), U.K. (Harmer and Alexander, 1986), Sweden (Alriksson and Eriksson, 1998) and the U.S. (Alban, 1982). This is augmented with information from other studies that compare forest floors under different tree species in natural stands.

The province of British Columbia (Figure 1) has a diversity of forest types as a result of its size (842,677 km^2), mountainous terrain and maritime influence. In coastal BC, lowland forests are primarily western hemlock (*Tsuga heterophylla* [Raf.] Sarg), western red cedar (*Thuja plicata* Donn ex. D. Don) and Sitka spruce (*Picea sitchensis* (Bong.) Carr.), with coastal Douglas-fir (*Pseudotsuga menziesii* (Mirb.) Franco) common on drier sites. At higher elevations (>700m), amabilis fir (*Abies amabilis* (Dougl.) Forbes, mountain hemlock (*Tsuga mertensiana* (Bong.) Carr.) and yellow cedar (*Chamaecyparis nootkatensis* (D. Don) Spach) dominate. The forests of interior BC range from ponderosa pine and interior Douglas-fir in dry zones to cedar, hemlock, western white pine (*Pinus monticola* Dougl. ex D. Don) and paper birch (*Betula papyrifera* Marsh.) in wet zones. Lodgepole pine (*Pinus contorta* Dougl.), trembling aspen (*Populus tremuloides* Michx.) and spruce hybrids (*Picea glauca* (Moench) Voss x *P. engelmannii* (Parry ex Engelm.) spruce are common throughout interior B.C.

QUESTION 1: WHAT DO WE KNOW ABOUT THE FOREST FLOORS CREATED BY BC TREE SPECIES?

Western Redcedar

Cedar has been considered to be an indicator of nutrient-rich sites, and is thought to produce N-rich forest floors. Cedar forest floors consistently have high concentrations of Ca and high pH relative to other species (Table 1, Alban, 1969;Turner and Franz, 1985; Prescott and Preston, 1994; Prescott et al., 2000a). Higher concentrations of nitrate or proportion of nitrate (relative to ammonium) are also commonly reported in cedar forest floors (Turner and Franz, 1985; Harmer and Alexander, 1986; Prescott et al., 2000a). However, rates of net N mineralization in cedar forests floors are usually not greater than other species (Prescott and Preston, 1994; Prescott et al., 1995); nor are N concentrations higher (Ovington, 1954; Alban, 1969; Prescott and Preston, 1994), indicating that cedar does not create particularly N-rich forest floors. In fact, at the Clonsast Bog trial in Ireland, cedar forest floors had the lowest rates of net N mineralization of 14 species, including *Calluna* (Table 1). This is consistent with cedar foliar litter consistently having high C:N ratio, high "lignin" content, and slow decay relative to other BC tree species (Figure 2) (Harmon et al., 1990; Prescott and Preston, 1994; Prescott et al., 2000a; Prescott et al., 2004). Cedar forest floors have been reported to have higher bacteria biomass (including ammonium-oxidizing bacteria) and lower fungal biomass (Turner and Franz, 1985; Grayston and Prescott, in press), probably as a consequence of the higher pH and Ca. Thus the high proportion of nitrate in cedar forest floors is more reflective of higher pH and associated changes in microbial populations, rather than an indication of high N availability. Similarly the tendency for cedar to naturally occur on nutrient-rich sites may instead be related to it preferring wet sites (which are often lower slope, rich sites). Cedar is more tolerant than other species of wet sites but actually has wide tolerance with respect to nutrient availability. Cedar grows better than other conifers on extremely N-poor cedar-salal sites and does not respond as much to fertilization (Bennett et al., 2003). Therefore, it seems that cedar, despite having higher nitrate in forest floors, neither prefers N-rich conditions, nor does it create them.

Western Hemlock

Forest floors under western hemlock consistently have lower pH (Table 1; Ovington, 1954; Alban, 1969; Turner and Franz, 1985; Prescott and Preston, 1994; Prescott et al., 2000a) and lower Ca concentrations (Prescott et al., 2000) than other species, as does hemlock needle litter (Prescott and

Table 1. pH and rates of net N mineralization during laboratory incubation of forest floors under 13 tree species at the Clonsast Bog in Ireland. Values followed by the same letter are not significantly different ($P<0.05$) based on one-way ANOVA and Bonferroni's multiple range test.

Species	pH	Net N mineralization $\mu g\ g^{-1}\ d^{-1}$
Sessile oak	4.7 bc	22.3 a
Sitka spruce	4.8 bc	15.1 b
Grand fir	5.0 b	15.1 bc
Douglas-fir	4.4 cde	13.9 bc
Norway spruce	4.9 b	12.7 bc
Lodgepole pine	3.9 e	12.6 bcd
Japanese larch	5.7 a	12.4 cde
Sitka spruce+Jap. Larch	4.3 cde	9.8 cde
Western hemlock	3.9 e	8.9 cdef
Scots pine	4.7 bc	7.4 def
Monterey pine	4.1 de	5.3 efg
Calluna	4.6 bcd	4.1 fg
Western red- cedar	6.1 a	1.2 g

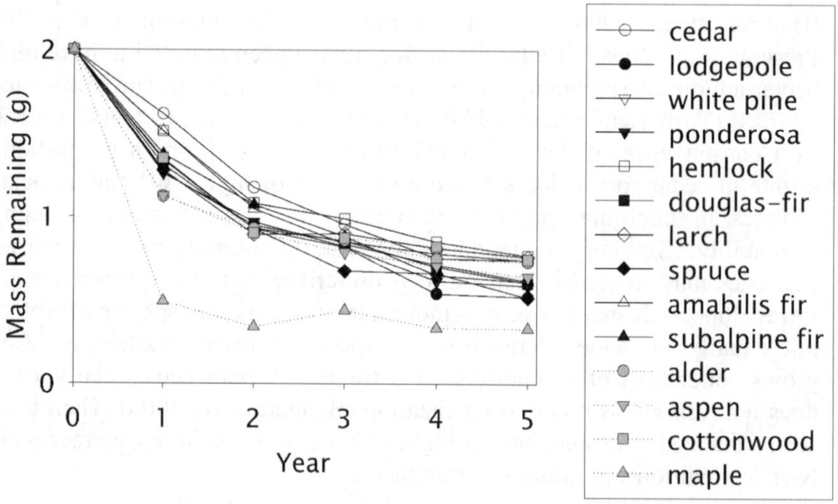

Figure 2. Mass remaining of foliar litter from 14 tree species native to British Columbia during a 4-year incubation in a coastal forest. Adapted from Prescott et al. (2004).

Preston, 1994). Western hemlock forest floors are intermediate (to douglas-fir and western red cedar) in rates of net N mineralization (Table 1; Prescott and Preston, 1994), N concentrations in forest floor and needle litter (Alban, 1969; Prescott and Preston, 1994) and rate of litter decay (Prescott et al., 2000a). Hemlock forest floors tend to be dominated by NH_4 rather than nitrate, and have higher fungal:bacteria ratio than cedar and Douglas-fir (Turner and Franz, 1985; Grayston and Prescott, in press). As with cedar, these characteristics are probably associated with the pH and Ca status rather than being indicative of N availability.

Sitka spruce

Sitka spruce forest floors have moderately high rates of net N mineralization (Table 1; Harmer and Alexander 1986). At the EP571 sites in coastal BC (Prescott et al 2000a), Sitka spruce litter and forest floors had high concentrations of N, P, Ca and K, but in other trials spruce was intermediate in these characteristics (Harmer and Alexander, 1986; Vesterdal and Raulund-Rasmussen, 1998). Harmon et al. (1990) reported fast decomposition of Sitka spruce needle litter relative to cedar, hemlock and Douglas-fir. Grayston and Prescott (in press) found high fungal:bacterial ratio in Sitka spruce forest floors (similar to hemlock). Thus Sitka spruce forest floors tend to be intermediate to high in most nutrients.

Douglas-fir

Douglas-fir forest floors have intermediate pH and concentrations of Ca and P (Prescott and Preston, 1994; Vesterdal and Raulund-Rasmussen, 1998; Prescott et al., 2000a). High rates of net N mineralization have been reported in Douglas-fir forest floors (Table 1, Prescott et al., 2000a; Prescott and Preston, 1994). In a 70-week laboratory incubation, net N mineralization from Douglas-fir forest floors even exceeded that released from forest floor of N-fixing red alder (*Alnus rubra* (Bong.) Carr.; Prescott, 1996; Figure 3), and concentrations of nitrate were also greatest in Douglas-fir after 40 weeks. Douglas-fir forest floors had substantially greater rates of net N mineralization than those of lodgepole pine and paper birch (Thomas and Prescott 2000; Figure 4). Nitrogen concentrations in Douglas-fir forest floors are higher than most conifers (Gloaguen and Touffet, 1982; Vesterdal and Raulund-Rasmussen, 1998, Thomas and Prescott, 2000; Prescott et al., 2000a). High N concentrations and low lignin:N ratio are also characteristic of needle litter of Douglas-fir relative to other conifers (Gloaguen and Touffet, 1982; Prescott and Preston, 1994; Prescott et al., 2000a; but see Thomas and Prescott, 2000). Decay rates of Douglas-fir needle litter are similar to other BC conifers (Figure 2; Harmon et al., 1990; Prescott and Preston, 1994; Prescott et al., 2000a, 2004). Although Douglas-fir forests had low fungal : bacterial ratios (like cedar; Grayston and Prescott, in press), they do not have the high pH and Ca usually found in cedar forest floors. Thus Douglas-fir forest floors are characterized by high N concentrations and net N mineralization rates, which are attributed to high N concentrations and low lignin:N in needle litter.

Lodgepole pine

Lodgepole pine forest floors have low pH, Ca and P (Table 1; Alriksson and Eriksson, 1998; Vesterdal and Raulund-Rasmussen, 1998; Thomas and Prescott, 2000). Lodgepole pine forest floors usually have low N concentrations and low rates of net N mineralization (Figure 4; Stump and Binkley, 1993; Prescott, 1996; Vesterdal and Raulund-Rasmussen, 1998; but see Table 1). In a laboratory incubation of lodgepole pine forest floor, negligible amounts of NH_4 and NO_3 were extracted during the first year, although half of the original forest floor mass decomposed (Prescott, 1996; Figure 3). Pine litter at the Skimikin trial had higher lignin concentration and lower concentrations of P, Ca and Mg but higher N concentrations than Douglas-fir and paper birch (Thomas and Prescott, 2000). In Rocky Mountain forests, lodgepole pine needle litter also had low P concentrations relative to white and Engelmann spruce and subalpine fir, but a wide range of

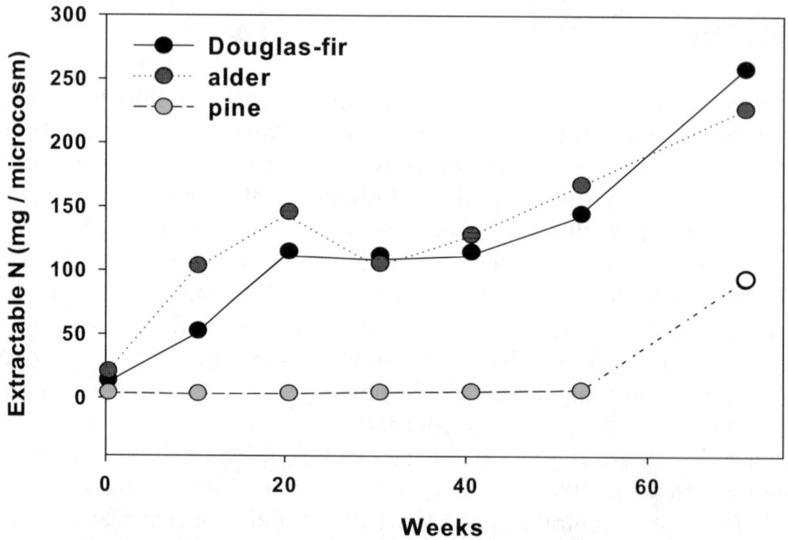

Figure 3. Concentrations of KCl-extractable N (NH_4-N + NO_3-N) in forest floors of Douglas-fir, red alder and lodgepole pine during a 70-week incubation in laboratory microcosms. Adapted from Prescott (1996).

N concentrations (Figure 5). Lodgepole pine litter had low P concentrations than other BC conifers (except western white pine; Prescott et al., 2004), but N concentrations were intermediate or high. High N concentrations in lodgepole pine litter were also reported by Thomas and Prescott (2000) and Moore et al. (in press). The decay rate of pine needle litter was similar to other BC tree species (Figure 2; Prescott et al., 2000b). Thus lodgepole pine forest floors, like its litter, tend to be low in pH, Ca and P, but variable in N concentration and net N mineralization rate.

Paper birch

Paper birch forest floors had N concentrations and net N mineralization rates that were greater than those of lodgepole pine, but less than Douglas-fir at the Skimikin trial (Figure 4). Birch forest floors had the highest pH and Ca and Mg concentrations and birch leaf litter had the highest Ca, Mg, K and P concentrations. The lignin content of birch leaf litter is intermediate to lodgepole pine and Douglas-fir (Thomas and Prescott, 2000; Prescott et al., 2004), and its rate of decay of was similar to that of needle litter (Figure 2). This is consistent with forest floors under European birch species which have high pH, Ca and Mg, but not higher rates of N or C mineralization (Saetre, 1998; Priha and Smolander, 1999; Saetre et al., 1999; Brandtberg et al., 2000).

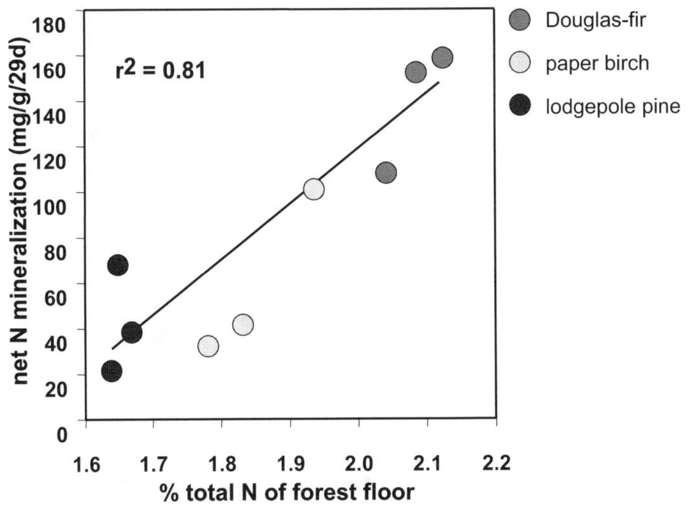

Figure 4. Relationship between net N mineralization rates and N concentrations in forest floors of lodgepole pine (P), paper birch (B) and Douglas-fir (F) growing in adjacent plots at the Skimikin species trial in interior British Columbia. Adapted from Thomas and Prescott (2000).

QUESTION 2: HOW IMPORTANT ARE SPECIES EFFECTS RELATIVE TO SITE INFLUENCES?

Understanding the effect that choice of tree species for planting will have on soil, or the potential effect of changes in species distribution in response to global climate change, requires that we understand how large species effects are relative to other factors such as soil type and slope position. This question cannot be satisfactorily addressed in natural forests, in which the influences of site factors on nutrient cycling are partly mediated through their influence on species composition of the vegetation. It is best addressed in common garden experiments that are replicated at several sites that differ in factors such as soil fertility, moisture availability or climate. Here we review several studies of this nature to address this question.

The pure effect of site (unsullied by species differences) on nitrogen cycling in litter and forest floors was assessed by comparing Douglas-fir forests along a productivity gradient (Figure 5). The productivity gradient was in turn related to soil N capital. (Prescott et al., 2000c). There was a clear effect of site on litter N content, forest floor net N mineralization and nitrification, all of which were positively correlated with soil N capital, despite all stands being Douglas-fir. The soil N capital appeared to be related to soil texture.

Harmer and Alexander (1986) compared forest floors of a) 16 tree species on the same site, and b) one species (Sitka spruce) on five sites of

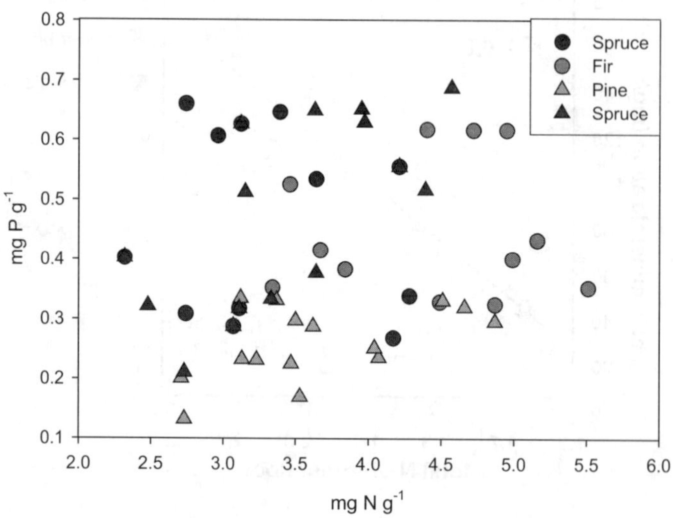

Figure 5. Concentrations of N and P in foliar litter of lodgepole pine, white and Engelmann spruce, and subalpine fir collected from 37 sites in the Rocky Mountains of Alberta.

varying Yield Class within the Banchory Forest in northeast Scotland. Differences in net N mineralization were greater among species (11 - 512 μg N g^{-1} organic matter) than among sites (25 - 71).

In the EP571 species trial in coastal British Columbia, Prescott et al. (2000) compared forest floors under cedar, hemlock, Sitka spruce and Douglas-fir at four sites. Rates of net N mineralization and nitrification were appreciative only at two of the four sites (Figure 6), and differences among tree species were apparent. At the other two sites, rates were too small for species differences to be detected. The low rates of N mineralization at two sites appeared to be related to cover of salal (*Gaultheria shallon*), which was high at those sites and negligible at the sites with high N mineralization. Like other ericaceous shrubs, salal is believed to produce tannins, which

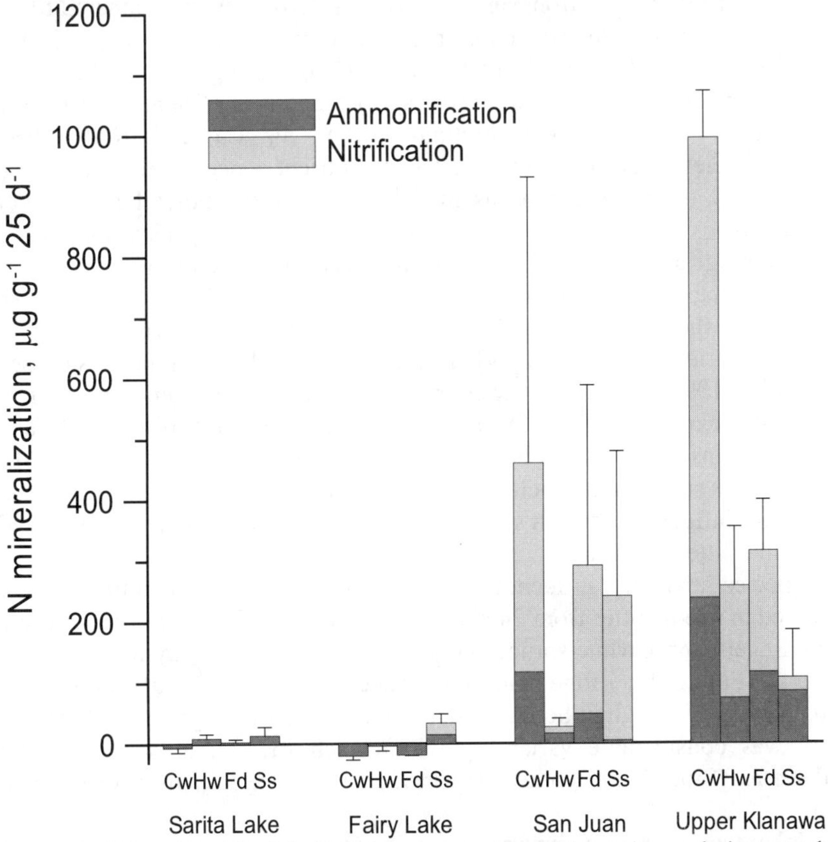

Figure 6. Rates of net N mineralization and proportion of ammonium and nitrate produced during a 25-day laboratory incubation of forest floors of four British Columbia tree species from four sites in coastal British Columbia. Tree species abbreviations: Cw = western redcedar; Hw = western hemlock; Fd = Douglas-fir; Ss = Sitka spruce. Adapted from Prescott et al.(2000a).

interfere with N mineralization. However, only mid-slope sites had salal; the other sites were in valleys and had understory vegetation such as *Rubus spectabilis* and *Polystichum munitum*, which are indicative of rich sites. Thus the differences in rates of N mineralization in these forest floors was most likely related to slope position hence moisture and nutrient status of the soil, and these influences appeared to overwhelm the effects of tree species on N mineralization. Tree species had consistent effects on other forest floor properties such as pH and nutrient concentrations at all sites, but again, differences among sites were greater than differences among tree species.

Vesterdal and Raulund-Rasmussen (1998) compared forest floor nutrient concentrations in a common garden experiment with seven conifer tree species (four native to BC) and seven sites along a soil fertility gradient in Denmark. Soils ranged from sandy Spodosols derived from weathered till to sandy-loam Alfisols of calcareous parent material. Mineral soil pH in the upper 50 cm ranged from 3.4 to 5.6, P from 17 to 175 mg Kg^{-1}), and Ca from 0.5 to 103 $mmol_c$ kg^{-1} soil. Tree species had a consistent influence on forest floor pH and concentrations of N, P, Ca and K. There was also a clear indication that forest floor nutrient concentrations were also modified by their concentrations in the mineral soil. For all forest floor characteristics measured, the range of values among tree species was greater than the range among sites (Figure 7). However, the greater range for species was attributable to the inclusion of lodgepole pine (*Pinus contorta* ssp *contorta*) in the trial; without pine, the effect of site would have been greater for most nutrients. This is supported by an earlier study (Raulund-Rasmussen and Vejre, 1995) in which only four of the species were studied at the two sites most extreme in soil fertility. The ranges in pH and most nutrient concentrations were greatest between sites when only beech, oak, Douglas-fir and Norway spruce were included. Bastrup-Birk et al. (2003) also found consistent influences of both site and species on concentrations of Ca, K and N in foliar litter in this trial.

Species and site influences on litter N and P concentrations were assessed in needle litter from 24 sites with various combinations of lodgepole pine, Engelmann x white spruce and subalpine fir in the Rocky Mountains of Alberta (Figure 5). Pine had low concentrations of P but variable N, subalpine fir was high in both nutrients, and spruce was high in P. However, there was considerable variability within each species, indicative of site influences. Litter P concentrations varied more among sites in spruce than in pine.

Johansson (1995) compared nutrient concentrations in needle litter of adjacent stands of Scots pine (*Pinus sylvestris*) and Norway spruce (*Picea abies*) at eight sites in Sweden. Spruce litter consistently had higher concentrations of N, P, K, Ca, Mg, Mn and lignin than pine litter. In both species, nutrient concentrations also varied considerably among sites. The average difference in %N was 0.11% between species and 0.09% among

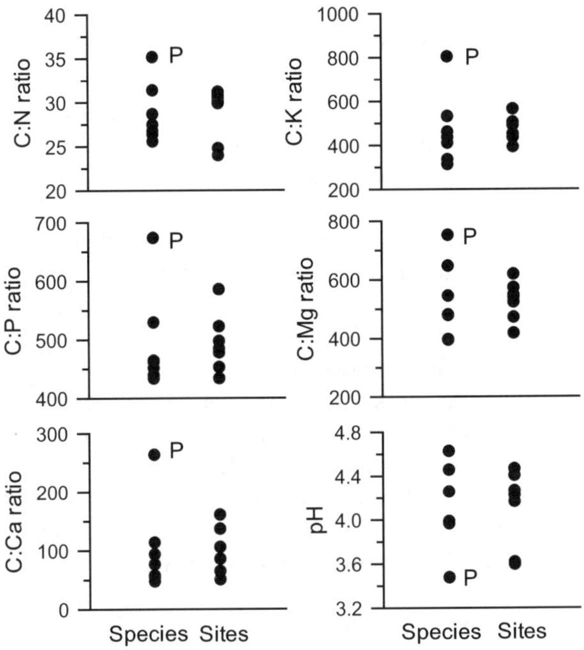

Figure 7. Range of mean values for pH and C:nutrient ratios in forest floors under each tree species and at each site in the Danish conifer tree species trial. Lodgepole pine is indicated with P. Adapted from Vesterdal and Raulund-Rasmussen (1998).

sites, indicating that site and species had similar influence. For %P, the average differences were 0.051% between species and 0.050% among sites. Spruce showed considerably more variation among sites than did pine (0.13 and 0.05, respectively for N; 0.084 and 0.017 for P).

These examples illustrate that both species and site influence forest floors, but the relative magnitude of the effects differs among studies. Of course, the relative magnitude of species and site influences detected in any study is largely determined by the sites and species selected for comparison (as demonstrated in the Danish trial), so the question is still largely unanswered. Nevertheless we may conclude from these examples that despite influences from other site factors, tree species have consistent and measurable effects on nutrient concentrations in litter and the resulting forest floors.

CONCLUSIONS

1. B.C. tree species have consistent effects on forest floor pH and nutrient concentrations. Species effects on forest floor net N mineralization rates are less consistent and predictable.

2. Cedar forest floors have high pH and Ca, high bacterial biomass and high proportion of NO_3, but do not have high N concentrations or rates of net N mineralization.
3. Hemlock forest floors have low pH and Ca, are dominated by NH_4 and have intermediate rates of net N mineralization.
4. Sitka spruce forest floors tend to be intermediate-to-high in most nutrients.
5. Douglas-fir forest floors have intermediate pH and concentrations of Ca and P but high N concentrations and net N mineralization rates.
6. Lodgepole pine forest floors tend to be low in pH, Ca and P, but variable in N concentration and net N mineralization rates.
7. Paper birch forest floors have high pH, Ca and Mg, but do not have particularly high rates of net N mineralization.
8. Forest floor nutrient concentrations are related to those in foliar litter but are not related to rate of decay of the litter.
9. Proportions of NH_4 and NO_3 in forest floors are more closely related to pH and Ca concentrations of the forest floors and their influence on bacterial and fungal populations than to N concentrations in litter and forest floor.
10. Actual nutrient concentrations in forest floors vary according to soil factors, but the nature of the effect (i.e. increase or decrease) of a given tree species does not change.

REFERENCES

Alban D H 1969 The influence of western hemlock and western redcedar on soil properties. Soil Sci. Soc. Amer. Proc. 33, 453-457.

Alban D H 1982 Effects of nutrient accumulation by aspen, spruce, and pine on soil properties. Soil Sci. Soc. Am. J. 46, 853-861.

Alriksson A and Eriksson H M 1998 Variations in mineral nutrient and C distribution in the soil and vegetation compartments of five temperate tree species in NE Sweden. For. Ecol. Manage. 108, 261-273.

Bastrup-Birk A, Hansen K, Ro-Poulsen H, Jørgensen B B, Mikkelsen T, Pilegaard K, Bille-Hansen J 2003. Biomasse og produktion. *In* Næringsstofkredsløb i skove - Ionbalanceprojektet. Ed. K Hansen. pp. 69-96. Forest and Landscape Research No 33-2003. Danish Forest and Landscape Research Institute, Hørsholm. [In Danish].

Bennett J N, Prescott C E, Barker J E, Blevins D P and Blevins L L. 2003. Long-term improvement in productivity and nutrient availability following fertilization and vegetation control on a cedar-hemlock cutover. Can. J. For. Res. 33:1516-1524.

Brandtberg P-O, Lundkvist H and Bengtsson J 2000 Changes in forest-floor chemistry caused by a birch admixture in Norway spruce stands. For. Ecol. Manage. 130, 253-264.

Gloaguen J C and Touffet J 1982 Evolution du rapport C/N dans les feuilles et au cours de la décomposition des litières sous climat atlantique. Le hêtre et quelques conifères. Ann. Sci. For. 39, 219-230.

Grayston S J and Prescott C E in press Microbial communities in forest floors under four tree species in coastal British Columbia. Soil Biol. Biochem.

Harmer R and Alexander I 1986 The effect of starch amendment on nitrogen mineralization from the forest floor beneath a range of conifers. Forestry 59, 39-46.

Harmon M E, Baker G A, Spycher G and Greene S E 1990 Leaf-litter decomposition in the *Picea/Tsuga* forests of Olympic National Park, Washington, U.S.A. For. Ecol. Manage. 31, 55-66.

Johansson M-B 1995 The chemical composition of needle and leaf litter from Scots pine, Norway spruce and white birch in Scandinavian forests. Forestry 68, 49-62.

Moore T R, Trofymow J A, Prescott C E, Fyles J, Titus B D and CIDET Working Group. In press. Patterns of carbon, nitrogen and phosphorus dynamics in decomposing foliar litter in Canadian forests. Ecosystems.

Ovington J D 1954 Studies of the development of woodland conditions under different trees. II. The forest floor. J. Ecol. 42, 71-80.

Prescott C E and Preston C M 1994 Nitrogen mineralization and decomposition in forest floors in adjacent plantations of western red cedar, western hemlock, and Douglas-fir. Can. J. For. Res. 24, 2424-2431.

Prescott C E, Thomas K D and Weetman G F 1995 The influence of tree species on nitrogen mineralization in the forest floor: Lessons from three retrospective studies. *In* Proceedings of the Trees and Soil Workshop. Eds. D J Mead and I S Cornforth. pp. 59-68. Lincoln University 28 February-2 March 1994. Lincoln University Press, Canterbury.

Prescott C E 1996 Influence of forest floor type on rates of litter decomposition in microcosms. Soil Biol. Biochem. 28, 1319-1325.

Prescott CE 2002 The influence of the forest canopy on nutrient cycling. Tree Physiol. 22, 1193-1200.

Prescott C E, Vesterdal L, Pratt J, Venner K H, de Montigny L M and Trofymow J A 2000a Nutrient concentrations and nitrogen mineralization in forest floors of single species conifer plantations in coastal British Columbia. Can. J. For. Res. 30, 1341-1352.

Prescott C E, Zabek L M, Staley C L and Kabzems R . 2000b Decomposition of broadleaf and needle litter in forests of British Columbia: influences of litter type, forest type, and litter mixtures. Can. J. For. Res. 30, 1742-1750.

Prescott C E, Chappell H N and Vesterdal L 2000c Nitrogen turnover in forest floors of coastal Douglas-fir at sites differing in soil nitrogen capital. Ecology 81, 1878-1886.

Prescott C E, Vesterdal L, Preston C M and Simard S W 2004 Influence of initial chemistry on decomposition of foliar litter in contrasting forest types in British Columbia. Can. J. For. Res. 34, 1714-1729.

Priha O and Smolander A 1999 Nitrogen transformations in soil under *Pinus sylvestris*, *Picea abies* and *Betula pendula* at two forest sites. Soil Biol. Biochem. 31, 965-977.

Raulund-Rasmussen K and Vejre H 1995 Effect of tree species and soil properties on nutrient-immobilization in the forest floor. Plant Soil 168-69, 345-352.

Saetre P 1998 Decomposition, microbial community structure, and earthworm effects along a spruce-birch soil gradient. Ecology 79, 834-836.

Saetre P, Brandtberg P-O and Lundkvist H 1999 Soil organisms and carbon, nitrogen and phosphorus mineralisation in Norway spruce and mixed Norway spruce - birch stands. Biol. Fertil. Soils 28, 382-388.

Stump L M and Binkley D 1993 Relationships between litter quality and nitrogen availability in Rocky Mountain forests. Can. J. For. Res. 23, 492-502.

Thomas K D and Prescott C E 2000 Nitrogen availability in forest floors of three tree species on the same site: the role of litter quality. Can. J. For. Res. 30, 1698-1706.

Turner D P and Franz E H 1985 The influence of western hemlock and western redcedar on microbial numbers, nitrogen mineralization, and nitrification. Plant Soil 88, 259-267.

Vesterdal L and Raulund-Rasmussen K 1998 Forest floor chemistry under seven tree species along a soil fertility gradient. Can. J. For. Res. 28, 1636-1647.

Chapter 3

NUTRIENT TURNOVER, GREENHOUSE GAS EXCHANGE AND BIODIVERSITY IN NATURAL FORESTS OF CENTRAL EUROPE

Sophie Zechmeister-Boltenstern[1], Evelyn Hackl[1], Gert Bachmann[*], Michael Pfeffer[1] & Michael Englisch[1]

[1]*Institut für Waldökologie und Boden, Bundesamt und Forschungszentrum für Wald, Seckendorff-Gudent Weg 8, A-1131 Wien, Austria.* [2]*Institut für Ökologie und Naturschutz, Althanstr. 14, A-1090 Wien, Austria.* E-mail: sophie.zechmeister@bfw.gv.at.

ABSTRACT

We measured microbial turnover of carbon (C) and nitrogen (N) in 12 natural forest reserves in Austria, and estimated potential emission rates of nitrous oxide (N_2O) and carbon dioxide (CO_2), and uptake rates of methane (CH_4). The community composition of soil microorganisms was investigated using PLFA (phospholipid fatty acid) analysis and molecular tools, and we examined the biodiversity of selected taxa of micro-, meso- and macrofauna. These characterizations of natural forests provide reference data for evaluating soil biology in managed, especially disturbed or damaged forests. Ecophysio-logical quotients were tested for their ability to make predictions about the carbon dynamics of forest soils. The 12 forests represented the six typical types in Central Europe: oak, beech, spruce-fir-beech, floodplain, and pine forests. Nitrogen turnover rates were high in moist soils with high pH. Nitrogen losses as nitrate or N_2O were small unless N deposition exceeded 30 kg ha^{-1} yr^{-1}. The fastest turnover of C and N occurred in the floodplain forests, based on microbial quotients, xylanase activity, the relative thickness of litter layer and ^{15}N abundance in the organic soil. Carbon turnover was slowest in the beech forests on acidic bedrock, and slow turnover may lead to the largest net C accumulation. Tree species had distinct effects on microbial communities, but high soil biodiversity in these natural forests may not be greater than in managed forests.

D. Binkley and O. Menyailo (eds.),
Tree Species Effects on Soils: Implications for Global Change, 31–49.

INTRODUCTION

Almost 50% of the total area of Austria is forested, and the forests are dominated by commercially valuable stands of Norway spruce (*Picea abies*). The few remaining forests that resemble the natural vegetation composition are located in forest reserves with restricted management. These natural forests are used as reference systems for evaluating silvicultural research on sustainable forest management. Natural forests are expected to have high biodiversity, where the structural richness of the habitat enables complex relationships between fauna, flora, and microflora. They also provide refugia for rare plants and animals found only in natural forest types. Austria had 180 of these forest reserves up to the year 2003. Most of these forests are privately owned, and owners are compensated by the government for loss of income associated with conservation status. The Ministerial Conference for the Protection of Forest Ecosystems (MCPFE) has launched a world-wide network of protected forest areas which should cover all major forest types (MCPFE and UNECE/FAO, 2003).

The sites selected for our investigation of soil conditions and communities were chosen by vegetation ecologists and soil scientists. The stands have developed under natural competition conditions with no management interventions. All sites were well documented with known forest history. Our set of sites spans gradients of environmental conditions as well as species composition, providing a realistic evaluation of the interactions of biotic and abiotic factors. This approach complements the "common garden" experiments, where the effects of tree species are compared in isolation from variations in environmental conditions. All investigated sites were within an area of 200 km in eastern Austria, representing a variety of vegetation types from subpannonic lowlands to the Bohemian massif.

Microbial nutrient turnover and greenhouse gas exchange of the forests provide key information about interactions between soils and trees. Soils support trees, and trees supply the growth substrate to the microbial populations in the soil via litter and root exudates. The microbial population developing under a certain forest type feeds back on tree growth via decomposition, nutrient release and immobilisation processes. Microbes mediate the turnover of greenhouse gases in the soil and hence determine whether a forest constitutes a net source or a net sink for greenhouse gases.

This chapter provides an overview of several studies of natural forests in Austria. We include some previously unpublished data, and draw conclusions in the context of tree and soil interactions and global climate change.

MATERIAL AND METHODS

Study sites and sampling design

Twelve natural forest stands were selected within the eastern part of Austria, featuring oak-hornbeam, woodruff-beech, acidic beech (beech forests on siliceous bedrocks with acidophilous ground vegetation), spruce-fir-beech, floodplain and Austrian pine forests. All stands were old growth forests, characterized by a natural tree species composition. An overview of the site characteristics and the geographical locations is given in Table 1. For the microbiological analyses, soils from each stand were sampled in spring and autumn of two successive years (Hackl et al., 2000; Hackl et al., 2004a) and for the DNA analysis another autumn sampling was performed (Hackl et al., 2004b).

Table 1. Site characteristics and soil chemical properties of 12 natural forest stands (data from Hackl 2004c). Soil chemical data represent means from 10 subsamples ± standard error.

Forest type	Site	Elevation (m)	Temperature (°C)	Precipitation (mm)	Soil type	Geology	Organic C (%)	Total N (%)	C:N	pH
Oak-hornbeam	JE	325	8.8	643	Dystric Planosol	Sandstone/ Claystone	5.04 ±0.34	0.22 ±0.01	23.4 ±0.6	4.5 ±0.2
	K	270	8.7	593	Calcaric Planosol	Micashist/ Limestone	4.23 ±0.21	0.20 ±0.01	21.0 ±0.4	5.4 ±0.3
Woodruff-beech	JB	320	8.8	643	Dystric Planosol	Sandstone/ Claystone	4.38 ±0.31	0.19 ±0.01	22.5 ±0.8	5.1 ±0.3
	Kl	510	7.6	768	Dystric Cambisol	Sandstone/ Claystone	4.36 ±0.32	0.33 ±0.02	13.1 ±0.4	4.1 ±0.0
Acidic beech	D	500	7.6	613	Dystric Cambisol	Gföhl gneiss	9.45 ±0.34	0.35 ±0.04	26.9 ±0.7	4.6 ±0.2
	S	550	7.4	631	Dystric Cambisol	Gföhl gneiss	7.03 ±0.56	0.30 ±0.02	23.5 ±0.5	4.0 ±0.0
Spruce-fir-beech	R	1035	5.5	1759	Chromic Cambisol	Dolomite	16.00 ±2.38	0.94 ±0.15	17.1 ±1.1	4.9 ±0.4
	N	995	5.8	1262	Stagnic Luvisol	Sandstone	6.46 ±0.63	0.38 ±0.03	16.9 ±0.7	4.0 ±0.1
Floodplain	M	160	9.7	582	Calcaric Fluvisol	Recent clay	5.46 ±0.34	0.47 ±0.02	11.7 ±0.4	7.2 ±0.0
	B	160	9.7	534	Calcaric Fluvisol	Recent clay	3.92 ±0.13	0.23 ±0.01	17.2 ±0.5	7.4 ±0.0
Austrian pine	St	640	7.0	668	Rendzic Leptosol	Dolomite	16.99 ±2.69	0.61 ±0.09	28.0 ±1.0	7.4 ±0.1
	Me	475	8.2	554	Rendzic Leptosol	Dolomite	9.64 ±0.63	0.26 ±0.02	37.0 ±2.0	7.4 ±0.0

Soil chemical and microbiological analyses

The soil samples were stored at -20 °C and prior to analysis were sieved to 2 mm. Soil pH was measured in $CaCl_2$-solution by glass electrode (soil: 0.01 M $CaCl_2$ = 1:5), and total soil carbon (C_t) and nitrogen (N_t) concentrations were analysed after dry combustion. Soil samples were weighed into tin capsules for isotope ratio mass spectrometry (IRMS); the continuous-flow IRMS system consisted of an elemental analyser interfaced to the gas isotope ratio mass spectrometer (Delta Plus, Finnigan MAT). Reference gas was calibrated to the atmospheric N_2 standard using IAEA (International Atomic Energy Agency, Vienna, Austria) reference material. The standard deviation of repeated measurements of a laboratory standard was 0.15‰ for $\delta^{15}N$ (Wanek and Arndt, 2002).

Extractable ammonium (NH_4^+-N) and nitrate (NO_3^--N) were determined in 2 *M* KCl-extracts as described in Hackl et al. (2004a). Low molecular weight organic compounds of the soil organic matter (sugars, amino acids) were analyzed in extracts of 60% acetone according to Hackl et al. (2000).

Nitrogen mineralization potential and urease activity were measured according to Kandeler (1996a,b). Microbial biomass was determined as ninhydrin–reactive N by a chloroform fumigation-extraction technique and was also measured by substrate-induced respiration (SIR) with glucose-amendment (Hackl et al., 2000). From SIR-data the microbial quotient as well as the metabolic quotient were calculated according to Insam (1996).

Nitrate translocation into deeper soil horizons was measured by applying a soil core-IER method (Binkley and Hart 1989). Resin bags containing ion exchange resins (IER, 4 mg Dowex 1X8 pract. and 4 mg Amberlite IR 120 pract.) were placed under soil cores (10.5 cm depth, 9 cm diameter) which were taken from each sampling point. The resin bags and the soil cores were enclosed in close-fitting PVC tubes and returned to the same holes from which the cores had been taken. Two months later the resin bags were removed and extracted two-fold with 1.59 *M* HCl. After neutralisation with NaOH, NO_3^--N was measured colorimetrically as NO_2^--N after reduction with copper sheathed granulated zinc (Kandeler 1996d).

Gas flux measurements

Nitrous oxide (N_2O) and carbon dioxide (CO_2) production and methane (CH_4) degradation were measured by incubating intact soil cores, which had been stored at 4 °C for not more than 4 days prior to analysis. Each soil core was enclosed in a 500 ml gas-tight glass-jar and kept at 25 °C for 24 hours in 1997 and for 6 hours in 1998, after ensuring that increases in concentration were linear. At the beginning and at the end of the incubation period gas samples were taken. Head space air of the vessels was transferred into evacuated 10 ml glass vials with a gas-tight syringe, and stored under water until analysis. Gas samples were analyzed by gas chromatography using a

^{63}Ni electron capture detector for N_2O (injector 120 °C, detector 330 °C, oven 30°C, carrier gas N_2; Zechmeister-Boltenstern, 1994), TCD for CO_2 (injector 120 °C, detector 150 °C, oven 80°C, carrier gas Helium) and a FID for CH_4 (Rigler and Zechmeister-Boltenstern, 1999).

Microbial communities were analyzed by phospholipid fatty acid (PLFA) extraction and analysis was done as described by Frostegård et al. (1996), with some modifications as described by (Hackl et al., 2004b). Bacterial communities in the forest soils were also analyzed by extracting bacterial community DNA from soil samples. After PCR amplification with 16S rRNA gene primers, amplicons were subjected to terminal restriction fragment length polymorphism (T-RFLP) analysis (Hackl et al., 2004b). For selected forest sites clone libraries were constructed using the same DNA preparations as for T-RFLP analysis (Hackl et al., 2004b).

Statistical analysis

The SAS System for Windows, Version 8 was used for statistical processing of data. Analysis of variance and Tukey test were applied to test for significant differences in ecophysiological parameters among forest types. Correlations between N_2O production rates and soil chemical and microbiological variables were determined using Spearman's rank correlations. Stepwise linear regression analyses were used to explore the relationships between nitrogen transformation rates. Cluster analysis of PLFAs were conducted using the program S-Plus after checking normal distribution of data. The average linkage procedure was selected. SPSS was used for the creation of Chernoff faces, which is a method of visualizing similarities and differences among multidimensional data. The program displays faces that are described by 15 facial characteristic parameters such as head eccentricity, eye eccentricity, pupil size, nose size, mouth shape, eye spacing, eye size, and mouth length.

RESULTS AND DISCUSSION

Nitrogen turnover

Plant roots release low molecular weight organic compounds, such as sugars and amino acids, which serve as carbon and nitrogen sources for microorganisms. The concentrations of 15 different amino acids in 12 natural forest soils are illustrated in Figure 1. Each parameter is represented by a number, namely the concentration of an amino acid, which is designed as a feature of the faces shown. For example the length of the nose indicates the amount of the amino acid glycine, whereas the shape of the mouth is determined by the amino acid citrullin. The figure shows forests with similar

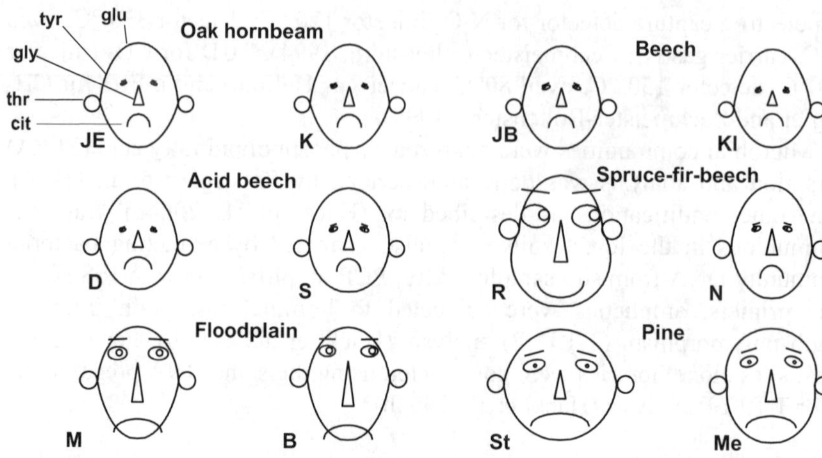

Figure 1: Chernoff faces reflecting concentrations of 15 amino acids (data from Hackl, 2000). Face breadth = aspartic acid, ear position = threonine, face length = serine, upper face eccentricity = glutamic acid, lower face eccentricity = proline, nose length = glycine, mouth center = alanine, mouth shape = citrulline, mouth width = valine, eye level = methionine, eye spacing = isoleucine, eye tilt = leucine, eye eccentricity = tyrosine, eye length = phenylalanine, pupil position = sum of amino acids.

amino acid composition. One forest stands out from the others, with a very different and well balanced amino acid composition: the spruce-fir-beech forest. This forest is called "Urwald Rothwald", and has never been managed in historical times. Microbial biomass and the concentration of sugar were twice the amounts of other soils (Hackl et al., 2000). This forest is situated in a remote valley in the limestone Alps and receives annually almost 1800 mm precipitation. The faces in Figure 1 show similarities in amino acid composition among the oak-hornbeam, beech and acidic beech forests, with high concentrations of leucine, serine and glutamine. In the pine forests traces of proline, methionine and tyrosine could be detected, which were not found in the other soils. These amino acids are produced by plants as osmoregulators in order to protect them from drought stress.

Three aspects of N-turnover are depicted in Figure 2: potential N-mineralization, N-storage in the microbial biomass, and activity of urease (an enzyme that releases ammonium (NH_4^+)). All three variables correlated positively with total soil N ($p<0.05$). Microbial biomass could be calculated from N mineralization potential and total soil N ($r^2=0.730$), urease activity was a function of N mineralization potential and total soil N ($r^2=0.653$) (Hackl et al., 2004). Turnover rates of N varied across forest types, partially in response to soil moisture and pH. Similar relationships have been found in forest soils close to New York (Templer et al., 2003). The highest rates of N turnover were measured in soils of the virgin forest "Rothwald", the virgin spruce-fir-beech forest that had the particular amino acid composition in Figure 1. The lowest N-turnover was

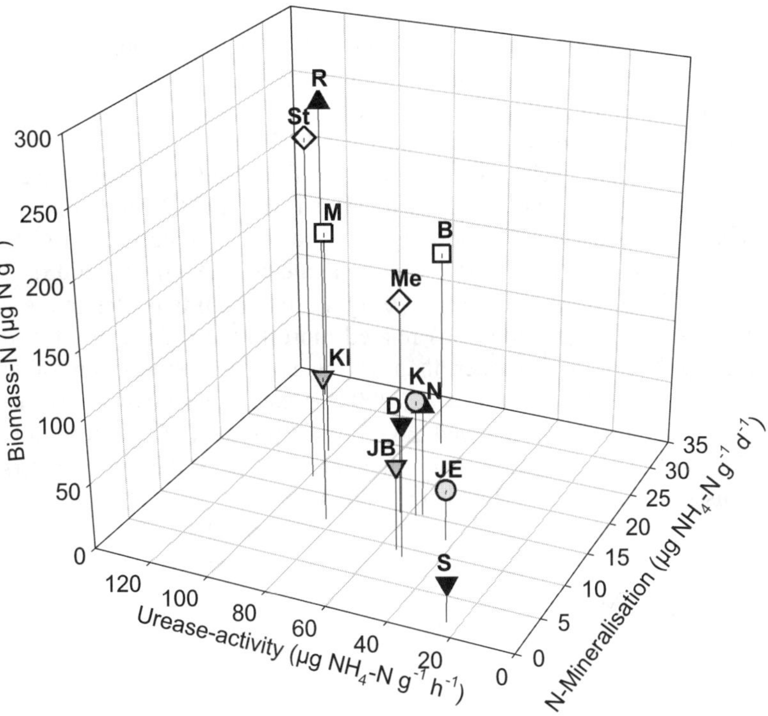

Figure 2: Potential nitrogen turnover at 12 natural forest sites (data from Hackl et al., 2004a).

measured in an acidic beech forest on silicate bedrock. As can be seen in Figure 3 total N, microbial N and extractable mineral N in the spruce-fir-beech forest "Rothwald" exceeded concentrations in the acidic beech forest "Saubrunn" by three- to fourfold. The ratio of nitrate to ammonium was very high (1:477) in the acidic beech forest, and moderate in the spruce-fir-beech forest with 1:5. Mineralisation rates differed strongly, whereas N_2O-emissions were low in both forests. Although N-turnover rates differed greatly between these forests, neither seemed prone to high losses of N_2O or nitrate, and both forests showed a fairly closed N-cycle. Other forests, including the floodplain forests, had slower but less closed N-cycles. This could be seen from intermediate N turnover rates but comparably high N_2O and nitrate leaching values (Hackl et al., 2004a; Zechmeister-Boltenstern et al., 2000).

Net greenhouse gas exchange

Carbon dioxide exchange of forests has received much attention although the soil compartment is still not fully understood (Giardina and Ryan, 2000; Trumbore et al., 1996; Valentini et al., 2000). Even less is known about nitrous oxide and methane turnover which can significantly affect the net greenhouse gas exchange. According to the Third Assessment Report of the IPCC (Intergovernmental Panel on Global Climate Change), one molecule of methane has the global warming potential of 23 molecules of CO_2, and one molecule of N_2O counts as 296 equivalents of CO_2 (IPCC, 2003). We found highest N_2O emissions (up to 170 μg m^{-2} h^{-1}) from two sites in the vicinity of Vienna; these high emissions related more to the high rates of N deposition (35 kg N ha^{-1} yr^{-1}) rather than to the species composition of the forests (Zechmeister-Boltenstern et al., 2002). Beside these two N-affected forests, the floodplain forests showed the highest emissions of N_2O (60 μg N_2O-N m^{-2} h^{-1}). Soil moisture and nitrate concentrations were the main drivers for N_2O emission within sites (Hackl et al., 2004a).

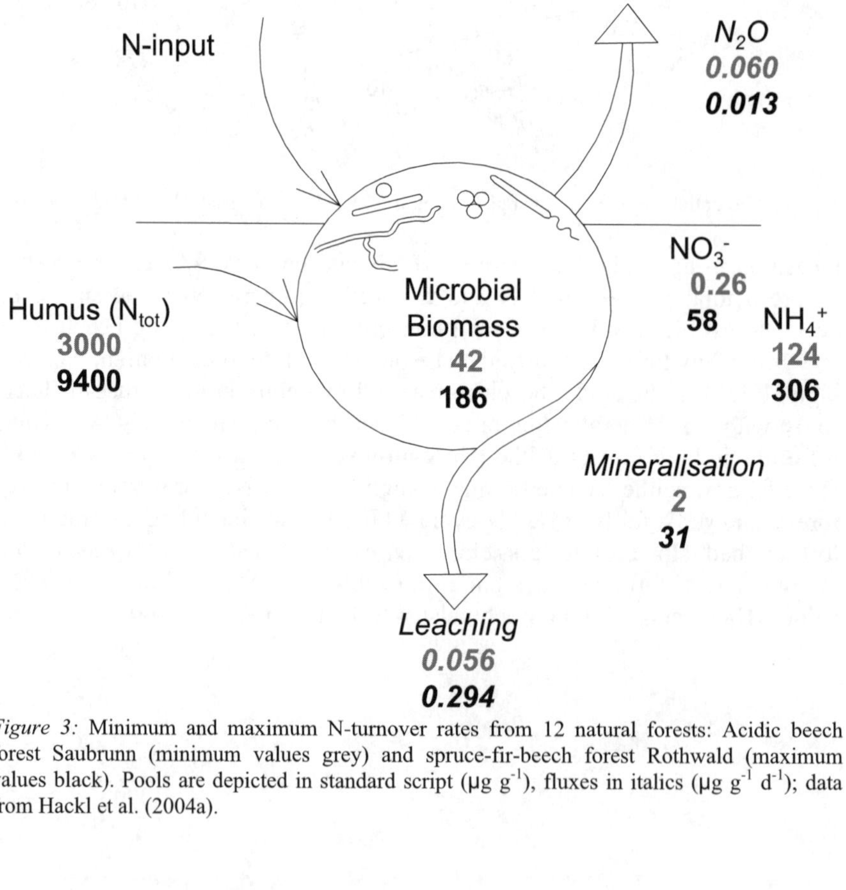

Figure 3: Minimum and maximum N-turnover rates from 12 natural forests: Acidic beech forest Saubrunn (minimum values grey) and spruce-fir-beech forest Rothwald (maximum values black). Pools are depicted in standard script (μg g^{-1}), fluxes in italics (μg g^{-1} d^{-1}); data from Hackl et al. (2004a).

Table 2: N_2O-emissions, CH_4-uptake (negative values), CO_2-emissions, and net greenhouse gas exchange calculated as CO_2-equivalents according to the Third Assessment Report (IPCC, 2001)

	N_2O	CH_4	CO_2	Greenhouse gas exchange
Site		$\mu g\ CO_2\text{-}E.m^{-2}.h^{-1}$		
JE	78891	-54	66159	144995
K	18392	-26	61558	79923
JB	38246	-71	57730	95905
Kl	10374	-95	35643	45922
D	1333	-37	46133	47428
S	685	-101	56984	57567
R	1980	-100	67513	69393
N	14239	-35	50570	64774
M	14000	-181	18180	31999
B	2121	-27	29072	31166
St	484	-16	24750	25218
Me	511	-61	11834	12284

Atmospheric methane is typically degraded (oxidation) in upland forest soils, and emitted from forests on peatland or swamps (Ambus and Christensen, 1995). Deciduous forests may take up more methane than conifer forests (Butterbach-Bahl and Papen, 2002). A comparison of 16 sites showed highest methane uptake (oxidation) rates in floodplain forests with high microbial biomass (Zechmeister-Boltenstern and Nikodim, 1999). Methane uptake in forest soils was reduced by increasing inputs of nitrogen and CO_2-concentrations (Rigler and Zechmeister-Boltenstern, 1999). Methane uptake rates in the natural forests ranged between 3 and 30 $\mu g\ CH_4\ m^{-2}\ h^{-1}$. Soil microbiological contribution to the global sink strength is about 6% for CH_4 (Crutzen, 1991). However, methane uptake rates had little effect on the net greenhouse gas exchange of our investigated forests. Greenhouse gas exchange of forest soils was mainly determined by N_2O and CO_2 release (Table 2). The lowest net greenhouse gas release occurred from floodplain and pine forests. These results are potential turnover rates as they result from laboratory incubation experiments. By investigating 30 individual cores for each measurement we may have covered some of the spatial variability, but temporal variability remained large and could not be accounted for (see Butterbach-Bahl and Kiese, this volume).

Decomposition and C storage

Ecophysiological quotients allow inferences about humus quality and decomposition processes, as well as the physiological state of the soil microflora. In Figure 4A the microbial quotients (Cmic:Corg) of the sampling sites are compared. The concept of the microbial quotient as an indicator of successional stage was introduced by Insam and Domsch (1988) and reviewed by Wardle et al. (1995). Here, the microbial quotient is an indicator for substrate quality on top of successional stage. A high microbial quotient means that the soil organic matter can sustain a large microbial community whereas a low microbial quotient would mean that the organic carbon is less palatable for the microbiota. Moreover, it is an indicator of carbon accumulation (Insam, 1988). The floodplain forests had a significantly higher microbial quotient than most other forest types. Floodplain forests, which are still affected by regular floods, may also be seen as forest types of an early successional stage, as compared to forests on higher terraces (Kaye et al., 2003). However this does not necessarily imply that these soils are accumulating organic matter. The high microbial quotient in our floodplain soils occurred with typical microflora of zymogenous or r-strategic organisms, which are able to quickly exploit easily decomposable carbon substrates (Hackl, 2004c; Killham, 1994; Paul and Clark, 1996). Fresh litter is decomposed immediately in the floodplain forests, which have a large fraction of bacteria (Figure 4B; Figure 5). In contrast, the acidic beech forests and the pine forests have a more recalcitrant carbon substrate sustaining a comparably small mass of microorganisms which are known as autochthonous or K-strategists. These microorganisms grow slowly, can survive for a long time under starving conditions and are able to break down recalcitrant substrate. The microbial community in the pine forest soils had a high fraction of fungi and actinomycetes (Figure 5; Hackl et al., 2004c). A high fungal to bacterial ratio in pine forests has also been measured in boreal ecosystems (Šantrůčková et al., 2003). A further indication for recalcitrant C substrate is the high xylanase activity in the pine soils (Figure 4C). Xylan is a component of plant cell walls, it is slowly decomposed and requires specialized microorganisms, such as basidiomycete fungi, for breakdown (Devlin and Witham, 1983; Paul and Clark, 1996).

The metabolic quotient ($qCO_2 = CO_2$ evolved g^{-1} microbial biomass) is an indicator of the respiration level of microbes (Insam, 1996). A low metabolic quotient means that the microbial biomass is in a “sleepy state” respiring at a low level. This behaviour is also said to be “efficient” in contrast to a high metabolic quotient which has been observed as a stress phenomenon (Klose et al, 2003; Šantrůčková and Straškraba, 1991). All natural forest soils had a low metabolic quotient of 2-4 except for the pine forests, which reached 18-19. Soils of the pine forests are shallow and dry. Soil moisture was usually at least 10% less than in the other forest soils. On top of a stress phenomenon a high metabolic quotient may also indicate a different microbial community (Šantrůčková and Straškraba, 1991). Indeed, the pine forests contained a high abundance of actinomycetes, which are known for their high respiration activity and occur in warm and dry soils (Figure 5; Hackl et al., 2004c).

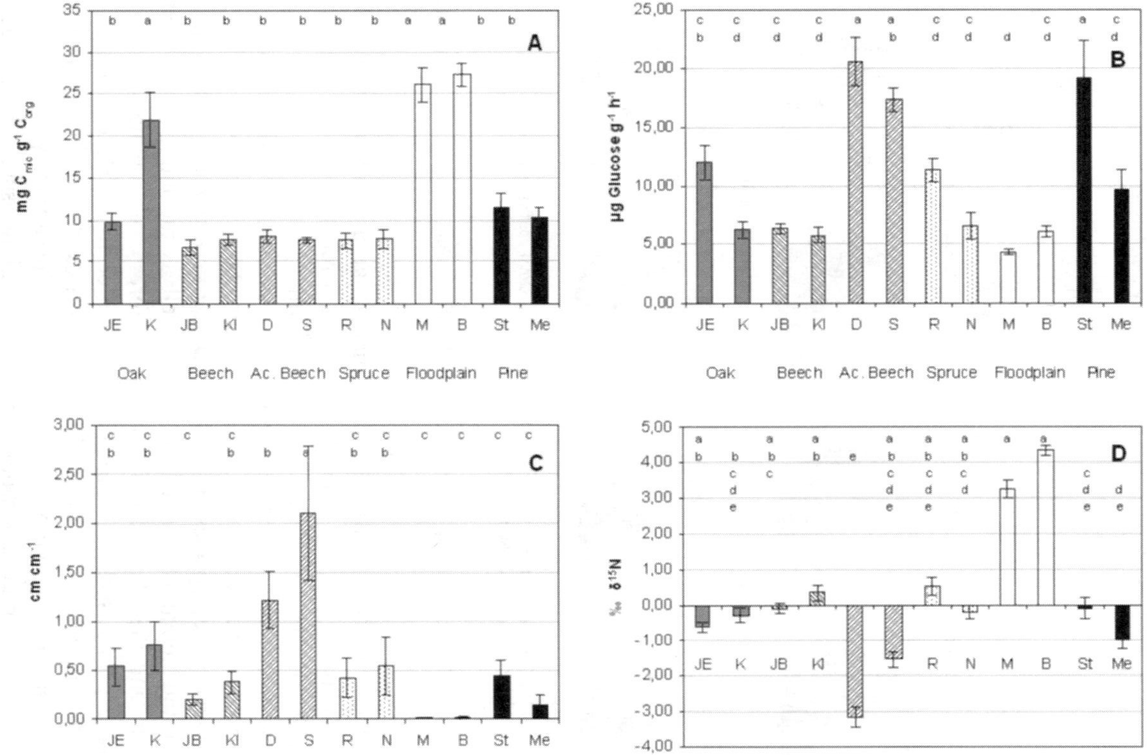

Figure 4: Ecophysiological Parameters providing information on humus quality and decomposition. A: Metabolic quotient B: Thickness forest floor/organic mineral horizon C: Xylanase activity D: $\delta^{15}N$ of total soil nitrogen.

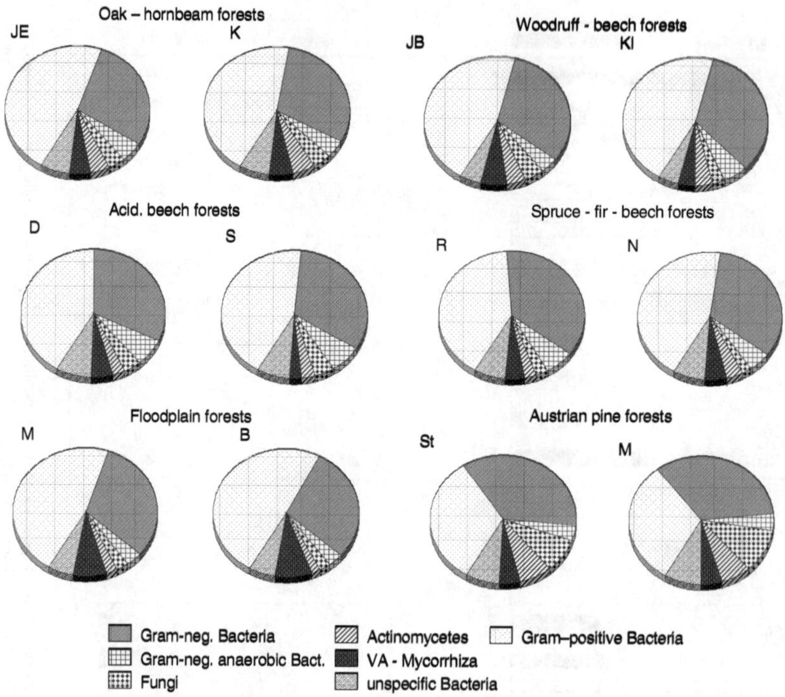

Figure 5: Microbial community composition of 12 natural forest sites according to PLFA analysis.

Another indication for organic matter accumulation rather than degradation is the ratio of forest floor thickness relative to organic mineral soil (Figure 4B; Nestroy et al., 2000). A thick litter layer was accumulating in the acidic beech forests, whereas all litter was comminuted (and perhaps decomposed) in the floodplain forests each year. This can be explained by the composition of tree species like ash (*Fraxinus excelsior* L.), willow (*Salix alba* L.) and poplar (*Populus alba* L.) as well as the nitrophilous ground vegetation and the likely differences in the forest soil communities. Litter from these plants is easily decomposed (Kaye et al., 2003; Prescott et al., 2000). We speculate that these floodplain forests do not accumulate soil carbon and nitrogen in the organic mineral soil. These forests can be seen as transient ecosystems, which would only reach a higher successional stage if they were cut off from riverine water. In this case different tree species would settle in and the soils might start to accumulate organic matter. Carbon accumulation might also happen in deeper horizons via deep roots, which have been shown to reach down to 2.9 m in floodplain forests (Kutschera and Lichtenegger, 2002).

The ideas described above are consistent with $\delta^{15}N$ natural abundance patterns (Figure 4D). Low $\delta^{15}N$ values in soil humus indicate the presence of undecomposed plant material whereas a high $\delta^{15}N$ typically indicate old, highly processed humus. Values of $\delta^{15}N$ typically increase with depth in soil (Haberbauer et al., 2002; Heil et al., 2000), as a long-term outcome of microbial discrimination and losses of ^{14}N. Microbial enzymes tend to process the lighter ^{14}N, leading to higher losses of this isotope (NO, NO_2, N_2O gases, or nitrate leaching) (Högberg, 1997; Koba et al., 1998). Our measurements showed highest $\delta^{15}N$ values in floodplain forests (Fig 4D), indicating higher rates of N loss. The acidic beech forests had negative $\delta^{15}N$, indicating more N derived from recent plant materials. These ecophysiological soil measurements are consistent with the acidic beech forest soils accumulating C and N, while this is not evident in the organic mineral soil from the floodplain forests.

Biodiversity

All the natural forests showed great diversity of soil organisms, microbes, as well as micro-, meso- and macrofauna (Drapela, 2003; Foissner et al., 2004; Waitzbauer and Zabransky, 2004; Zechmeister-Boltenstern, 2003). More than one thousand species of soil organisms have been identified, including 35 new species (DIANA project: http://bfw.ac.at/300/2197.html). These were protozoa (Foissner et al., 2004), as well as one springtail (Pomorski et al., 2003) and one spider (Milasowszky personal communication). Earthworms occurred only in the floodplain forests, where the annual litterfall was comminuted into the mineral soil each year. Oribatid mites were most abundant in acidic beech forest soils (Laibl and Bruckner, personal communication). The floodplain forests had very high species richness of many soil taxa, perhaps resulting from the recurring floods. Many rare and new species were found in pine forests. The spruce-fir-beech forests contained carabid beetles and spiders which are specialized on old-growth forests. This high biodiversity in all the natural forests may not be greater than that of managed forests. For example, spider diversity was significantly lower in the natural forests than in 89 managed forests in Central Europe (Milasowszky personal communication). In managed forests many common and ubiquitous spiders are found due to edge effects and disturbances. The species found in natural forests were specialized forest spiders, which may only survive under the conditions provided in protected areas.

Microbes contribute the largest fraction to soil biomass. Microbial biodiversity and community composition can be determined by PLFAs (Phospholipid fatty acids) analysis. Cluster analysis groups forests according to similiarity in the overall PLFA profiles (Figure 6). The PLFA profile in the pine forests differed the most from the other forest types, and the floodplain forests also showed a distinct microbial community composition. The microbiota of oak-hornbeam forests were similar to those of the beech forests, whereas the spruce-fir-beech forests and acidic beech forests were less similar to each other.

Pine forests were rich in actinomycetes and the fungal biomass (Figure 5), perhaps resulting from the recalcitrant pine needle litter which fungi can utilize better than other microorganisms. An investigation of 13 forests across Europe revealed a close similarity of microbial communities in mediterranean pine forests (*Pinus pinaster*, Italy) to those in boreal pine forests (*Pinus sylvestris*, Finland) (Zechmeister-Boltenstern et al., 2004). In spite of the climatic differences pine needles supported similar functional groups of microorganisms which distinguished them from other forest types.

The floodplain forests contained a large portion of gram positive bacteria, as well as arbuscular mycorrhizal fungi. Bacteria are known to dominate at neutral pH (Paul and Clark, 1996). Arbuscular mycorrhizae occur in symbiosis with herbal vegetation, which is abundant in floodplain forests, as well as in symbiosis with tree species such as ash and poplar. The total amount of PLFAs correlated well with microbial biomass determined by chloroform fumigation extraction, indicating both methods provide the similar estimates of microbial pool sizes. The measurements of spring and autumn samples revealed similar results, indicating that the microbial community had not changed significantly between seasons (Hackl et al., 2004c).

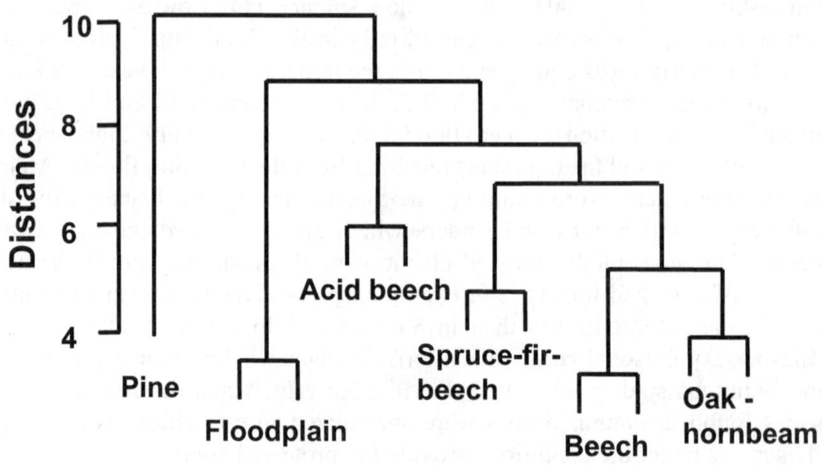

Figure 6: Dendrogram of individual sampling sites according to microbial communities as determined by PLFA analysis (Cluster analysis in S-Plus, average linkage procedure).

Soil DNA was extracted for a closer view on bacterial populations; 16S rRNA genes were amplified and subjected to terminal restriction fragment length polymorphism analysis. Clone libraries were created and taxonomic assignments were made by comparison with existing databases (Hackl et al., 2004b). The bacterial communities in pine forests were compositionally distinct from those in oak and spruce-fir-beech forests, as we found for the overall microflora. High –G+C gram positive bacteria (mainly belonging to the *Actinobacteria* group) prevailed in the pine forest. High –G+C gram positives were also most prevalent in 16S rRNA clone libraries from mineral forest soil samples in British Columbia, where lodgepole pine (*Pinus contorta* Douglas) was a major tree species (Axelrood et al., 2002). In oak and spruce-fir-beech forests, representatives of the *Holophaga-Acidobacterium* group were most abundant. Only a few cultivated representatives of the *Holophaga-Acidobacterium* group exist, and these bacteria have never been detected by classical methods. Only recently and after the introduction of molecular methods have they been acknowledged as an important bacterial group. Among forest soils, those under pinyon pine-juniper woodlands have been reported to contain acidobacteria in highest numbers (Dunbar et al., 1999; Hackl et al., 2004b).

CONCLUSIONS

Our investigations showed an influence of forest type on the composition and function of the soil microbial communities. Abiotic factors such as bedrock, moisture and nitrogen deposition were also important in nutrient turnover rates and the emission of greenhouse gases. Ecophysiological quotients could reveal information about humus decomposition and carbon storage, which was not evident from primary data.

When reviewing tree-soil interactions in the light of global climate change, two viewpoints have to be considered. First, how do different forest types affect global climate change? Should management schemes aim to maximize wood production or to conserve the soil carbon? In the first case fast growing tree species with easily decomposing litter and a narrow C:N ratio have to be favoured. This could be species of floodplain forests, such as poplar. Poplar plantations are already used as so called "Kyoto forests" planted for maximum C sequestration in the tree biomass (Leip et al., 2003). In the second case tree species with high lignin concentrations in the litter, and slow microbial decomposition rates (Lovett et al., 2004; Thomas and Prescott, 2000) may maximise carbon conservation. Drainage of forests should then be avoided especially on highly organic soils (Roulet, 2000). In the long term, C sequestration and storage in old-growth forests may be considered more important than short time carbon turnover. This potential must not be underestimated, for example it has been calculated that Siberian forests reassimilate 90% of European fossil fuel emissions (Schimel et al., 2001).

Second, how will global climate change affect forests? Forest types will clearly react differently to changes in climate, and we can offer some speculation in the absence of strong experimentation. Drought probably hampers decomposition in some forests (such as Austrian pine forests), and increased precipitation might increase decomposition. Decomposition in other forests (such as the acidic beech forests) may or may not be limited by low pH, with less response to changes in precipitation. A rise in temperature is most likely to affect soils with nutrient rich litter and organic mineral soils, as found in spruce-fir-beech forests in limestone Alps. These forests have a short vegetation period and longer growing seasons might therefore enhance decomposition rates and losses of soil carbon. The pattern of climate change would also be important, with different responses to slow or rapid changes.

We recommend that afforestation programs should take into account the potential natural vegetation of a site (Bohn et al., 2003; Mayer, 1984; Ricotta et al., 2002). Many people would assume that indigenous tree species will be supported by the native soil community, finding compatible symbiotic partners (including mycorrhizal fungi). Hence their connection to the soil community could be considered as tightly woven, where the fitness of the tree is enhanced by its long-term-effect on soils. This view is a slight adaptation of the concept by Binkley and Giardina (1998), who considered tree-microbe interactions more frayed. In addition, indigenous trees are adapted to prevalent climate and soil conditions and may therefore be resilient to various stress effects (Grabherr et al., 1995). Our study suggests that natural forest soils have relatively closed nutrient cycles with small losses of nitrate and nitrous oxide. They may be considered as beneficial for the environment and the conservation of site-specific biodiversity

ACKNOWLEDGEMENTS

This work was financially supported by the Federal Ministry for Agriculture and Forestry, Environment and Water management. We thank R. Jandl for helpful comments on the manuscript and the Kitzler sisters for editing.

REFERENCES

Ambus P and Christensen S 1995 Spatial and seasonal nitrous oxide and methane fluxes in Danish forest- , grassland-, and agroecosystems. J.Environ. Qual. 24, 993-1001.

Axelrood P E, Chow M L, Radomski C C, McDermott J M and Davies J 2002 Molecular characterization of bacterial diversity from British Columbia forest soils subjected to disturbance. Can. J. Microbiol. 48, 655-674.

Binkley D and Giardina C 1998 Why do tree species affect soils? The warp and woof of tree-soil interactions. Biogeochemistry 42, 89–106.

Binkley D and Hart S C 1989 The components of nitrogen availability assessments in forest soils. Adv. Soil Sci. 10, 57-112.

Bohn U, Gollub G, Hettwer C, Neuhauslova Z, Schlueter H and Weber H 2003 Karte der natuerlichen Vegetation Europas. Map of the natural vegetation of Europe. Federal Agency for Nature Conservation, Bonn.

Butterbach-Bahl K and Papen H 2002 Four years continuous record of CH_4-exchange between the athmosphere and untreated and limed soil of a N-saturated spruce and beech forest ecosystem in Germany. Plant Soil 240, 77-90.

Crutzen P J 1991 Methane's sinks and sources. Nature 350, 380-381.

Devlin R M and Witham F H 1983 Plant Physiology. PWS Publishers, California. 577 pp.

Drapela T 2003 Ground beetle communities of two floodplain forests in eastern Austria: indicating natural and dynamic conditions? 33. Jahrestagung Ges. f. Ökologie 33, 333 pp.

Dunbar J, Takala S, Barns S M, Davies J A and Kuske C R 1999 Levels of bacterial community diversity in four arid soils compared by cultivation and 16S rRNA gene cloning. Appl. Environ. Microbiol. 65, 1662-1669.

Foissner W, Berger H, Xu K and Zechmeister-Boltenstern S 2004 A huge, undescribed soil ciliate (Protozoa: Ciliophora) diversity in natural forest stands of Central Europe. Biodiversity and Conservation. In press.

Frostegård Å, Tunlid A and Bååth E 1996 Changes in microbial community structure during long-term incubation in two soils experimentally contamined with metals. Soil Biol. Biochem. 28, 55-63.

Giardina C P and Ryan M G 2000 Evidence that decomposition rates of organic carbon in mineral soil do not vary with temperature. Nature 404, 858-861.

Grabherr G, Koch G, Kirchmeir H and Reiter K 1995 Hemerobie österreichischer Waldökosysteme – Vorstellung eines Forschungsvorhabens im Rahmen des österreichischen Beitrags zum MAB- Programm der UNESCO. Zeitschrift für Ökologie und Naturschutz 4, 131-136.

Haberbauer G, Gerzabek M H and Krenn A 2002 Nitrate dynamics in an alpine forest site (Mühleggerköpfl). O and N stable isotope analysis in natural water samples. Environmental Science and Pollution Research. Special Issue 2, 37-41.

Hackl E, Zechmeister-Boltenstern S and Bachmann G 1999 Mikrobielle Umsatzraten in Böden von Naturwäldern. Mittlgn. Dtsch. Bodenkundl. Gesellsch. 89, 241-244.

Hackl E, Bachmann G and Zechmeister-Boltenstern S 2000 Soil microbial biomass and rhizosphere effects in natural forest stands. Phyton 40, 83-90.

Hackl E, Pfeffer M, Donat C, Bachmann G and Zechmeister-Boltenstern S 2004a Microbial nitrogen turnover in soils under different types of natural forest. For. Ecol. Manag. 188, 101-112.

Hackl E, Zechmeister-Boltenstern S, Bodrossy L and Sessitsch A 2004b Comparison of diversities and compositions of bacterial populations inhabiting natural forest soils. Appl. Environ. Microbiol. 70, 5057-5065.

Hackl E, Pfeffer M, Donat C, Bachmann G, Zechmeister-Boltenstern S 2004c Composition of the microbial communities in the mineral soil under different types of natural forest. Soil Biol. Biochem. In press.

Haertel E, Zechmeister-Boltenstern S and Gerzabek M 2002 Gaseous nitrogen losses from a forest site in the north Tyrolean limestone alps. Environmental Science and Pollution Research. Special Issue 2, 23 – 30.

Heil B, Ludwig B, Flessa H and Beese F 2000 ^{13}C and ^{15}N distributions in three spodic dystric cambisols under beech and spruce. Isotopes Environ. Health Stud. 36, 35-47.

Högberg P 1997 ^{15}N natural abundance in soil-plant systems. New Phytol. 137, 179-203.

Insam H and Domsch K H 1988 Relationship between soil organic carbon and microbial biomass on chronosequences of reclamation sites. Microbial Ecol. 15, 177-188.

Insam H and Öhlinger R 1996 Ecophysiological parameters. *In* Methods in Soil Biology. Eds. F Schinner, R Öhlinger, E Kandeler, R Margesin. pp. 306-309. Springer Verlag, Berlin.

IPCC 2001 Climate Change 2001: The Scientific Basis. Contribution of Working Group I to the Third Assessment Report of the Intergovernmental Panel on Climate Change Eds. J T Houghton, Y Ding, D J Griggs, M Noguer, P J van der Linden, X Dai, K Maskell, and C A Johnson. pp. 881. Cambridge University Press, Cambridge.

Jones R J A, Hiederer R, Rusco E, Loveland P J and Montanarella L 2004 The map of organic carbon in top soils in Europe, Version 1.2 September 2003: Explanation of Special Publication Ispra 2004 No.72 (S.P.I.04.72). European Soil Bureau Research Report No.17, EUR 21209 EN, 26pp. and 1 map in ISO B1 format. Office for Official Publications of the European Communities, Luxembourg.

Kandeler E 1996a Ammonium. *In* Methods in Soil Biology. Eds. F Schinner, R Öhlinger, E Kandeler and R Margesin. pp. 406-408. Springer, Berlin.

Kandeler E 1996b N-Mineralisation under waterlogged conditions. *In* Methods in Soil Biology. Eds. F Schinner, R Öhlinger, E Kandeler and R Margesin. pp. 141-143. Springer, Berlin.

Kandeler E 1996c Urease activity by colorimetric technique. *In* Methods in Soil Biology. Eds. F Schinner, R Öhlinger, E Kandeler and R Margesin. pp. 171-174. Springer, Berlin.

Kandeler E 1996d Nitrate. *In* Methods in Soil Biology. Eds. F Schinner, R Öhlinger, E Kandeler and R Margesin. pp. 408-410. Springer, Berlin.

Kaye J P, Binkley D and Rhoades C 2003 Stable soil nitrogen accumulation and flexible organic matter stoichiometry during primary floodplain succession. Biogeochemistry 63, 1-22.

Killham K 1994 Soil Ecology, Cambridge University Press, Cambridge.

Klose S, Wernecke K D and Makeschin F 2003 Microbial biomass and enzyme activities in coniferous forest soils as affected by lignite-derived deposition. Biol. Fertil Soils 38, 32-44.

Knoepp J D, Coleman D C, Crossley Jr D A and Clark J S 2000 Biological indices of soil quality: an ecosystem case study of their use. For. Ecol. Manag. 138, 357-368.

Koba K, Tokuchi N, Yoshioka T, Hobbie E A and Iwatsubo G 1998 Natural abundance of nitrogen-15 in a forest soil. Soil Sci. Soc. Am. J. 62, 778-781.

Kutschera L and Lichtenegger E 2002 Wurzelatlas mitteleuropäischer Waldbäume und Sträucher. Stocker Verlag, Graz. 604 pp.

Leip A, Matteucci G, Seufert G and Meroni M 2003 Greenhouse gas fluxes in a poplar plantation compared to a natural forest in Parco Ticino, Italy: the JCR Kyoto experiment. Geophysical Research Abstracts 11594.

Lovett G M, Weathers K C, Arthur M A and Schultz J C 2004 Nitrogen cycling in a northern hardwood forest: Do species matter? Biogeochemistry 67, 289-308.

Mayer H 1984 Waldbau auf soziologisch-ökologischer Grundlage, 3. Auflage. Gustav Fischer, Stuttgart-New York.

MCPFE Liaison Unit Vienna and UNECE/FAO (Eds) 2003 State of Europe's Forests 2003 - The MCPFE Report on Sustainable Forest Management in Europe, Liaison Unit Vienna, Austria; 115pp. (www.mcpfe.org).

Nestroy O, Danneberg O, Englisch M, Gessl H, Hager H, Kilian W, Nelhiebl P, Pecina O Schneider W, Pehamberger A and Wagner J 2000 Österreichische Bodensystematik 2000. Mitt. OEBG 60, Wien, 92 pp.

Paul E A and Clark F E 1996 Soil Microbiology and Biochemistry, 2nd Ed. Academic Press, London.

Pomorski R J, Leithner C and Bruckner A 2003 Protaphorura christiani n. sp. from Austria, with remarks on the generic status of Protaphorura stiriaca (Stach, 1946) (Collembola: Onychiuridae). Genus 14, 307-312.

Prescott C E, Zabek L M, Staley C L and Kabzems R 2000 Decomposition of broadleaf and needle litter in forests of British Columbia: influences of litter type, forest type and litter mixtures. Can. J. For. Res. 30, 1742-1750.

Ricotta C, Carranza M J, Avena G and Blasi C 2002 Are potential natural vegetation maps a meaningful alternative to neutral landscape models and quest? Appl. Veg. Sci. 5, 271-275.

Rigler E and Zechmeister-Boltenstern S 1999 Oxidation of ethylene and methane in forest soils - effect of CO_2 and mineral nitrogen. Geoderma 90, 147-159.

Roulet N T 2000 Peatlands, carbon storage, greenhouse gases and the Kyoto-Protocol: prospects and significance for Canada. Wetland 20, Issue 4, 605-615.

Schimel D S, House J I, Hibbard K A, Bousquet P, Ciais P, Peylin P, Braswell B H, Apps M J, Baker D, Bondeau A, Canadell J, Churkinal G, Cramer W, Denning A S, Field C B, Friedlingstein P, Goodale C, Heimann M, Houghton R A, Melillo J M, Moore III B, Murdiyarso D, Noble I, Pacala S W, Prentice I C, Raupach M R, Rayner P J, Scholes R J, Steffen W L and Wirth C 2001 Recent patterns and mechanisms of carbon exchange by terrestrial ecosystems. Nature 414, 169-172.

Swift M J, Andrén O, Brussaard L, Briones M, Couteaux M.-M., Ekschmitt K, Kjoller A, Loiseau P and Smith P 1998 Global change, soil biodiversity, and nitrogen cycling in terrestrial ecosystems: three case studies. Glob. Change Biol. 4, 729-743.

Šantrůčková H, Bird M I, Kalaschnikov Y N, Grund M, Elhottová D, Šimek M, Grigoryev S, Gleixner G, Arneth A, Schulze E D and Lloyd J 2003 Microbial characteristics of soil on a latitudinal transect in Siberia. Glob. Change Biol. 9, 1106-1117.

Šantrůčková H and Straškraba M 1991 On the relationship between specific respiration activity and microbial biomass in soils. Soil Biol. Biochem. . 23, .525-532.

Templer P, Findlay S and Lovett G 2003 Soil microbial biomass and nitrogen transformations among five tree species of the Catskill Mountains, New York, USA. Soil Biol. and Biochem. 35, 607-613

Thomas K D and Prescott C E 2000 Nitrogen availability in forest floors of three tree species on the same site: the role of litter quality. Can. J. For. Res. 30, 1698-1706.

Trumbore S E, Chadwick O A and Amundson R 1996 Rapid exchange of soil carbon and atmospheric CO_2 driven by temperature change. Science 272, 393-396.

Valentini R, Matteucci A J, Dolman A J, Schulze E D, Rebmann C, Moors E J, Granier A, Gross P, Jensen N O, Pilegaard K, Lindroth A, Grelle A, Bernhofer C, Grunwald T, Aubinet M, Ceulemans R, Kowalski A S, Vesala T, Rannik U, Berbigier P and Loustau D 2000 Respiration as the main determinant of carbon balance in European forests. Nature 404, 861-865.

Waitzbauer W and Zabransky P 2003 Der Urwald "Rothwald" (niederösterreichische Kalkalpen), ein Zentrum entomologischer Biodiversität. 33. Jahrestagung Ges. f. Ökologie 33, 98 pp.

Wanek W and Arndt S K 2002 The $\delta^{15}N$ variation between nodulated roots and shoots of soybean reflects the relative contribution of symbiotic N_2 fixation to plant N acquisition. J. Exp. Bot. 53, 1109-1118.

Wardle D A and Ghani A 1995 A critique of the microbial quotient (qCO2) as a bioindicator of disturbance and ecosystem development. Soil Biol. Biochem. 12, 1601-1610.

Zechmeister-Boltenstern S 1994 Measurement of actual and potential denitrification and soil respiration with an automated gas chromatographic system. Austrian Journal of Agricultural Research 45, 219-226.

Zechmeister-Boltenstern S and Nikodim L 1999 Effect of water tension on ethylene production and consumption in montane and lowland soils. Eur. J. Soil Sci. 50, 1-8.

Zechmeister-Boltenstern S, Hackl E, Bachmann G, Donat C and Pfeffer M 2000 Microbial nutrient turnover in forests with natural tree species composition. Proceedings of the International Conference on Forest Ecosystem Restoration, April 2000, Universität für Bodenkultur, Vienna, pp. 296-299.

Zechmeister-Boltenstern S, Hahn M, Meger S and Jandl R 2002 N_2O emission and nitrateleaching in relation to microbial dynamics in a beech forest soil. Soil Biol. and Biochem. 116, 823-832.

Zechmeister-Boltenstern S, Schindlbacher A, Ambus P and Butterbach-Bahl K 2004 Nitrogen oxides emissions in relation to microbial parameters at 13 European forest sites. Geophysical Research Abstracts 6, 5151.

Chapter 4

TREE SPECIES EFFECTS ON NITROGEN CYCLING AND RETENTION: A SYNTHESIS OF STUDIES USING ^{15}N TRACERS

Pamela H. Templer
Department of Environmental Science, Policy and Management, University of California, Berkeley, CA 94720, USA, ptempler@nature.berkeley.edu

ABSTRACT

The forests of the Catskill Mountains in New York receive some of the highest rates of nitrogen (N) deposition in the northeastern United States, and many watersheds are beginning to show signs of N saturation. The watershed export of N is variable, despite relatively uniform N deposition. I tested the hypothesis that tree species vary in their influence on forest N retention and loss. A laboratory study showed that tree species influenced soil microbial biomass and organic soil C:N, but that variation in these properties cannot explain differences in potential net mineralization, net nitrification, or microbial uptake of ^{15}N. A greenhouse experiment showed that sugar maple seedlings take up more NH_4^+ than NO_3^-, while beech seedlings take up more NO_3^- than NH_4^+. Results from a 300-day ^{15}N experiment in the field showed that most of the N deposited onto forests is retained within the forest floor, and the magnitude of N retention varied among tree species. Sugar maple stands retained the least N, and red oak stands retain the most. A fertilization experiment showed that red oak stands are most likely to have the greatest decrease in N retention if availability increases in the future. Sugar maple stands currently export substantially more N than the other forests, in part due to a limited capacity to take up NO_3^-. Tree species composition can be a strong regulator of forest N retention, but differences among species may depend on N inputs. Future research needs to address the interacting effects of increased N inputs and changes in tree species composition, as non-linear effects may likely occur in biogeochemical cycles.

D. Binkley and O. Menyailo (eds.),
Tree Species Effects on Soils: Implications for Global Change, 51–69.

INTRODUCTION

Humans have dramatically increased the amount of fixed nitrogen (N) in the terrestrial biosphere through the production of N fertilizers, fossil fuel combustion, and the use of N-fixing species in agriculture and forestry (Galloway et al. 1995). This has led to increasing amounts of N being deposited onto terrestrial ecosystems. Increased N inputs to forest ecosystems can lead to N saturation, the syndrome of responses in which excess N supply to forests leads to nitrate (NO_3^-) leaching into groundwater and streams and other alterations of forest nutrient cycling (Agren and Bosatta 1988; Aber et al. 1989; Stoddard 1994; Peterjohn et al. 1996). Symptoms of N saturation and excess N leaching have been observed in temperate forests of the United States (Johnson et al. 1991; Driscoll and Van Dreason 1993; Gilliam et al. 1996) and Europe (Gundersen et al 1998, MacDonald et al. 2002). Despite these signs of nitrogen saturation syndrome, most forests in the United States continue to retain a large proportion of N deposition, often 90% (Peterjohn et al. 1996; Lovett et al. 2000; see chapters in this volume by Rothe, Butterbach-Bahl and Kiese, and Papen and coauthors for truly N saturated forests in Bavaria.)

Forest fertilization and ^{15}N tracer experiments in the United Stated have shown that of the N deposited onto forests, the majority is retained within the soil, with the vegetation accounting for a minority of the N retention (Johnson 1992, Fenn et al. 1998; Nadelhoffer 1999; Magill et al. 1997). Even if forests retain a large amount of deposited N, N leached in the form of NO_3^- can cause essential cations such as calcium (Ca) and potassium (K) to be leached out of the forest soil, and some have suggested this could lead to nutrient imbalances in trees (Friedland et al. 1988; Schulze 1989). Nitrogen leaching can also lead to acidification of stream water (Vitousek et al. 1997), eutrophication of estuaries and coastal areas (Howarth 1988) and changes in species composition (Vitousek et al. 1997).

The Catskill Mountains in southeastern New York receive among the highest inputs of N deposition in the northeastern United States (Ollinger et al. 1993; Stoddard 1994), with wet N deposition rates of 4 to 9 kg N ha-1 yr^{-1} from 1986 through 2003 (National Atmospheric Deposition Program 2004). The inclusion of dry deposition of N raises the total N deposition estimate to about 11 kg N ha^{-1} yr^{-1} (Lovett and Reuth 1999). Concentrations of nitrate (NO_3^-) in stream water have increased in the past 25 years (Murdoch and Stoddard 1993), but these increases have not been consistent across all watersheds in the Catskill Mountains. Nitrogen retention ranges between 49-90% of atmospheric input, and stream NO_3^- concentrations vary by 17-fold, even among watersheds that are completely forested and have similar rates of N deposition (Lovett et al. 2000).

The mechanisms behind the variation in stream NO_3^- concentration remain unclear despite extensive research on potential factors such as hydrology (Burns et al 1998; West et al. 2001), N deposition and topography

(Lovett et al. 2000; Weathers et al. 2000). Some of the variation may relate to tree species composition; NO_3^- loss relates inversely to soil C:N, which in turn relates with species composition of the forests (Lovett et al. 2002). Mixed-species stands dominated by sugar maple (*Acer saccharum* Marsh) have low C:N and high nitrate leaching, whereas forests dominated by red oak (*Quercus rubra* L.) have higher C:N and lower nitrate losses (Lovett et al. 2002). These observations led to the hypothesis that tree species composition may influence N retention and thereby influence the amount of NO_3^- reaching streams (Lovett et al. 2000, Lovett et al. 2002).

Plant species can affect the localized movement of N through ecosystems via indirect effects on soil chemical properties and microbial activity (Vitousek et al. 1982; Zak et al. 1986; Finzi et al. 1998). Tree species differ in leaf litter quality (e.g. lignin:N or C:N ratio of litter; see Prescott and Vesterdal, this volume), providing a range of organic matter quality for microbial communities (Pastor and Post 1986). Therefore, soil microbes associated with different tree species may have variable rates of organic matter decomposition (Melillo et al. 1982), N mineralization (Vitousek et al. 1982; Zak and Pregitzer 1990; Zak et al. 1986), and nitrification (Lovett and Rueth 1999; Robertson 1982; Zak et al. 1986).

Plant species can also affect forest N cycling rates through differences in uptake and sequestration of N (Gharbi and Hipkin 1984; Horsley 1988; Crabtree and Bazazz 1993; Nadelhoffer et al. 1995). For example, plant species differ in their capacity to take up NH_4^+ or NO_3^- as their primary N source in part because of physiological trade-offs in taking up either form of N (Crabtree and Bazazz 1993; Horsley 1988; Stewart et al. 1992). While NO_3^- is readily available because of its high solubility in soil water, plants must use energy to reduce the NO_3^- prior to incorporation into their amino acids. With this constraint, it may be more efficient for plants to take up NH_4^+ because it can be immediately incorporated into amino acids. However, NH_4^+ is not always available to plants for uptake because it is tightly associated with soil exchange sites and does not move readily through the soil in solution.

This research tested the hypothesis that N cycling and retention varies significantly among forested stands dominated by different tree species in the Catskill Mountains. I expected N retention to be low in stands dominated by sugar maple, high in stands dominated by red oak, and intermediate in beech (*Fagus grandifolia* Ehrh.) stands. This pattern is based on differences in litter quality and soil chemistry characteristics associated with each species. For example, laboratory (Lovett and Rueth 1999; Templer et al. 2003; Lovett et al. 2004) and field rates (Finzi et al. 1998; Lawrence 2000) of net nitrification were higher in stands of sugar maple compared to the other three tree species. Higher net nitrification rates in sugar maple stands may increase NO_3^- leaching and lower forest N retention compared to forest stands dominated by other tree species. Lower rates of net nitrification in red oak stands may lead to greater N retention within these stands compared to other tree species.

This paper summarizes three experiments that used ^{15}N isotope methods to:

(1) examine the effect of tree species on soil microbial biomass, determining if microbial biomass strongly influences microbial processes that affect retention or loss of N;

(2) compare NH_4^+ and NO_3^- uptake and sequestration by dominant tree species, and

(3) quantify the fate of deposited N within stands of the dominant tree species, and determine the impact of N fertilization on forest N retention.

MATERIALS AND METHODS

Site Description

The Catskill Mountains are a range of low, mostly flat-topped mountains in southeastern New York. The bedrock is composed mostly of sandstone, shale, and conglomerate (Stoddard and Murdoch 1991), covered by glacial till that ranges from 0 to > 30 m in depth (Kudish 2000). The Inceptisol soils have moderate to high acidity (Stoddard and Murdoch 1991) and are well drained and moderately steep. Soils average 60% sand, 30% silt and 10% clay content (Kudish 2000). Vegetation between 500-1100 m elevation is dominated by northern hardwood forests common throughout the northeastern United States, including sugar maple, American beech, red oak, and eastern hemlock (*Tsuga Canadensis* L.; Kudish 1971; McIntosh 1972). About 80-90% of the original forest of the Catskill Mountains was subject to some level of harvesting by the end of the nineteenth century, although most of the cutting was selective harvest rather than clear-cutting. The Catskill Forest Preserve was created in 1885 with most of the land within its current borders added by the 1930's (Kudish 2000). The Catskill Park covers approximately 285,500 hectares, with about 41% of that protected in the Catskill Forest Preserve (Van Valkenburgh 1996). Mean annual temperature is 4.3 °C and mean annual precipitation is 153 cm at the 808 m elevation weather station on Slide Mountain.

Objective 1: Tree species effects

A laboratory study examined microbial biomass, potential net mineralization, potential net nitrification, and microbial uptake of added $^{15}NH_4$ in soils collected in August 1997 from plots of the five dominant tree species inhabiting watersheds of the Catskill Mountains, NY. We chose two 12 m diameter plots that were dominated by mature individuals of each of the

4 target species: American beech, red oak, sugar maple, and yellow birch (*Betula alleghaniensis* Britton). These stands were similar in age, had no recent disturbance, and were surrounded by mixed species stands. Litter of the target tree species represented 67-84% of the total litter at each plot. Soils under beech, sugar maple and yellow birch were collected from the upper Rondout watershed, and red oak soils were collected from the Kanape Brook watershed (see Lovett et al. 2000 for watershed characteristics). After brushing aside the litter layer (Oi), we collected 4 samples of the rest of the O horizon (two samples from each plot, including Oe and Oa where present). Each sample was a composite of four soil cores (6 cm diameter; 12 cm depth maximum each). Composited samples were sieved (8 mm sieve) to remove larger roots, woody fragments and stones, then analyzed for total C and N (Carlo-Erba NA 1500 C-N analyzer), soil organic matter content (loss on ignition), extractable N (KCl extraction), microbial biomass N (MBN, chloroform fumigation-extraction), microbial uptake of added ^{15}N, and potential net mineralization and nitrification for 5 incubation periods (0, 1, 3, 10 or 28 days; see Templer et al. 2003 for more detail).

Two-way analyses of variance (ANOVA) tested for the effects of tree species and incubation period for MBN, soil NO_3^- and NH_4^+ using SAS JMP software (Version 3.2.5, 1999). One-way ANOVA tested for the main effects of species on other variables. We also examined correlations between soil C, N, C:N, field condition soil moisture and microbial properties and processes to examine potential relationships among factors. The effects of species were confounded with any differences that existed between the sites in other factors, but for simplicity the effects will be referred to simply as species effects. In all experiments, a *P* value of 0.05 was used to avoid Type I errors in our hypothesis tests.

Objective 2: Ammonium versus nitrate uptake

A greenhouse study in 1998 examined N uptake by American beech, red oak, and sugar maple seedlings. We collected 14 seedlings of each species, and their native Oe and Oa horizons, which were randomly selected from mixed species stands of the Kanape Brook watershed in the Catskill Mountains, NY in November 1997. Seedlings retained their own soil and were kept moist and cool while being transported from the field. To maintain the dormancy normally experienced in the field, we over-wintered them at 2.2 °C. The plants were moved into a greenhouse (25-28 °C) on April 3, 1998 to break dormancy. Most leaves reached full expansion by May 5, 1998, when we added 99 atom% ^{15}N-NH_4 (5 ml of 12 mg N/L as $^{15}NH_4$-NO_3) to six of the seedlings of oak and maple, and 99 atom% enriched ^{15}N-NO_3 (5 ml of 12 mg N/L as NH_4-$^{15}NO_3$) to six others to examine uptake of NH_4^+ and NO_3^-. Two individuals of each tree species served as controls, receiving no N addition and providing the natural abundance of ^{15}N for each tree species. The same experimental design was used for beech, except ^{15}N tracer was added to only 10 individuals because 2 died during the winter. Plants were harvested 4 days after ^{15}N addition to minimize the impact of microbial activity on N transformations of NH_4^+ and NO_3^- in the soil; this was a long enough time to detect enriched ^{15}N in the plant biomass. We analyzed C, N and ^{15}N on roots, stem and leaves of each seedling and extractable NH_4^+ and NO_3- in the soil of each pot using the methods described in Templer and Dawson (2004).

We calculated plant N uptake as:

$$N_{up} = B^{-1} * \frac{(N_{plt})(atom\ \%^{15}N_{plt}) - (N_{plt})(atom\ \%^{15}N_{NA})}{atom\ \%^{15}N_S}$$

where N_{up} is the amount of N taken up by the plant; B= biomass of the plant at the end of the experiment; atom % $^{15}N_{plt}$ = atom percent ^{15}N of the plant at the end of the experiment; N_{plt} = mass of N in the plant at the end of the experiment; atom %$^{15}N_{NA}$ = atom percent ^{15}N of the reference plants (seedlings that did not receive ^{15}N tracer); atom % $^{15}N_S$ = atom %^{15}N of soil NO_3^- or NH_4^+ at the end of the experiment. We calculated uptake of N per unit plant biomass for a measure of plant demand for N that accounts for differences in plant biomass among tree species. We also calculated uptake of N per total soil NH_4 and NO_3 to account for differences in availability of the two forms of N.

We conducted one-way analyses of variance (ANOVA) with tree species as the main effect to examine plant biomass, plant N, as well as soil NH_4^+ and NO_3^-. To examine preference for NH_4^+ or NO_3^-, we conducted an ANOVA within each tree species with form of N (NH_4^+ vs NO_3^-) as the main effect. All of the data were log-transformed prior to statistical analysis. A Kruskal-Wallis test was performed if the data were not normally distributed after the log transformation. The effects of species were again confounded with any differences that covaried with the locations where the seedlings had established.

Objective 3: Fate of N deposition

A 300-day field ^{15}N experiment examined soil and plant pools in the field, tracing N movement and retention times within various pools. If retention within different pools differed among tree species, changes in tree species composition in the future could alter the location and retention of N within forests. A fertilization treatment was used to examine the potential effect of higher N inputs on N retention. Methods are described in detail in Templer et al. (2005).

In 1999-2000, we examined the fate of N in 6 sets of 12-m diameter paired plots for sugar maple, American beech and red oak (12 plots per species total). The plots were located within mixed-species forest and were composed of clusters of the target tree species. The subsequent measurements showed that litter of the target tree species represented 67-84% of the total litter at each plot, with non-target litter resulting primarily from understory trees and trees outside the plot. The 6-paired plots were distributed across three watersheds for each of the tree species. Because of the distribution of tree species within the Catskill Mountains, we could not easily locate all three species within each watershed. One plot of each pair received only ambient levels of N deposition (approximately 11.2 kg N/ha/yr), while the other had been fertilized with an additional 50 kg N/ha/year (as granular NH_4NO_3 in four doses per year-- June, July, August and November of each year) for the 1.5 years prior to and during this experiment; thus the total N enhancement in the fertilized plots by the end of this experiment was 112.5 kg N / ha. This enabled us to compare the movement of N within plots that received ambient levels of N to those that received a higher level of N. To each of the paired plots, we added trace amounts of 99 atom % enriched $^{15}NH_4Cl$ to the inner 8 m of each plot during July, August and October 1999 (3 additions of 10 mg $^{15}NH_4Cl$-N / m^2 each; dissolved in 5 L deionized water for each plot). We added ^{15}N three times, including one dose in the fall after leaf drop, rather than as a single pulse, for a more natural simulation of N availability throughout the year. In July and August of 1999, ^{15}N was added approximately 1.5 weeks following the N fertilization additions. We added the ^{15}N isotope as NH_4 to trace the large flux of N that is mineralized from organic matter, as opposed to the smaller flux of N received from ambient deposition, which occurs primarily as NO_3. The tracer solution was applied to an 8-m diameter subplot using a backpack sprayer to ensure even distribution. For this experiment, the plot size was reduced to 8 m diameter because that area was adequate to detect ^{15}N tracer additions and because of the significant cost of ^{15}N addition.

To determine natural abundance background ^{15}N, roots and soil (Oe and Oa horizons) samples were collected from each plot in June 1999. Surface litter layer (Oi) samples were collected from areas outside, and directly adjacent to, each plot during August 1999. Wood and bark samples were collected from areas outside and directly adjacent to each plot during May 2000. To track the fate of the added ^{15}N in the soil, we collected 4 samples of the litter layer (Oi), organic (Oe and Oa horizons) and mineral soil and fine root samples from each plot on three dates. The first sampling occurred 2 days following the first ^{15}N addition in July 1999. The second sampling (day 90 after initial ^{15}N addition) occurred just prior to the third ^{15}N addition in October 1999. The third sampling (day 300) occurred during May 2000, prior to budbreak of the deciduous tree species. In the calculations of ^{15}N recovery percentages, we used only ^{15}N inputs occurring before the sampling date (see Templer et al. 2005 for details).

We also measured total N content and ^{15}N concentrations of aboveground plant pools as described in Templer et al. (2005). We collected the outer 2 cm of wood and bark from three individuals of the target tree species within each plot during May 2000 (prior to budbreak) to examine how much of the N was moved to long-term sinks within the trees. We used allometric equations for each tree species (Tritton and Hornbeck 1982) to calculate total aboveground tree biomass of each plot (4 m radius). Litterfall samples were collected biweekly during leaf fall in 1998 and 1999. Sun-lit foliage samples were collected from each plot during the peak-growing season (August) of 1998, 1999 and 2000 and were composited by plot and year prior to ^{15}N analysis.

We calculated ^{15}N recovery using N mass, the amount of ^{15}N added, and the atom % ^{15}N enrichment of the various ecosystem pools:

$$\%^{15}N_{rec} = 100 * \frac{m_{pool} * \left(atom\%^{15}N_{pool} - atom\%^{15}N_{ref}\right)/100}{^{15}N_{tracer}}$$

where % $^{15}N_{rec}$ = percent of ^{15}N tracer recovered in the labeled N pool; m_{pool} = N mass of the labeled pool; atom % $^{15}N_{pool}$ = atom percent ^{15}N in the labeled pool; atom % $^{15}N_{ref}$ = atom percent ^{15}N of the reference (non-labeled) plots; $^{15}N_{tracer}$ = amount of ^{15}N added to each plot prior to sample collection (Buchmann et al. 1996). We use ^{15}N recovery as an estimate of net retention of N in a given pool at a given point in time.

We averaged the mass and N pool size for soil and plant samples over time for each plot, and conducted two-way analyses of variance (ANOVA) testing the effects of tree species and N treatment (ambient vs. fertilized). Each plot was nested within tree species because of the paired-plot design. Data that were not normally distributed were log-transformed prior to statistical analysis. We conducted linear contrasts of the means with Tukey-Kramer post hoc tests to examine the effect of tree species on forest ^{15}N recovery and to compare ambient and fertilized plots. Only the results from day 300 (May 2000) are presented. The effect of tree species was again confounded with any other factors that may have differed among the plots.

Enriched ^{15}N samples were analyzed on a Europa Integra, which is a combined sample combustion unit and isotope ratio mass spectrometer. Natural abundance ^{15}N samples were analyzed on a Europa 20-20 mass spectrometer after combustion in a Europa ANCA-GSL combustion unit. The standard used was 0.36679 atom% ^{15}N calibrated against IAEA N1, an International Atomic Energy Agency standard.

RESULTS

Objective 1: Tree species effects

The N content of microbial biomass and rates of N transformation in the laboratory clearly differed among soils collected from sites dominated by

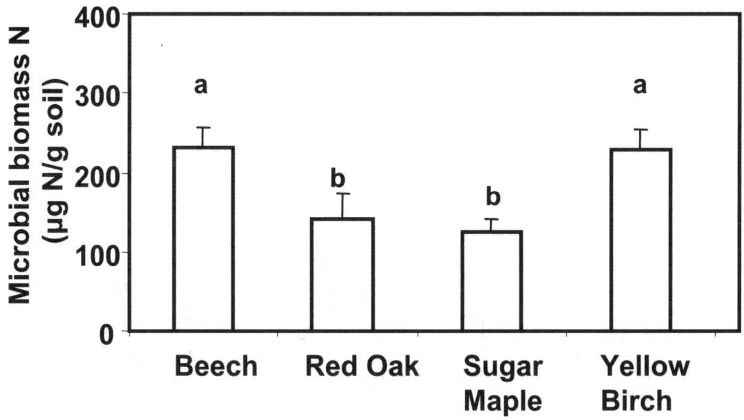

Figure 1. Microbial biomass N among tree species (error bars represent standard error; n=4 for each tree species; each bar represents the mean of four sample replicates averaged across the following 5 sampling periods: 0, 1, 3, 10, 28 days). From Templer et al. 2003.

different tree species, but the patterns among parameters was not consistent. Soil MBN ranged 2-fold, with lower values in stands of red oak and sugar maple than beech and yellow birch (Figure 1). Sugar maple soils consistently had larger soil NO_3^- pools (Table 1) and greater rates of potential net nitrification (Figure 2) than red oak soils, despite similar soil MBN. Short-term recovery (day 3 minus day 0) of ^{15}N recovery per unit MBN varied among tree species with greater values in soils of red oak (0.023) stands compared to sugar maple stands (-0.024). The uptake of ^{15}N per unit MBN in 28 days was similar to the 3-day trend, but the differences among tree species were not significant (p=0.6). Soil moisture accounted for 22% of the variation in MBN and 36% of the variation in net mineralization. The C:N ratio of the soil samples differed among species, with lowest values in sugar maple soils and the greatest values in red oak soils (Table 1). However, C:N did not correlate with rates of net N mineralization, net nitrification, or MBN.

Objective 2: Ammonium versus nitrate uptake

Total plant uptake of NH_4-N was 5-fold greater (per unit plant biomass) than uptake of NO_3-N in sugar maple seedlings (Figures 3 and 4). The red oak seedlings contained about 3 times more N derived from $^{15}NH_4^+$ than $^{15}NO_3^-$, but the power of the experiment could not show that this difference was significant. In contrast to the other three tree species, beech took up 5 times more N as $^{15}NO_3^-$ than $^{15}NH_4^+$.

When total soil NH_4-N and NO_3-N were taken into account, beech seedlings still took up 10 times more N as NO_3-N than NH_4-N (p=0.022; Figure 4). Sugar maple seedlings again took up almost 5 times more N as

Table 1. Organic horizon properties by tree species with standard error in parentheses (n=4 for each tree species). NH_4^+, NO_3^-, dissolved organic N (DON) and total dissolved N (TDN) were averaged across days 0, 1, 3, 10 and 28 for each tree species. Different letters above values represent statistically significant differences at $p<0.05$. From data in Templer et al. 2003.

Species	Moisture	Organic Matter	Carbon on	Nitrogen Nitroge	C:N	NH_4-N	$N0_3$-N	DON	TDN
	(%)					(µg/g soil)			
Beech	47 (15)	79 (7)	34.3 (7.6)	1.88 (0.41)	18.4^{ac} (0.9)	69.4 (15.6)	11.0^{a} (2.5)	67.7 (8.4)	147 (18)
Red Oak	18 (6)	78 (5)	35.9 (3.5)	1.74 (0.20)	20.7^{b} (0.7)	56.4 (11.1)	1.3^{a} (0.3)	59.1 (9.4)	113 (15)
Sugar Maple	32 (9)	84 (2)	42.7 (1.4)	2.49 (0.09)	17.3^{c} (1.0)	51.2 (14.0)	53.6^{b} (19.1)	62.4 (27.6)	148 (37)
Yellow Birch	48 (8)	86 (5)	43.5 (3.8)	2.23 (0.19)	19.5^{abc} (0.1)	81.0 (13.4)	8.0^{a} (2.3)	71.2 (8.3)	157 (13)

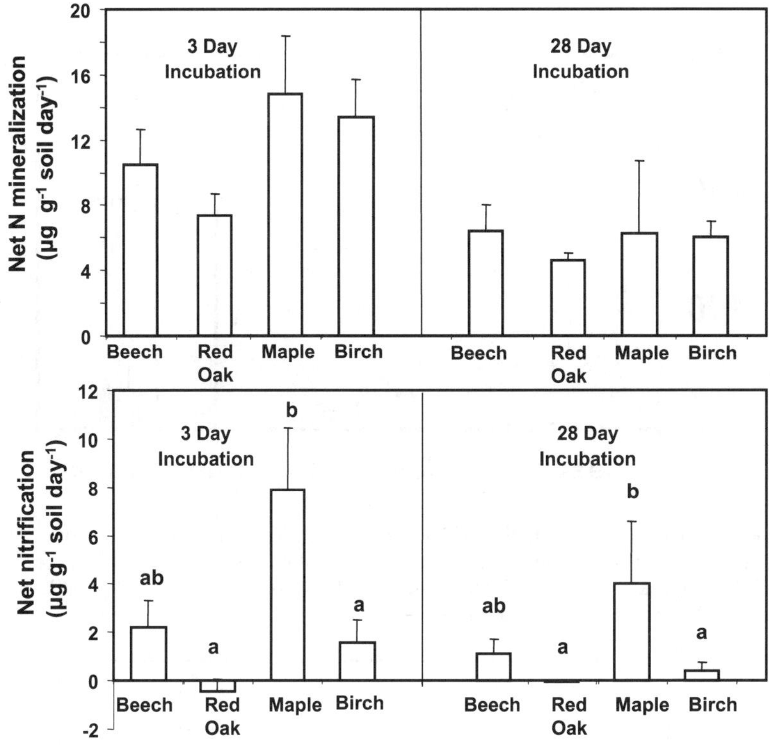

Figure 2. Potential net mineralization (upper) and potential net nitrification (lower) measured after 3 and 28 day laboratory incubations (error bars represent standard error; n=4 for each tree species). Different letters above bars indicated differences at $p<0.05$ (after Templer et al. 2003).

NH_4-N than NO_3-N (p=0.0005). Red oak seedlings averaged one-third higher uptake of NO_3 than NH_4, but this difference was again not significant.

In addition to variation in composition of N uptake, tree species also varied in the magnitude of NH_4-N and NO_3-N uptake per unit plant biomass. For example, sugar maple took up over 6 times more NH_4-N than red oak, and over 180-times more NH_4-N than beech seedlings ($p<0.05$; Figure 3). Sugar maple and red oak seedlings had similarly high amounts of NO_3-N uptake into their biomass.

Objective 3: Fate of N deposition

Total plot ^{15}N recovery (sum of the O horizon, fine roots, and any mineral soil to a depth of 12 cm, and aboveground biomass) did not vary among tree species (p=0.7), but the addition of extra N in fertilizer reduced the quantitative recovery of the added $_{15}N$ (Figure 5). Recovery of added $_{15}N$

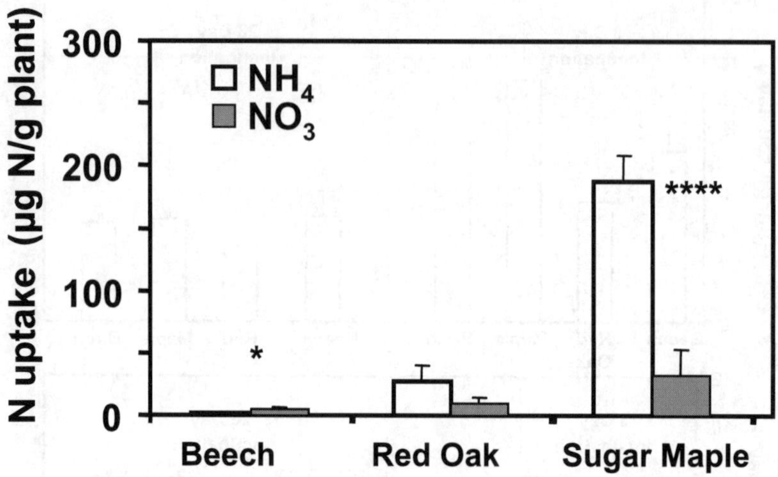

Figure 3. Plant uptake and sequestration of NH_4-N and NO_3-N per g plant biomass (n=4 for beech and n=5 for red oak, and sugar maple seedlings). Within a tree species, differences in NH_4-N vs NO_3-N uptake are denoted by the following: * $p<0.10$, ** $p<0.05$, *** $p<0.01$, and **** $p<0.005$, after Templer and Dawson 2004.

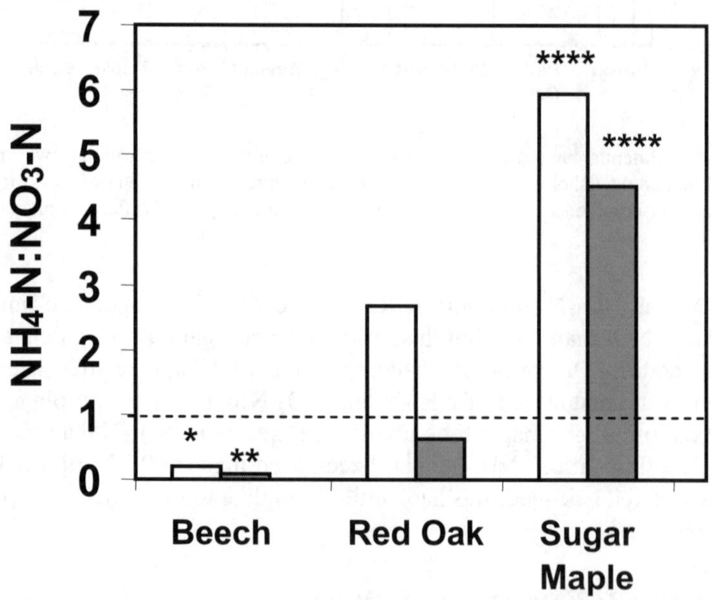

Figure 4. Ratio of N taken up as NH_4 vs NO_3 by plants. Open bars are the ratio of total plant uptake of NH_4-N:NO_3-N per g plant biomass (n=4 for beech and n=5 for red oak and sugar maple seedlings). Solid black bars show ratio of total plant uptake of NH_4-N:NO_3-N per g plant biomass per total µg NH_4-N/NO_3-N in pot. Within a tree species, differences in NH_4 vs NO_3 uptake are denoted by the following: *$p<0.1$, **$p<0.05$, ***$p<0.01$, and ****$p<0.005$. Dashed line denotes equal uptake of NH_4-N and NO_3-N, after Templer and Dawson 2004.

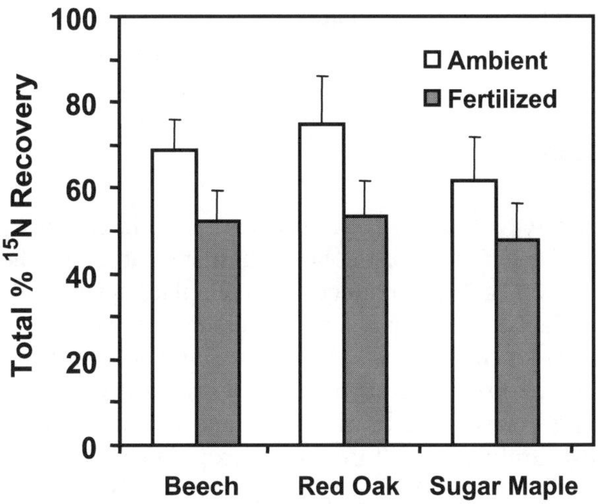

Figure 5. Fertilization reduced total %^{15}N retention ($p=0.02$) at day 300. Includes surface O horizon, fine roots, any mineral soil with 12 cm depth and aboveground biomass (means with standard errors, after Templer et al. 2005).

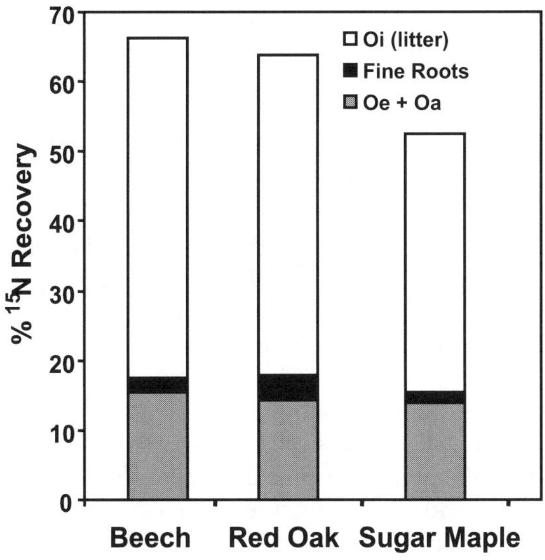

Figure 6. Recovery of ^{15}N within the forest floor (Oi, Oe+Oa, and fine roots) of ambient plots (no N fertilizer) at day 300 (means May 2000). Beech and red oak differed from sugar maple ($p<0.05$, after Templer et al. 2005).

ranged from 62-75% in the unfertilized ambient plots to 48-61% in the fertilized plots.

The O horizon (Oi+Oe+Oa, with fine roots included) was the dominant sink for $_{15}$N in all plots, and recovery of $_{15}$N in the O horizon differed among the species (Figure 6). Sugar maple plots retained 53% of the tracer N, compared with 63 to 66% retention in soils from the other species.

DISCUSSION

Soils dominated by different tree species varied in capacity to retain N in these Catskill Mountain forests, both indirectly through their effect on microbial processing of N, and directly through plant uptake and sequestration (Figure 7). Soils beneath sugar maple cycle soil N rapidly (Frehlich et al. 1993, Templer et al. 2003, Lovett et al. 2004), with a substantial proportion of N leaching from the soil (Fitzhugh et al., *in preparation*). Some of this loss results from high rates of net nitrification, and low rates of plant uptake of NO_3^-. The most surprising result from these experiments was that although red oak forest stands retained a large portion of added N, they had the lowest capacity to retain larger additions of N (from fertilization). We need to understand not only how tree species currently retain N, but how they may respond to increases in N availability in the future.

Tree species influenced the size of the soil microbial biomass pool and soil C:N, but the patterns in these properties did not account for patterns in microbial processes and N transformations. Sugar maple and red oak had low microbial biomass compared to the other beech and birch, but this similarity in microbial biomass did not translate into similar rates of N transformation. Indeed, soils from under sugar maple had lower C:N, higher pools of NO_3^- and rates of potential net nitrification than the other three species. Microbes in sugar maple soils took up the highest amount of ^{15}N immediately and released it, whereas soil microbes from beneath the other four species continued to immobilize added ^{15}N throughout the experimental period. Properties other than soil C:N and microbial biomass need to be examined to better understand the factors controlling microbial N transformations in these forest soils.

The greenhouse experiment showed that sugar maple seedlings took up much more NH_4^+ than NO_3^-, and beech seedlings showed the opposite pattern. The results of the laboratory and greenhouse studies suggest that there is a combination of soil and plant processes that can help explain why sugar maple stands may export more N than other forested stands such as red oak. Net nitrification and soil pools of NO_3^- are highest within sugar maple soils. However, sugar maples are not able to take up a significant amount of NO_3^-, the form of N that is produced in the largest amount in their stands and is most likely to be leached. We also note that both sugar maple and beech are

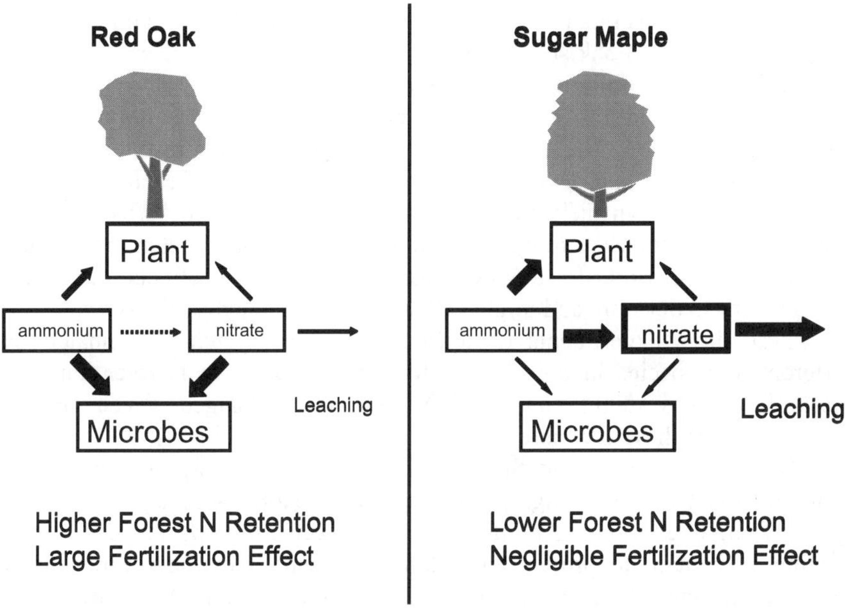

Figure 7. Conceptual model demonstrating potential mechanisms leading to greater forest N retention within red oak than sugar maple stands. Thicker arrows (indicating fluxes), boxes (indicating pools) and fonts indicate larger pool size or fluxes. Dashed lines represent non-detectable amounts of a particular flux. Sugar maple stands had higher nitrification rates, larger soil nitrate pools, less microbial uptake of N, six times greater plant uptake of ammonium relative to nitrate, and larger leaching losses of N (summary from data presented in this chapter, and cited references).

considered dominant components of mature (climax) northern hardwoods forests, yet they each rely primarily on a different form of N; generalizations about patterns in form of N usage with succession would not be supported.

These differences among species in uptake of NO_3^- versus NH_4^+ may also be important in other forests in this region. For example, clearcutting northern hardwood forests in New Hampshire increases leaching losses of NO_3^-, and these losses decline sharply as forest regeneration develops (Likens et al. 1970, Pardo et al. 1995). Regeneration often includes a dominant component of pin cherry (*Prunus pennsylvanica* L.; Marks 1974, Bormann and Likens 1979), with rapid rates of N accumulation. Intriguingly, pin cherry has very low rates of NO_3^- reductase activity, indicating that its role in retaining ecosystem N probably comes from rapid uptake of NH_4^+ and not NO_3^- (Truax et al. 1994a, 1994b).

Similar to the laboratory and greenhouse experiments, tree species varied in forest retention of added N in a 300-day field ^{15}N experiment. Most of the added N was retained within the forest floor, but the magnitude of N retention varied among the soils from beneath different tree species. Sugar maple stands retained the least amount of N with current levels of N deposition, while red oak stands retain the most. However, the fertilized treatment of this study shows that red oak stands may be more likely to have decreased forest N retention if availability increases in the future.

How long would it take for soils to reflect a change in overstory from red oak to sugar maple? The differences in these experiments resulted from an unknown period of soil development under the influence of each species, as well as any unknown differences in other site factors. If the effects of tree species are particularly strong, the change in soils (and N retention) should develop over a period of a few decades. Weaker species effects might take much longer to develop, and might differ substantially across site types.

These three experiments demonstrated that ecosystems dominated by different tree species in the Catskill Mountains varied in N retention, and these differences might change if N deposition changed. The species composition of these forests changed dramatically over the last 100 years in the Catskill Mountains, and these changes continue in response to climate changes, selective harvesting, and the spread of exotic pests and pathogens. Each of these environmental disturbances affects tree species differently, making it difficult to predict the future composition of these forests. For example, introduced beech bark disease (Houston et al. 1979; Griffin et al. 2003) and the hemlock wooly adelgid (Orwig and Foster 1998) have increased mortality of beech and hemlock across the northeastern United States. The dominance of sugar maple appears to be increasing in response to the decline of the other species (Griffin et al. 2003). This shift in species composition could increase N losses from forests, as beech and hemlock currently take up and retain a larger amount of N than the sugar maple stands that replace them. Other environmental changes could impair the increase in sugar maple, including climate change, acid rain and other introduced pests could (Lovett and Mitchell 2004). Future research is needed to address the interactions of multiple anthropogenic disturbances so that we can better predict which tree species will remain as dominants of northeastern United States forests. This information is necessary if we want to better manage these forests to retain the maximum amount of N and to understand the biogeochemical effects of changes in tree species composition.

REFERENCES CITED

Aber J, Nadelhoffer K J, Steudler P and Melillo J M 1989 Nitrogen saturation in northern forest ecosystems. BioScience 39:378-386.

Agren G I and Bosatta E 1988 Nitrogen saturation of terrestrial ecosystems. Environ. Pollut. 54:185-197.

Bormann F H and Likens G E 1979 Pattern and Process in a Forested Ecosystem: Disturbance, Development and the Steady State Based on the Hubbard-Brook Ecosystem Study. Springer-Verlag, New York.

Buchman N, Gerbauer G and Schulze E D 1996 Partitioning of ^{15}N-labeled ammonium and nitrate among soil, litter, below- and aboveground biomass of trees and understory in a 15-year old Picea abies plantation. Biogeochemistry 33:1-23.

Burns D A, Murdoch P S, Lawrence G B and Michel R L 1998 Effect of groundwater springs on N03- concentrations during summer in Catskill Mountain streams. Water Resour. Res. 34:1987-1996.

Crabtree R C and Bazazz F A 1993 Tree seedling response of four birch species to simulated nitrogen deposition: ammonium vs nitrate. Ecol. Appl. 3:315-321.

Driscoll C and Van Dreason R 1993 Seasonal and long-term temporal patterns in the chemistry of Adirondack lakes. Water Air Soil Poll. 67:319-344.

Fenn M, Poth M, Aber J, Baron J, Bormann B, Johnson D, Lemly A, McNulty S, Ryan D and Stottlemyer R 1998 Nitrogen excess in North American ecosystems: predisposing factors, ecosystem responses, and management strategies. Ecol. Appl. 8:706-733.

Finzi A C, Van Breemen N and Canham C D 1998 Canopy tree-soil interactions within temperate forests: species effects on soil carbon and nitrogen. Ecol. Appl. 8:440-446.

Fitzhugh R, Lovett G M, Weathers K C and Arthur M A. Canopy tree species and soil solution C, N, and P chemistry in northern hardwood forests of the Catskill Mountains, New York, USA. *In Prep.*

Frelich L E, Calcote R R, Davis M B and Pastor J 1993 Patch formation and maintenance in an old-growth hemlock-hardwood forest. Ecology 74:513-527.

Friedland A J, Hawley G J and Gregory R A 1988 Red spruce *Picea-rubens sarg*. Foliar chemistry in northern Vermont and New York USA. Plant Soil 105:189-194.

Galloway J N, Schlesinger W H, Levy II H, Michaels A and Schnoor J L 1995 Nitrogen fixation: atmospheric enhancement-environmental response. Global Biogeochem. Cy. 9:235-252.

Gharbi A and Hipkin C 1984 Studies on nitrate reductase in British angiosperms. I. A comparison of nitrate reductase activity in ruderal, woodland-edge and woody species. New Phytol. 97:629-639.

Gilliam F S, Adams M B and Yurish B M 1996 Ecosystem nutrient responses to chronic nitrogen inputs at Fernow Experimental Forest, West Virginia. Can. J. Forest Res. 26:196-205.

Griffin J M, Lovett G M, Arthur M A and Weathers K C 2003 The distribution and severity of beech bark disease in the Catskill Mountains, NY. Can. J. Forest Res. 33:1754-1760.

Gundersen P, Emmett B A, Kjonaas O J, Koopmans C J and Tietema A 1998 Impact of nitrogen deposition on nitrogen cycling in forests: a synthesis of NITREX data. Forest Ecol. Manag. 101:37-55.

Horsley S B 1988 Nitrogen species preference of Prunus-serotina Ehrh. And Betula-Alleghaniensis Britt. Seedlings. Am. J. Bot. Annual Meeting of the Botanical Society of America, Davis, California, USA, August 14-18, 1988.

Houston D R, Parker E J and Lonsdale D 1979 Beech bark disease: patterns of spread and development of the initiating agent *Crptococcus fagisuga*. Can. J. Forest Res. 9:336-344.

Howarth R W 1988 Nutrient limitation of net primary production in marine ecosystems. Annu. Rev. Ecol. Syst. 19:89-110.

Johnson D, Van Miegroet H, Lindberg S, Harrison R and Todd D 1991 Nutrient cycling in red spruce forests of the Great Smoky Mountains. Can. J. Forest Res. 21:769-787.

Johnson D W 1992 Nitrogen retention in forest soils. J. Environ. Qual. 21:1-12.

Kudish M 2000 The Catskill Forest *A History*. Purple Mountain Press, New York. 217p.

Lawrence G B, Lovett G M and Baevsky Y H 2000 Atmospheric deposition and watershed nitrogen export along an elevational gradient in the Catskill Mountains, New York. Biogeochemistry 50:21-43.

Likens G E, Bormann F H, Johnson N M, Fisher D W and Pierce R S 1970 Effects of Forest Cutting and Herbicide Treatment on Nutrient Budgets in the Hubbard Brook Watershed-Ecosystem. Ecol. Mon. 40, 23-47.

Lovett G M and Rueth H 1999 Soil nitrogen transformations in beech and maple stands along a nitrogen deposition gradient. Ecol. Appl. 9:1330-1344.

Lovett G M, Weathers K C and Sobczak W V 2000 Nitrogen saturation and retention in forested watersheds of the Catskill Mountains, New York. Ecol. Appl. 10:73-84.

Lovett G M, Weathers K C and Arthur M A 2002 Control of nitrogen loss from forested watersheds by soil C:N ratio and tree species composition. Ecosystems 5:712-718.

Lovett G M, Weathers K C, Arthur M A and Schultz J C 2004 Nitrogen cycling in a northern hardwood forest: do species matter? Biogeochemistry 67:289-308.

Lovett G M and Mitchell M J 2004 Sugar maple and nitrogen cycling in forests of eastern North America. Front. Ecol. Environ. 2:81-88.

MacDonald J A, Dise N B, Matzner E, Armbruster M, Gundersen P and Forsius M 2002 Nitrogen input together with ecosystem nitrogen enrichment predict nitrate leaching from European forests. Global Change Biol. 8:1028-1033.

Magill A H, Aber J D, Hendricks J J, Bowden R D, Melillo J M and Steudler P A 1997 Biogeochemical response of forest ecosystems to simulated chronic nitrogen deposition. Ecol. Appl. 7:402-415.

McIntosh R P 1972 Forests of the Catskill Mountains, New York. Ecol. Monogr. 42:143-161.

Melillo J M, Aber J D and Muratore J F 1982 Nitrogen and lignin control of hardwood leaf litter decomposition dynamics. Ecology 63: 621-626.

Murdoch P S and Stoddard J L 1993 Chemical characteristics and temporal trends in eight streams of the Catskill Mountains, New York. Water Air Soil Poll. 67:367-395.

Nadelhoffer K J, Downs M R, Fry B, Aber J D, Magill A H and Melillo J M 1995 The fate of ^{15}N-labelled nitrate additions to a northern hardwood forest in eastern Maine, USA. Oecologia 103:292-301.

Nadelhoffer K J, Downs M R and Fry B 1999 Sinks for ^{15}N-enriched additions to an oak forest and a red pine plantation. Ecol. Appl. 9:72-86.

National Atmospheric Deposition Program 2004. http://nadp.sws.uiuc.edu/ (accessed December 10, 2004).

Ollinger S V, Aber J D, Lovett G M, Millham S E and Lathrop R G 1993 A spatial model of atmospheric deposition for the northeastern United States. Ecol. Appl. 3:459-472.

Orwig D A and Foster D R 1998 Forest response to the introduced hemlock woolly adelgid in southern New England, USA. J. Torrey Bot. Soc. 125:60-73.

Pardo L H, Driscoll C T and Likens G E 1995 Patterns of nitrate loss from a chronosequence of clear-cut watersheds. Water Air Soil Poll. 85, 1659-1664.

Pastor, J and Post W M 1986 Influence of climate, soil moisture, and succession on forest carbon and nitrogen cycles. Biogeochemistry 2: 3-27.

Peterjohn W T, Adams M B and Gilliam F S 1996 Symptoms of nitrogen saturation in two central Appalachian hardwood forest ecosystems. Biogeochemistry 35: 507-522.

Robertson G P 1982 Factors regulating nitrification in primary and secondary succession. Ecology 63: 1561-1573.

Schulze E D 1989 Air pollution and forest decline in a spruce (*Picea abies*) forest. Science 244:776-783.

Stewart G R, Joly C A, and Smirnoff N 1992 Partitioning of inorganic nitrogen assimilation between the roots and shoots of cerrado and forest trees of contrasting plant communities of South East Brasil. Oecologia 91:511-517.

Stoddard J L and Murdoch P S 1991 Catskill Mountains. *In* Acidic Deposition and Aquatic Ecosystems: Regional Case Studies. Ed. D F Charles. pp. 237-271. New York: Springer-Verlag New York.

Stoddard J L 1994 Long-term changes in watershed retention of nitrogen. In *Environmental Chemistry of Lakes and Reservoirs*, Advance in Chemistry Series, Volume 237. Ed. L A Baker. pp. 223-284. American Chemical Society, Washington, D.C.

Templer P, Findlay S and Lovett G 2003 Soil microbial biomass and nitrogen transformations among five tree species of the Catskill Mountains, New York, USA. Soil Biol. Biochem. 35:607-613.

Templer P H and Dawson T E 2004 Nitrogen uptake by four tree species of the Catskill Mountains, New York: Implications for forest N dynamics. Plant Soil 262:251-261.

Templer P H, Lovett G, Weathers K, Findlay S and Dawson T 2005 Tree species influence on forest nitrogen retention in the Catskill Mountains, New York, USA. Ecosystems, *in press.*

Tritton L M and Hornbeck J W 1982 Biomass equations for major tree species of the northeast. General Technical Report NE-69. United States Department of Agriculture. Forest Service. Northeastern Forest Experiment Station.

Truax B, Gagnon D, Lambert F and Chevrier N 1994a Nitrate assimilation of raspberry and in cherry in a recent clear-cut. Can J Bot 72, 1343-1348.

Truax B, Lambert F, Gagnon D and Chevrier N 1994b Nitrate reductase and glutamine-synthetase activities in relation to growth and nitrogen assimilation in red oak and red ash seedlings – effects of N-forms, N-concentration and light-intensity. Trees-Struct. Funct. 9, 12-18.

Van Valkenburgh N J 1996 Forest Preserve of New York State in the Adirondack and Catskill Mountains: A Short History. Purple Mountain Press, New York.

Vitousek P M, Gosz J R, Grier C C, Melillo J M and Reiners W 1982 A comparative analysis of potential nitrification and nitrate mobility in forest ecosystems. Ecol. Monogr. 52:155-177.

Vitousek P M, Aber J, Howarth R, Likens G, Matson P, Schindler D, Schlesinger W and Tilman D 1997 Human alteration of the global nitrogen cycle: source and consequences. Ecol. Appl. 7:737-750.

Weathers K C, Lovett G M, Likens G E and Lathrop R 2000 Effect of landscape features on deposition to Hunter Mountain, Catskill Mountains, New York. Ecol. Appl. 10:528-540.

West J A, Findlay S E G, Burns D A, Weathers K C and Lovett G M 2001 Catchment-scale variation in the nitrate concentrations of groundwater seeps in the Catskill Mountains, NY. Water Air Soil Poll. 132:389-400.

Zak D R, Pregitzer K S and Host G E 1986 Landscape variation in nitrogen mineralization and nitrification. Can. J. Forest Res. 16:1258-1263.

Zak D R and Pregitzer K S 1990 Spatial and temporal variability of nitrogen cycling in Northern lower Michigan. Forest Sci. 36:367-380.

Chapter 5

TREE SPECIES MANAGEMENT AND NITRATE CONTAMINATION OF GROUNDWATER: A CENTRAL EUROPEAN PERSPECTIVE

Andreas Rothe

Bayerisches Staatsministerium für Landwirtschaft und Forsten Ludwigstr. 2 80539 München Germany Tel: +49 89 2182 2534 e-mail: andreas.rothe@stmlf.bayern.de

ABSTRACT

Nitrate concentrations in seepage water from pristine forests usually are very low, regardless of species composition. In most parts of Europe, anthropogenic nitrogen deposition has led to increased nitrate leaching from forests, and the rates of nitrate leaching in seepage water may be influenced by the dominant tree species. Spruce forests have a higher leaf area than beech forests, and also maintain high leaf area throughout the year. This leads to higher rates of N deposition from the atmosphere and to higher interception losses (resulting in a less deep seepage) compared to beech stands. For these reasons, spruce-dominated forests tend to have more than double the concentrations of nitrate in seepage water compared with broadleaved forests. These species effects depend in part on the age of the forests and on the fate of nitrate in deeper soil layers. In Bavaria, the Southernmost State of Germany active plantation management shifted the balance from the original broadleaved forests towards spruce dominated forests in the 18th and 19th century. Planting of forests in the late 20th Century shifted to an emphasis on broadleaved species, and this shift contributes to reductions in nitrate concentrations of seepage water in areas with high N deposition. For sensitive areas like ground water protection zones mixed-species forests with higher proportions of broadleaved species are recommended in order to reduce the risk of excessive nitrate leaching.

INTRODUCTION

Most forests are naturally nitrogen-limited ecosystems. Without human impact both coniferous and deciduous forests are characterized by a tight

D. Binkley and O. Menyailo (eds.),
Tree Species Effects on Soils: Implications for Global Change, 71–83.

nitrogen (N) cycle and N-losses are low except during occasional disturbances like wind throw or fire (Gosz 1981, Melillo 1981). Nitrogen concentrations of forest streams in remote forests, and forests in areas with low rates of atmospheric N deposition (<5 kg N ha^{-1} yr^{-1}), are very low and dominated by organic-N forms (Hedin et al. 1995, Binkley et al. 2004). Increased atmospheric deposition of N across much of Europe has altered biogeochemical cycles, increased rates of N leaching and shifted the balance from dominantly organic-N in leachates to nitrate. Across 181 European forests, N deposition averaged 16.8 kg ha^{-1} yr^{-1}, with maximum N-inputs up to 70 kg ha^{-1} yr^{-1} (MacDonald et al., 2002). Nitrogen supply may exceed the uptake capacity of the soil microbial community and the vegetation, increasing rates of N loss; about 2/3 of European forests show increased leaching losses of N, though few are N saturated with outputs matching inputs (MacDonald et al., 2002). The increasing N-saturation of European forests has raised concerns about a contamination of ground waters and stream waters with nitrate, which prevails in soil water output.

Tree species greatly influence the biogeochemistry of forests (Stone 1975, Kreutzer 1989, Binkley and Giardina 1998), so the choice of tree species in forestry management systems may also influence nitrate leaching. In this paper I examine the question of species influence on nitrate leaching rates, considering how large a shift in species composition would be needed to influence N leaching. I focus on the temperate forests of Central Europe for three reasons. Nitrate contamination of ground water is an important water-quality issue in this heavily populated area, and an intensive database on nitrate leaching in this region is available. From the broader perspective, forest management in the 18th and 19th Centuries shifted the original beech-dominated hardwood forests in the area to spruce plantations, and current management schemes are increasing the emphasis on hardwood forests.

NITRATE CONCENTRATIONS ARE HIGHER IN SPRUCE FORESTS THAN IN BEECH FORESTS

The concentrations of nitrate in seepage water in spruce forests are typically higher then in beech forests for comparable sites (Figure 1). The maximum nitrate concentrations are up to 11 mg N/L under spruce, more than three times the maximum for beech. Many mechanisms may influence how spruce and beech differ in N cycling (e.g. rooting patterns, litter quality, mineralization), the dominant factor influencing nitrate leaching may be differences in the rate of throughfall input of N to the soil. Spruce forests sustain a higher leaf area than beech, and the spruce canopies are maintained throughout the year. These canopy differences lead to higher dry deposition of N in spruce stands (Fig. 1). Across Europe, rates of N outputs correspond reasonably well with rates of N deposition (Callesen et al. 1999, MacDonald

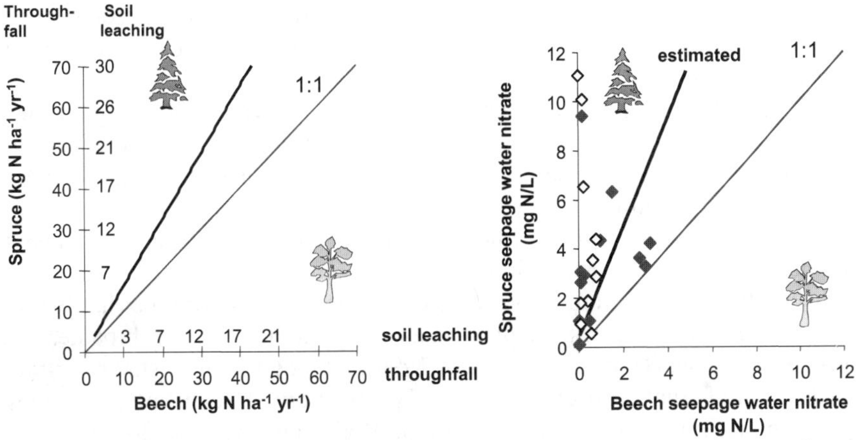

Figure 1. The flux of N in throughfall and soil leaching in spruce stands versus beech stands (left); throughfall N_{beech} = throughfall N_{spruce}/1,62 (Rothe et al. 2002), and soil leaching N = -1.87 + 0.46*Throughfall N (McDonald et al. 2002). Concentrations of nitrate-N in seepage water below the rooting zone of beech-spruce pairs (right); filled symbols are from European Case studies (Rothe et al., 2002), and open symbols are data from three model areas in Southern Germany (Rothe and Mellert, 2004). The "estimated" line for seepage water nitrate was calculated from the leaching flux (left graph) assuming a water yield of 270 mm for spruce and 370 mm for beech (precipitation 800 mm, interception 35 % for spruce and 22 % for beech, transpiration 250 mm for both species according to Augusto et al. 2002).

et al. 2002), indicating that the effect of species on deposition rates are more important than other species-related processes. Other processes may be important under special circumstances. For example, all three spruce stands in Figure 1 with nitrate-N concentrations higher than 9 mg/L were located 50 km northwest of Munich in a region characterized by high N inputs (throughfall N in spruce stands about 30 kg N ha^{-1} yr^{-1}). Despite the high N inputs nitrate concentrations in beech forests are very low and high gaseous losses (mainly as N_2 and N_2O) are more important for total N losses than leaching (Butterbach-Bahl et al. 2002).

Fewer data are available for species other than Norway spruce and beech. Nitrate concentrations in Scots pine stands often are very low (Gensior et al. 2003), but this may reflect the poor sites that are commonly occupied by pine, rather than a property of pine per se. High rates of N deposition could lead to increased N loading of these pine forests, and high nitrate concentrations have been observed in pine stands exposed to high N inputs for a long time (De Schrijver et al. 2000, Meesenburg et al. 2003). Available data for fir indicate a lower nitrate leaching compared to spruce (Heitz and Rehfuess 1999, Rothe and Mellert 2004). Since deposition rates are similar for both stand types, the species effect may indicate a more effective internal N cycling in fir stands. Due to its deeper rooting system, fir stands take up N more efficiently and store more N in woody biomass. Lower nitrate concentrations compared to spruce can also be expected in deciduous larch stands as a consequence of lower N inputs due to a reduced N interception in the dormant season (Adamson et al. 1993).

Strong effects on nitrate leaching have been shown for N-fixing species like alder (Kreutzer 1981, Binkley et al. 1992, Brown 1992). The effect on groundwater quality is small, however, due to their limited regional distribution in Central Europe (< 1% of forest cover). Additionally alders often occur on wet sites, where denitrification in deeper soil layers compensates nitrate formation in the topsoil (Kreutzer 1989).

Very little information is available on nitrate leaching for other hardwood species such as birch, oak, maple and ash. Differences among broadleaved species may exist (Heitz and Rehfuess 1999), although in most studies nitrate leaching did not differ significantly between broadleaved species (Falkengren-Grerup and Bergkvist 1995, Rothe and Mellert 2004). This indicates that differences between non N-fixing broadleaved species may not be very important for water quality, although they may be interesting biogeochemically. The key issues for forest management appear to center on the differences between broadleaves and conifers.

TREE SPECIES DIFFERENCES SCALE WITH POLLUTION LOAD

Atmospheric N deposition often is the crucial factor for nitrate leaching (Johnson and Lindberg 1992, Gundersen et al. 1998). For 181 European forests, throughfall N input explained over 60 % of the variation of leached nitrate (Mac Donald et al. 2002), and species may influence deposition rates. A spruce stand will tend to capture 60% more N deposition than a beech forest, and will have nearly twice the leaching loss of nitrate (Figure 1). The difference is even more pronounced for nitrate concentrations due to a higher seepaage water yield in beech stands. On average nitrate concentrations in beech stand will be only about 35-40 % of those of spruce stands (Fig. 1). Certainly variability for individual sites is high due to factors like stand history, soil type and decomposition rates.

Species-related differences in nitrate leaching are small under low rates of N deposition. As long as nitrate leaching is below 5 kg N ha^{-1} yr^{-1} (corresponding to throughfall N of about 15 kg N ha^{-1} yr^{-1}) nitrate concentrations even in spruce stands normally remain below about 2 mg N/L for a situation with average water balance. Figure 1 may even overestimate species differences for low atmospheric inputs assuming a linear relationship N input spruce vs. beech for the whole range of N input. However, the difference between spruce and beech was significantly smaller for throughfall N (spruce) < 20 kg N ha^{-1} yr^{-1} (Rothe et al. 2002). This corresponds well to the observation that throughfall N inputs are independent of vegetation cover in unpolluted areas (Hedin et al. 1995). In unsaturated N ecosystems with N leaching smaller than 5 kg N ha^{-1} yr^{-1} even major changes in tree species composition will have little effect on nitrate concentrations of soil drainage, and water quality of forest water resources is generally high. With increasing levels of N input and decreasing N retention, nitrate leaching will increase especially in spruce stands. Average nitrate concentrations of about 5 mg N/L can be expected in spruce stands vs. only about 2 mg N/L in beech stands when throughfall N input in spruce stands is about 30 kg N ha^{-1} yr^{-1} (Figure 1).

NITRATE CONCENTRATIONS ARE NEGATIVELY CORRELATED WITH PRECIPITATION INPUT

The concentration of N in seepage water depends on the total quantity of N leached, as well as the volume of water. Quantities of N loss may be most interesting in terms of biogeochemistry, whereas concentrations are important for water quality issues (including human health). The quantity of water leaching from beech stands is higher than from spruce stands due to lower interception losses from deciduous canopies. The effect of precipitation amount on nitrate concentration is illustrated in Table 1, with a constant output of 10 kg N ha^{-1} yr^{-1}. For high precipitation rates, nitrate concentrations are lower than 2 mg/L even for spruce, and the differences between the species are comparably low. The species effect is more pronounced when water fluxes are low. This means that in areas with low precipitation deciduous species will not only increase water yield but also enhance water quality significantly.

Table 1. Calculated nitrate concentrations in soil water output for a flux of 10 kg N ha^{-1} yr^{-1}. Water fluxes were calculated assuming canopy interception rates of 35 % for spruce and 22 % for beech, and 250 mm transpiration for both species (according to Augusto et al. 2002).

Precipitation mm	Spruce	Beech
	Nitrate (mg N/L)	
500	13.3	7.1
600	7.1	4.6
700	4.9	3.4
800	3.7	2.7
900	3.0	2.2
1000	2.5	1.9
1200	1.9	1.5
1400	1.5	1.2

STUDIES IN MATURE FORESTS MAY OVERESTIMATE NITRATE LEACHING.

Most investigations concerning nitrate leaching have been carried out in mature stands, even though rates of nitrogen transformations vary during stand development. The importance of stand development is illustrated in Figure 2 for a spruce stand exposed to high N deposition load. Elevated nitrate leaching occurs for 2 to 5 years after clearcutting, followed by a period of low nitrate concentrations in soil drainage. At this stage N demand usually exceeds N input mainly due to the establishment of nutrient rich foliage (Miller 1979; Jacobsen et al. 2002). As stands mature, foliage biomass remains constant on a stand level and N demand may decline even on productive sites. N input may significantly exceed N uptake at this stage and nitrate leaching occurs in stands with a limited retention capacity. As a

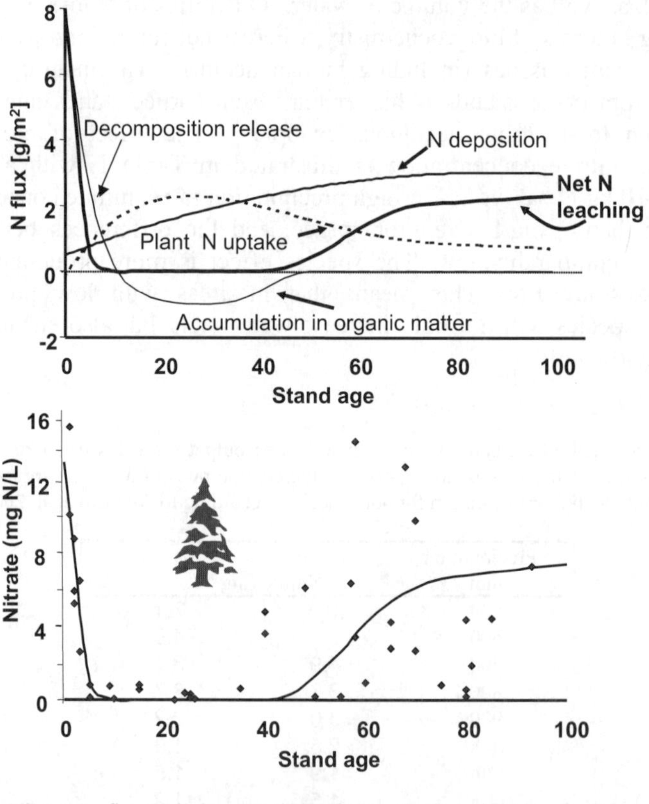

Figure 2. Nitrogen fluxes during stand development after clearcut for a spruce stand in Southern Germany supposed to high N deposition (upper, modified from Rothe and Mellert, 2004), and nitrate concentrations with stand age for spruce stands in three study areas in Southern Germany (lower, data from Rothe and Mellert 2004 and Weis 2003). The bold line in the lower graph was derived from the net leaching values in the upper graph, assuming a seepage water flux of 300 mm for mature stands and 600 mm after stand removal.

consequence nitrate concentrations usually are significantly lower in younger stands compared to mature forests (Leak and Martin 1975, Stevens et al. 1994).

Younger age classes commonly form an important fraction of forest cover in managed forests, and regional nitrate concentrations will be significantly lower than those of mature stands. The effect will be more important for forests close to N saturation and lower in cases where forest N retention remains high. This corresponds well with our findings in three research areas in Southern Germany, where the difference between nitrate concentrations of larger forest areas and those of mature spruce stands depended on N deposition. In the research area with the highest N deposition, the regional nitrate concentration of drainage water was "only" 4.7 mg N/L, compared with nitrate concentrations of some mature spruce stands in excess of the EU drinking water standard of 11.3 mg N/L. In the research area with the lowest N depostion nitrate conentrations in mature spruce stands were only 0.5 mg N/L higher compared to the regional nitrate concentrations of 1.6 mg N/L (Rothe and Mellert 2004).

NITRATE CONCENTRATIONS IN SOIL DRAINAGE WATER ARE HIGHER COMPARED TO GROUND OR STREAM WATER VALUES

Most studies concerning tree species effects on nitrate leaching have investigated soil water drainage using tension lysimeter cups. This method is well suited to study forest related effects, although rapid water movement through preferential flow pathways is not sampled well by these lysimeters (e.g. Hansen and Harris 1975, Hatch et al. 1997). An even bigger uncertainty arises from the fact that nitrate concentrations can change significantly during solute transfer below the rooting zone. Nitrate concentrations in stream and groundwater are typically lower than concentrations measured in soil drainage below the rooting zone. The reductions in concentration reflect mixing with other water in some cases, or nitrate removal. A famous example is the "Fuhrberger Feld" in Northern Germany, where the public water supply of the city of Hannover is located. Nitrate concentrations in the rooting zone are about 22 mg N/L for agricultural fields, but nitrate concentrations are near zero 8 meters below the soil surface, as a result of denitrification in deeper soil layers (Strebel et al. 1993). Denitrification effects in deeper soil layers have also been reported in forested areas (Rothe et al. 2000). However, information on the fate of nitrate as water moves from the rooting zone to deeper levels is too incomplete for any quantitative assessment.

NITRATE LEACHING IN FORESTS IS SMALL COMPARED TO AGRICULTURAL AREAS

In densely populated Central Europe, land use patterns are complex and most watersheds containing forests also contain other land uses. Large, completely forested watersheds are rare. Fertilization is not a common practice in forests in Central Europe, whereas agricultural fields typically receive 100 to 200 kg N ha^{-1} yr^{-1}. Although nitrate concentrations in agricultural areas vary widely depending on crop type and management, water draining from agricultural areas usually shows significantly higher nitrate concentrations compared to those from forests (e.g., Callesen et al. 1999, Binkley et al. 2004). Unfortunately few studies have analyzed the importance of individual land use types for nitrate leaching in a special catchment. Usually such studies find that forest contribution to overall nitrate leaching is small (Strebel et al. 1993). This indicates that the impact of forest management often will be negligible compared to the impacts of differences in agricultural management. Additionally agriculture can change crop type and fertilizer input annually in contrast to long term forestry operations. Nevertheless forest management may play role in catchments where “clean” water from forest dilutes contaminated water form agricultural areas. In this case even a comparably small change of nitrate leaching from the forest area may threaten water quality from the whole catchment.

SOME QUANTITATIVE ESTIMATES FOR BAVARIA

Bavaria is the southernmost state of Germany, and 36% of its 7.5 million hectares is covered by forests. Norway spruce is the major species in the coniferous 70% of the forest landscape, and European beech dominates the broadleaved areas. The atmospheric deposition of N ranges from 5 to 40 kg ha^{-1} yr^{-1}, and throughfall N input to the soil surfaces commonly ranges between 15-20 kg ha^{-1} yr^{-1} for conifer forests, and 10-15 kg ha^{-1} yr^{-1} for broadleaved forests (Bayerisches Staatsministerium für Landwirtschaft und Forsten 2004). Nitrate concentrations of soil water in forests have been examined in detail within a statewide inventory in the years 2001/02, covering 399 sampling points in a regular grid (Gensior et al. 2003). Nitrate concentrations below the rooting zone were below 2.2 mg N/L for 65% of all forests and 8% of all forests exceeded 11 mg N/L. The distribution of values was skewed to lower concentrations, so the median nitrate concentration for Bavarian forests was about 1.2 mg N/L. Water quality of forested areas is generally high, and most forests continue to accumulate N (outputs less than inputs). However, this general pattern has the notable exception of 35% of all sites showing elevated nitrate concentrations higher than 2.3 mg N/L.

The historic species composition of the state-owned forests (about 1/3 of total forest area) has been well characterized for a long time (Figure 3). The original natural forest cover was dominated by beech, and most broadleaf stands were converted to conifer plantations in the18th and 19th century. By 1900 the percentage of broadleaved species had declined to only 20% where it remained for the next 50 years. The planting of broadleaved species increased after the Second World War, but then declined between 1965 and 1980 because of the greater market value of spruce (Figure 3). In 1982 the Bavarian State Forest Service introduced new silvicultural guidelines emphasizing "close to nature forestry," with major objectives of increasing the proportion of broadleaved and mixed-species stands. The proportion of broadleaves in artificial regeneration (planting and seeding) has increased since then, rising to about 70% currently. These regeneration efforts are clearly reflected in the increased proportion of the forest landscape dominated by broadleaved species. The extent of broadleaved forests has increased by 40,000 hectares in 30 years, corresponding to about 5 football fields per day. The data, however, also give an impression about the long time

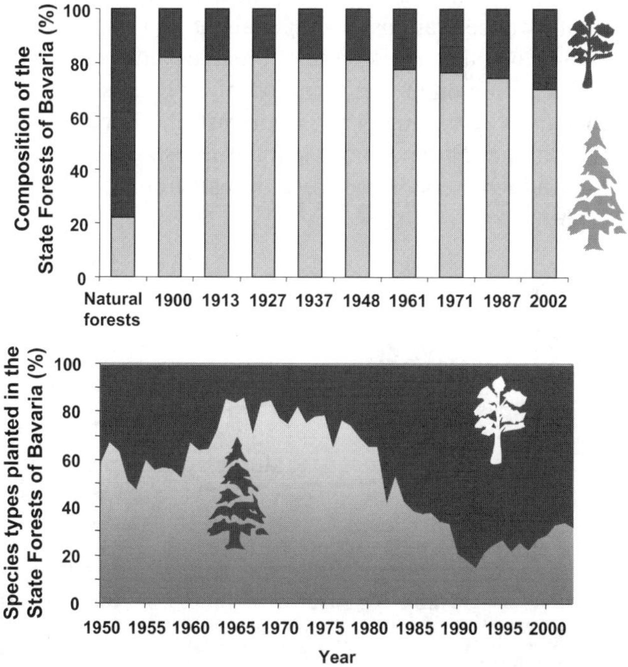

Figure 3. Composition of stands in the State Forests of Bavaria. (upper; modified from Rothe and Borchert 2003, data for the natural forest cover from Walentowski et al. 2004), and types of species planted (modified from Rothe and Borchert 2003).

spans inherent to forest management. Due to relatively high rotation periods (on the average about 100 years) and due to a high proportion of natural regeneration (which is mainly coniferous according to the present species distribution) even the strong efforts of the last 20 years in favor of broadleaved species “only” resulted in a yearly conversion speed from conifers to broadleaves of 0.3% of forest areas per year. This means that it will take at least another 60 to 100 years to achieve the long-term goal of a balanced proportion of conifers and broadleaves in Bavaria’s state forests.

As noted above, the quality of forest water resources in Bavaria remains generally high, with median nitrate concentrations in the rooting zone of only 1.2 mg N/L despite the prevalence of spruce. A rough calculation indicates that the 30-year increase of broadleaved forests of 40,000 hectares (corresponding to about 6 % of total states forests) would have had a minimal impact on average nitrate concentrations (< 0.2 mg N/L) in seepage water. Such general calculations may overlook local effects, so we analyzed a scenario for alternative species composition for two study areas assuming all other factors remaining constant (Figure 4). The forests in the “Munich gravel plain” study area are still accumulating most of the incoming nitrogen, and regional nitrate concentrations average about 1.5 mg N/L. Our model calculations indicated that an increase of broadleaved species from the current 15% up to 60% within one rotation period of 100 years would reduce nitrate concentrations only by 0.5 mg N/L. In contrast, the Eurasburger Forest study area has much higher N deposition, and the forests retain little if any of the deposited N. An increase in broadleaved forests from 15% of the area to 60%

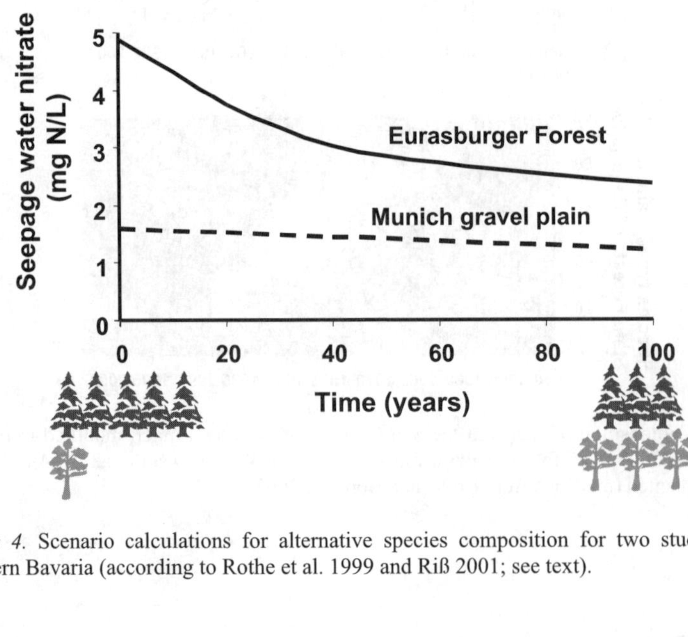

Figure 4. Scenario calculations for alternative species composition for two study areas in Southern Bavaria (according to Rothe et al. 1999 and Riß 2001; see text).

might reduce nitrate concentrations in seepage water by 2.4 mg N/L. The effect of tree species management clearly depends on the ecosystem's stage concerning N saturation. The Munich gravel plain currently is more representative of the overall region, but high N inputs across the region enhance the importance of tree species in moderating a possible future increase of N concentrations in soil water. Mixed-species forests with higher proportions of broadleaved species may be less prone towards nitrate leaching even with increasing nitrogen saturation. The policy of increasing the coverage of broadleaved species in Bavaria also serves the goal of reducing risk of nitrate contamination of forest groundwater resources.

CONCLUSIONS

1. Nitrate leaching in forests is generally low as long as nitrogen deposition is small and forest continue to accumulate N. At this stage tree species management is of minor importance for ground water quality.
2. Tree species likely influence N leaching in areas with high N deposition. Nitrate concentrations in soil drainage water in conifer forests are typically higher than for broadleaved forests. Broadleaved species would be recommended for planting in sensitive areas such as ground water protection zones.
3. Substantial shifts in the species composition of forests take long periods of time, so foresters can help to reduce the risk of ground water contamination only in the long run.
4. In catchments with mixed land uses, the importance of forests for nitrate leaching usually is small compared to agricultural lands. Changes in agricultural management will have a much stronger and faster impact than any forestry activities.
5. The problem of elevated nitrate concentrations in forest water resources is not a problem of the forests per se but the consequence of anthropogenic N deposition exceeding the accumulation capacity of forest ecosystems. The fundamental solution to protecting ground water resources would be a reduction of N emissions.
6. While there is a good knowledge about the influence of tree species on soil water, information about the effects on groundwater is limited. Studies following the fate of nitrate from the rooting zone into the groundwater are necessary to close this knowledge gap.
7. Nearly all studies on nitrate leaching concentrated either on forests or agriculture. Studies incorporating different land use types within common watersheds would be very helpful in order to identify the relevance of the land use types for nitrate contamination of groundwater.

REFERENCES

Adamson J K, Hornung M, Kennedy V H, Norris D A, Paterson I S and Stevens P A 1993 Soil solution chemistry and throughfall under adjacent stands of Japanese Larch and Sitka Spruce at three contrasting locations in Britain. Forestry 66 (1), 51-68.

Augusto L, Ranger J, Binkley D and Rothe A 2002 Impact of several common tree species of European temperate forests on soil fertility. Ann. For. Sci. 59, 233-253.

Bayerisches Staatsministerium für Landwirtschaft und Forsten 2004 Waldzustandsbericht 2004. Bayerisches Staatsministerium für Landwirtschaft und Forsten, Munich, Germany, http://www.forst.bayern.de.

Binkley D and Giardina C 1998 Why do tree species affect soils? The warp and woof of tree-soil interactions. Biogeochemistry 42, 89-106.

Binkley D, Burnham H, and Allen H L 1999 Water quality impacts of forest fertilization with nitrogen and phosphorus. For. Ecol. Manage. 121, 191-213.

Binkley D, Ice G G, Kaye J and Williams C A 2004 Patterns of variation in nitrogen and phosphorus concentrations in forest streams of the United States. Journal of the American Water Association 2004:1277-1291.

Binkley D, Sollins P, Bell R, Sachs D and Myrold D 1992 Biochemistry of adjacent conifer and alder conifer stands. Ecology 73, 2022-2033.

Brown A H F 1992 Functioning of mixed-species stands at Gisburn, NW-England. In: Cannell MGR, Malcolm DC and Robertson PA (eds.): The ecology of mixed-species stands of trees. Special Publication 11 of the Britisch Ecological Society, Blackwell Scientific Publications, Oxford, 125-150.

Butterbach-Bahl K, Gasche R, Willibald G and Papen H 2002 Exchange of N-gases at the Höglwald forest - A summary. Plant and Soil 240, 117-123.

Callesen I, Raulund-Rasmussen K, Gundersen P and Stryhn H 1999 Nitrate concentrations in soil solution below Danish forests. For. Ecol. Manage. 114, 71-82.

De Schrijver A, Van Hoydonck G, Nachtergale L, De Keersmaeker L, Mussche S and Lust N 2000 Comparison of nitrate leaching under Silver Birch (Betula pendula) and Corsican Pine (Pinus Nigra SSP. Laricio) in Flanders (Belgium). Water, Air and Soil Pollution 122, 77-91.

Falkengren-Grerup U and Bergkvist B 1995 Effects of acidifying air pollutants on soil/soil solution chemistry of forest ecosystems. Annali di chimica 85 (7-8), 317-327.

Gensior A, Kölling C and Mellert K H 2003 Die Nitratinventur in Bayern. Methodik und Ergebnisse. Berichte Freiburger Forstliche Forschung, Heft 49, 101-113.

Gosz J R 1981 Nitrogen cycling in coniferous forests. In: Clark FE and Rosswall T (eds.): Terrestrial Nitrogen Cycles. Ecol. Bull. (Stockholm) 33, 405-425.

Gundersen P, Emmett B A, Kjonaas O J, Koopmans C J and Tietmema A 1998 Impact of nitrogen deposition on nitrogen cycling in forests: a synthesis of NITREX data. For.Ecol.Manage. 101, 37-56.

Hansen E A and Harris A R 1975 Validity of soil water samples collected with porous ceramic cups. Soil Science Society of America Proceedings 39, 528-536.

Hatch D J, Jarvis S C, Rook A J and Bristow A W 1997 Ionic contents of leachate from grassland soils, a comparison between ceramic suction cup samples and drainage. Soil Use and Management 13, 68-74.

Hedin L O, Armesto J J and Johnson A H 1995 Patterns of nutrient loss from unpolluted, old-growth temperate forests: Evaluation of the biogeochemical theory. Ecology 76 (2), 493-509.

Heitz R and Rehfuess K E 1999 Reconversion of Norway spruce (Picea abies (L.) Karst.) stands into mixed forests: effects on soil properties and nutrient fluxes. in Olsthoorn et al. (eds.) Mangement of mixed-species forest: silviculture and economics. IBN Scientific contributions 15, IBN-DLO, Wageningen, NL., 37-45.

Jacobsen C, Rademacher P, Meesenburg H and Meiwes K J 2002 Gehalte chemischer Elemente in Baumkompartimenten – Literaturstudie und Datensammlung. Niedersächsische Forstliche Forschungs- und Versuchsanstalt, Göttingen, http://www.nfv.gwdg.de.

Johnson D W and Lindberg S E (Eds.) 1992 Atmospheric deposition and forest nutrient cycling. Ecological Studies 91, Springer Verlag, New York.

Kreutzer K 1981 Die Stoffbefrachtung des Sickerwassers in Waldbeständen. Mitteilungen Deutsche Bodenkundliche Gesellschaft 32, 273-286.

Kreutzer K 1989 The impact of forest management practices on the soil acidification in established forests. Air Pollution Research Report 13, 75-90. Comission of the European Communities, Brussels, Belgium.

Leak W B and Martin C W 1975 Relationship of stand age to streamwater nitrate in New Hampshire. USDA Forest Service Research Note NE-211 Upper Darby, Pennsylvania.

MacDonald J A, Dise N B, Matzner E, Armbruster P, Gundersen P and Forsius M 2002 Nitrogen input together with ecosystem nitrogen enrichment predict nitrate leaching from European forests. Global Change Biology 8, 1028-1033.

Meesenburg H, Horvath B and Meiwes KJ 2003 Stoffhaushalt von Waldökosystemen NW-Deutschlands unter hoher Stickstoffbelastung. Berichte Freiburger Forstliche Forschung, Heft 49, 57-69.

Melillo J M 1981 Nitrogen cycling in deciduous forests. In: Clark FE and Rosswall T (eds.): Terrestrial Nitrogen Cycles. Ecol. Bull. (Stockholm) 33, 427-442.

Miller H G 1979 The nutrient budget of even-aged forests. In Ford E D, Malcolm D C and Atterson J (eds.) The ecology of even-aged forest plantations. Institute of Terrestrial Ecology, Cambridge.

Riß M 2001 Untersuchungen zur Nitratbelastung in den Wäldern der südlichen Münchner Schotterebene 2. Teil: Flächige Abschätzung und Prognose. Master Thesis, Forest Faculty, Technische Universität München.

Rothe A and Borchert H 2003 Der Wald für morgen – eine Naturalbilanz über 25 Jahre naturnahe Forstwirtschaft im Bayerischen Staatswald. LWF-Bericht 39, Bayersiche Landesanstalt für Wald und Forstwirtschaft, Freising, Germany.

Rothe A and Mellert K H 2004 Effects of forest management on nitrate concentrations in seepage water: Results from three model areas in Southern Bavaria, Germany. Water, Air and Soil Pollution, in press.

Rothe A, Brand S and Hurler R 1999 Waldbewirtschaftung und Nitratbelastung des Grundwassers. AFZ/Der Wald 10, 531-533.

Rothe A, Englschall M, Hurler R, Wittfoth J and Butterbach K 2000 Nitratverlagerung in tieferen Bodenschichten eines süddeutschen Waldgebietes. Wasser und Boden 52 (11), 52-56.

Rothe A, Huber C, Kreutzer K and Weis W 2002 Deposition and soil leaching in stands of Norway spruce and European beech: Results from the Höglwald in comparison with other European case studies. Plant and Soil 240, 33-45.

Stevens P A, Norris D A, Sparks T H and Hodgson A L 1994 Soil and Stream water interactions for different aged forest and moorland catchnments in Wales. Water, Air, and Soil Pollution 73, 297-317.

Stone E 1975 Effects of species on nutrient cycles and soil change. Philosophical Transactions of the Royal Society, London (B) 27, 149-162.

Strebel O, Böttcher J and Duynisveld W H M 1993 Ermittlung von Stoffeinträgen und deren Verbleib im Grundwasserleiter eines norddeutschen Wassergewinnungsgebietes. Texte des Umweltbundesamtes 46, 86 S. Federal Environmental Agency of Germany, Berlin.

Walentowski H, Ewald J, Fischer A, Kölling C and Türk W 2004 Handbuch der natürlichen Waldgesellschaften Bayerns. Verlag Geobotanica, Freising, Germany, 441pp.

Weis W 2003 Vergleichende Untersuchungen zum Stoffverlust in Waldökosystemen bei Verjüngung über Gruppenschirmstellung und Kleinkahlschlag. Abschlussbericht zum Forschungsbericht B59, Bayerische Landesanstalt für Wald und Forstwirtschaft, Freising, Germany.

Chapter 6

PLANT EFFECTS ON SOILS IN DRYLANDS: IMPLICATIONS FOR COMMUNITY DYNAMICS AND ECOSYSTEM RESTORATION

Jordi Cortina[1] and Fernando T. Maestre[2]
[1] Departament d'Ecologia, Universitat d'Alacant Ap. 99 03080 Alacant, Spain; [2] Department of Biology, Duke University, Phytotron Building, Box 90348, Durham, NC 27708 USA.

INTRODUCTION

Almost 50% of the emerged land (6.15 10^6 ha) are considered drylands (Reynolds and Stafford Smith, 2002). They are distributed in four continents, covering from 31% (South America) to 75% (Australia) of the continental land area. Drylands encompass areas with a wide range of conditions in relation to the ratio Precipitation/Evapotranspiration (P/E), from hyperarid (P/E<0.03) to Dry Subhumid (0.5<P/E<0.75) (UNESCO, 1977). Throughout the text, we will use the term drylands in a broad sense, meaning areas where precipitation is so scarce that it becomes the main factor controlling biological processes. This definition corresponds to what Noy Meir (1973) refers to as arid ecosystems

In areas like the Mediterranean, drylands have been subjected to a long history of land use (Hillel, 1992; Grove and Rackham, 2001). But this is not always the case, as different dryland areas have supported extremely contrasted human population densities. The boundary between subhumid to semiarid conditions may be particularly sensitive to disturbances, and population buildup in subhumid and semiarid zones have probably favoured the expansion of areas characterized by discontinuous vegetation cover. This process is not recent, as population density was relatively high in areas such as the Mediterranean basin 2000 years ago (Grove and Rackham, 2001). For example, Butzer (1990) suggests that population density in the Iberian Peninsula was close to the carrying capacity for agricultural systems several times since 2,000 BP. Phases of economic welfare and social integration correspond to higher demands for resources, including land, leading to defor-

D. Binkley and O. Menyailo (eds.),
Tree Species Effects on Soils: Implications for Global Change, 85–118.

estation and, eventually, land degradation. The 20th century has shown an unprecedented increase in population density in many drylands (Le Houérou, 1992). But, unlike what happened in the past, population density has mainly concentrated in coastal areas. As a result, recent population increase has been paralleled by a reduction in the intensity of inland land use (and an increased demand for resources overseas).

Changes in land use are relevant in the context of species effect on soil properties for two reasons. First, land use, together with natural factors, have favoured land degradation in vast areas. For example, alpha grass (*Stipa tenacissima*) steppes, once covering more than 8 million ha in the Maghrib, are being destroyed at a rate of 1% per year (Aïdoud, 1989; Le Houérou, 2001). Past disturbances may interact with plant species to generate unexpected effects. For example, litterfall inputs may not be the same if individuals are established in a slope that had either been terraced or left aside. Second, with few exceptions, most individual plants may have established in their present locations not long ago. Examples of changes in land use dating less than 50 years are common (Grove and Rackham, 2001; Bonet, 2004). In contrast, the effect of land uses such as agriculture or grazing on ecosystem structure and functioning may last for centuries (Bruun et al., 2001; Maestre 2004). Changes in soil phosphorus content in the surroundings of Roman farms have been detected up to 2,000 years after abandonment (Dupouey et al., 2002). These changes were large enough to affect P concentration in oak leaves. Thus, information accumulated in soils for centuries adds noise to the interpretation of species effect on soil properties (Cortina, 1992).

There are many ways in which plants may affect soils. They can modify soil properties directly, e.g. through inputs of organic matter and nutrients, and indirectly, by affecting abiotic and biotic conditions that influence soil properties. For example, symbiosis with specific strains of mycorrhizal fungi can be relevant for the establishment of their own progeny or for the colonization by other species (Palenzuela et al., 2002; Azcón-Aguilar et al., 2003). Under dry conditions, shade may affect litter decomposition directly (Duguay and Klironomos, 2000; Verhoef et al., 2000), and as a result of changes in microclimatic conditions (Jackson and Caldwell, 1992; Cortina and Vallejo, 1994). Sometimes, changes in soil properties affect plant composition and growth, generating positive feedbacks (Northrup et al., 1995; Morehead et al., 1998). On the other hand, plants may affect soil properties in ways that may appear subtle to us, but may be rather evident to other organisms (Bais et al., 2003, 2004). This complex network of interactions hampers the interpretation of plant-soil relations. Unfortunately, our knowledge is still quite broad, and surprises arise too often.

Despite the intrinsic variability in soil properties and the complexity of soil-plant interactions, plant effects on soils play a major role in population, community and ecosystem dynamics in drylands. Several features of drylands may contribute to this. By definition, climatic conditions are harsh and soil conditions are unfavourable for plant growth due to salinity, low organic matter, P immobilisation, etc. Thus, any slight modification of microclimate or soil properties may have disproportionate effects on other organisms (e.g., Pugnaire et al., 1996a; Moro et al., 1997; Maestre, 2003). Furthermore, xerophytes are commonly small and isolated, as compared to plants from temperate and tropical areas. This may favour litterfall, throughfall, stemflow and fine root inputs accumulating underneath the canopy, and thus intensifying plant effects. Finally, isolated plants act as obstacles for runoff carrying organic matter, nutrients and sediments (Greene et al., 2001). Thus, it is not surprising that certain plants and vegetation patches have higher infiltration rates, improved soil structure and nutrient content, and higher biological activity, creating "hotspots" of favorable soil conditions that have been referred to as "fertile islands" or "resource islands" (Whitford, 2002).

As we just mentioned, plant alteration of soil and microclimate conditions may affect other organisms to the point of controlling the composition and function of the whole ecosystem (West, 1989; Schlesinger and Pilmanis, 1998; Montaña et al., 2001). Identifying these interactions is crucial to understand and to manage dryland ecosystems. In this review, we will describe the way plants affect soil properties in drylands, and how these feedbacks, together with changes in microclimatic conditions, affect plant-plant interactions and community dynamics. In the last section, we will discuss the use of this knowledge for the restoration of degraded dryland ecosystems.

PLANT EFFECTS ON WATER DYNAMICS AND USE

In drylands, water scarcity controls plant effects on soils in several ways. Biological processes, including plant productivity, are strongly limited by water. Water limitation results in discontinuous plant cover, which is often arranged as a two-phase mosaic of vegetated and bare ground patches (Valentin et al., 1999). Spatial patterns of vegetation, together with morphological and physiological attributes of plants greatly affect water fluxes and availability (Bromley et al., 1997, Cerdà, 1997). Indeed, the spatial pattern of plant patches is essential to maintain ecosystem composition and function in drylands (Noy Meir, 1973; Tongway et al., 2001), and can be used as an indicator of the degradation status of arid and semiarid lands (Wu et al., 2000; Bastin et al, 2002; Maestre and Cortina, 2004a).

At the patch level, canopies differ in their capacity to intercept water. For small rains, which normaly account for a large number of rain events, interception can be high (Martínez-Meza and Whitford, 1996; González-Hidalgo and Bellot, 1997; Bellot et al., 2001), and may promote the formation of dry shadows (Valladares and Pearcy, 2002). Stemflow and throughfall are strongly dependent on plant architecture (Martínez-Meza and Whitford, 1996, Domingo et al., 1998). Funnel-like structures, as found in many shrubs, favour moisture concentration around the base of the stems (West, 1989; Martínez-Meza and Whitford, 1996), whereas tussock grasses may concentrate throughfall inputs around the canopy edge (Puigdefábregas et al., 1999). Rainfall redistribution may affect deep soil moisture content as well, as water fluxes at the base of the stem may circulate down the soil profile following the roots and root pores (West, 1989; Martínez-Meza and Whitford, 1996; Ryel et al., 2003). These fast tracks for water fluxes may explain why increases in moisture content following a rainfall event are not necessarily sequential, from top to bottom, but water may saturate deeper soil profiles earlier than shallow ones (González-Hidalgo et al.,2003). Species with the highest stemflow, and with deep roots may favour the recharge of deep soil horizons (Martínez-Meza and Whitford, 1996). These traits are common in some desert and Mediterranean plants (Cannon, 1911; Kummerov, 1981). On the other hand, hydraulic lift –water translocation to upper or lower parts of the soil profile through the rooting system – has been described in drylands worldwide (Caldwell et al., 1998; Burguess et al., 1998; Filella and Peñuelas, 2003; Gutiérrez and Squeo, 2004). This movement of water may affect N mineralization and microbial activity as well (Caldwell et al., 1998). The

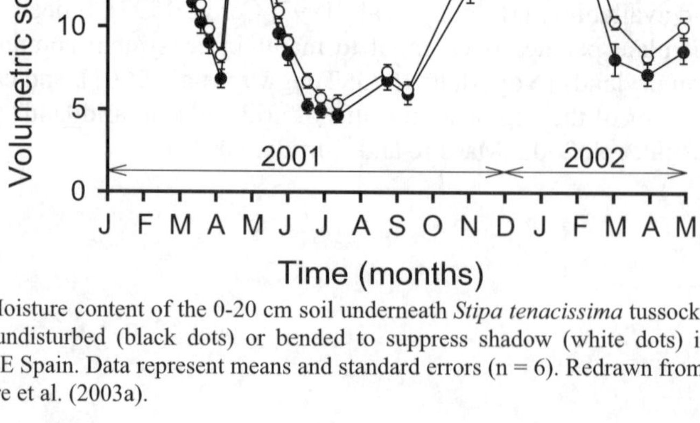

Figure 1. Moisture content of the 0-20 cm soil underneath *Stipa tenacissima* tussocks that were either left undisturbed (black dots) or bended to suppress shadow (white dots) in a steppe located in SE Spain. Data represent means and standard errors (n = 6). Redrawn from Appendix A in Maestre et al. (2003a).

shade provided by plant canopies, together with litter accumulation, influence the soil water balance in the surroundings of plant patches (Fig. 1). It is important to note that shade may not only affect soil moisture content, but may also influence individual plant performance and plant community composition by modifying evaporative demand, and light quantity and quality (Valladares, 2003).

Stormy rains are common in some drylands, such as the Mediterranean and West Peru (De Luís et al., 1997; Arntz and Fahrbach, 1996). For example, in Spain, more than 100 rain events with intensity higher than 100 mm in 24 h, including seven rain events higher than 500 mm in 24 h, were recorded between 1901 and 1989 (Olcina, 1994). There is increasing evidence that the amount and frequency of large vs. small rainfall events may control ecosystem function and composition in drylands (Ehleringer et al., 1999). One of the consequences of rainfall concentration is that part of the water does not accumulate in soils, but is either drained or exported as runoff (Bellot et al., 2001; De Luís et al., 2001). Surface runoff can have strong effects on plant-soil interactions because it may transport sediments and organic matter within or beyond the slope, and because it favours the increase in soil fertility and water availability under plant canopies (Cerdà, 1997; Puigdefábregas et al., 1999). Structures in contact with the surface soil (lower branches, multiple stems, litter, tussock plants, etc.) favour runoff retention (Rostagno, 1989; Bromley et al., 1997), and attributes of plant patches like size, width parallel to the slope, and spatial pattern are critical to define the ability of an ecosystem to retain and use those resources transported by runoff (Ludwig et al., 1999).

Most of the water inputs in drylands are lost by evapotranspiration. Thus, in Mediterranean watersheds, annual evapotranspiration, and not drainage, is well correlated with precipitation (Piñol et al., 1991). Vegetation management, involving shifts from woody to herbaceous vegetation, partial clearing and formation of stone pavements, etc. has been used to collect more drainage water (Hillel, 1992; Lavee et al., 1997; Burch et al., 1987). In general, evapotranspiration is proportional to LAI, and suppression of woody vegetation commonly results in increasing soil moisture availability, seepage drainage or watershed runoff (Burch et al., 1987; Bellot et al., 2001), although exceptions exist (Dodd et al., 1998). On the contrary, increases in woody cover (e.g. such as shrub encroatchment) may lead to a reduction in surface water availability or groundwater levels (Puigdefábregas and Mendizábal 1998; Ohte et al., 2003; Bellot et al., 2004) (Fig. 2). The increase in the cover of woody invasive species can have important economic consequences (Le Maitre et al., 2002). Species identity may also be relevant for water losses. Obvious examples are summer-deciduous or semi-deciduous species (e.g. *Cistus salviifolius*, *Cistus albidus*, *Euphorbia dendroides*, etc.), that reduce evapotranspiration losses by leaf shedding (Ne'eman and Goubitz, 2000). The difference between drought escapers, drought avoiders and drought tolerant species (Levitt, 1980), could be helpful in this context. Comparisons in water use between pairs of species are frequent. For example, pines commonly close stomata at relatively high water potential, and thus may reduce water losses during drought as compared to other species. *Pistacia lentiscus* and *Quercus coccifera*, two coexisting shrubs common in the western Mediterranean basin,

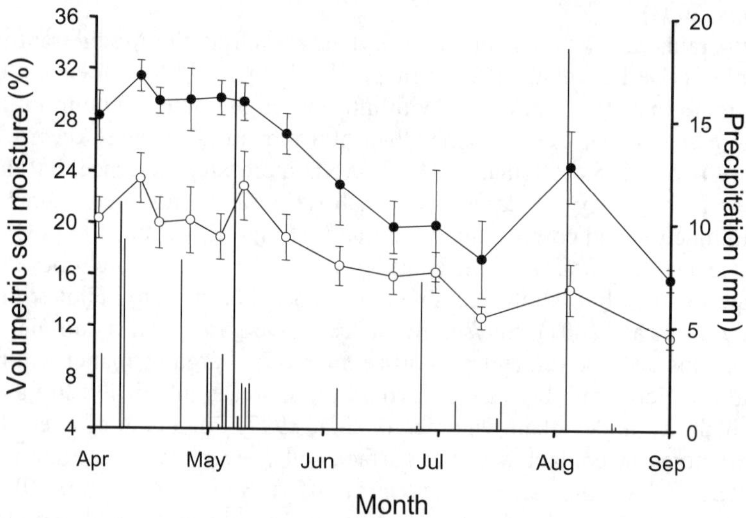

Figure 2. Changes in moisture content of the surface soil (0-10 cm depth) of adjacent Mediterranean semiarid grasslands with (white circles) and without (black circles) a dense cover of *Pinus halepensis* (3470 trees ha^{-1}). Bars correspond to rainfall events recorded during the period of study (April-September 1996). Soil moisture data represent means and standard errors (n = 5). Redrawn from Bellot et al. (2004).

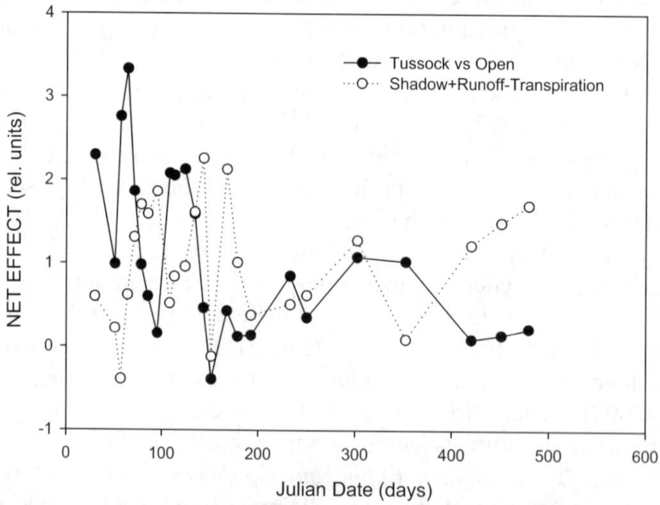

Figure 3. Net effect of the tussock grass *Stipa tenacissima* on soil moisture content (0-20 cm depth) as compared to the arithmetic addition of the net effects of transpiration, runoff and shadow in a Mediterranean semiarid steppe of SE Spain. Net effects are calculated from the standardized differences in moisture content between tussock and open microsites (net tussock effect), tussock and herbicided tussock (net transpiration effect), tussock with and without runoff (net runoff effect) and standing and bended tussock (net shadow effect). From original data in Maestre et al. (2003a).

show contrasted water use strategies (Vilagrosa et al., 2003). Other factors being equal, the water spender *P. lentiscus* is likely to deplete soil moisture sooner after a rain that *Q. coccifera*. But, to our knowledge, no attempt to systematically relate plant strategy to withstand drought, and water dynamics at a plot and catchment scale has been made so far.

Finally, plants may affect water availability by favouring the formation of calcite and laterites, and thus modifying soil volume (Viles, 1990). Water uptake from the vadose zone may lead rhizoconcretion precipitation around particular roots, whereas phreatophytes may favour different forms of calcite precipitation in the zone of capillary rise (Thomas, 1988). These processes result in the reduction of root absorptive capacity, root death, and finally the decrease in available soil.

Given the diversity of effects of plants on water fluxes, it is not surprising that the effect of plants on water availability, a critical soil property in drylands, may substantially vary according to plant community composition and structure. Furthermore, it is difficult to explain variations in moisture content as the additive effect of changes in single fluxes (Fig. 3). On the other hand, plant effects on soil moisture content may be short-lived, being restricted to the lapse between a rainfall event (and, eventually, homogeneous high moisture content), and soil dessication, or to particular periods of the year (Belsky et al., 1993) (Fig. 4). This is particularly true for shallow soils and surface soil horizons.

ORGANIC MATTER AND NUTRIENTS: ISLANDS OF FERTILITY

Litterfall inputs are relatively low in drylands due to constrains in plant productivity (Berg et al., 1999; Breckle, 2002), but they may be substantially higher immediately underneath the canopy. Variability in litter decomposition rates is very high, ranging from some of the lowest rates recorded (0.07 yr^{-1}; Hart et al., 1992), to relatively high values (Gallardo and Merino, 1992). For Mediterranean ecosystems, Aerts (1997) found that short term decay rates were largely variable, averaging 0.35 $years^{-1}$. These values were very close to those of temperate ecosystems (0.36 $years^{-1}$), and substantially lower than decay rates in tropical areas (2.33 $years^{-1}$). Within a given site, variability in litter decomposition rates can be very high (Gallardo and Merino, 1992), suggesting that species identity may control soil organic matter (SOM) dynamics and nutrient availability. In Mediterranean ecosystems litter decay rates are related to actual evapotranspiration (AET) and litter quality (Aerts, 1997), but the amount of variability explained by single climatic or chemistry parameters is very low. In desert ecosystems, AET underestimates decomposition rates, probably due the activity of soil fauna (Whitford et al., 1981). The relationship between litter decay rates and a widely used index of

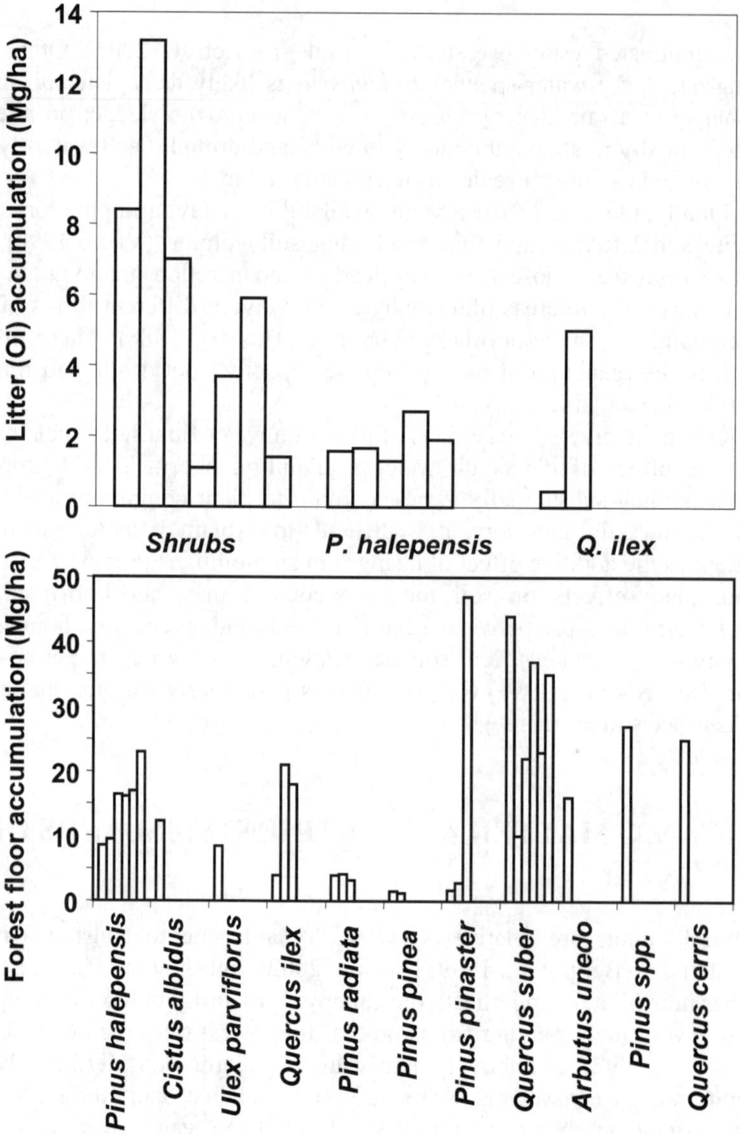

Figure 5. Litter (upper graph) and forest floor (lower graph) accumulation in Mediterranean forests and shrublands dominated by different woody species. Data from Fons (1995), Ferran (1996), Huesca et al. (1998), Serrasolsas and Vallejo (1999), Cortina (1992), Berg et al. (1993), Sevink et al. (1989), and Van Wesemael and Veer (1992). Only maximum values were taken from studies reporting measurements at several locations.

recalcitrance, the lignin to nitrogen ratio, is often poor in drylands (Gallardo and Merino, 1992; Aerts, 1997). The resistance of the external layers of leaves and needles may be more important drivers of decomposition rates in these environments than the quality of the whole tissue (Gallardo and Merino, 1992; Cortina and Vallejo, 1994).

Forest floor accumulation in drylands is commonly lower than in more mesic environments because of low litterfall, and relatively high decomposition rates and vertical transfers (Vallejo et al., 1998; Fig. 5). However, particularly dry conditions and surface accumulation of rock fragments may favour the formation of xeromoder type duff layers (Sevink et al., 1989; Fons, 1995). This is probably due to reduced transfer to lower soil horizons, as CO_2 efflux is promoted by the presence of a stone layer (Casals et al., 2000). As previously mentioned, small height and isolation may increase the potential of dryland vegetation to concentrate litterfall, and locally promote forest floor build-up. Plant species differ in forest floor accumulation, and morphology (Peltier et al., 2001), although other factors

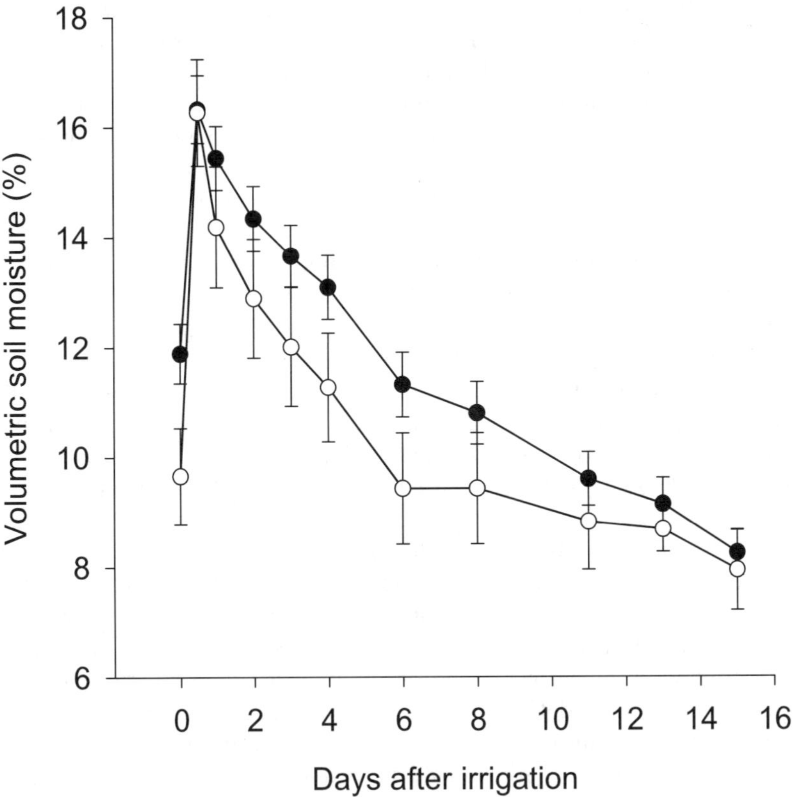

Figure 4. Changes in volumetric soil moisture content in the surface soil (0-20 cm) underneath *Stipa tenacissima* tussocks (black circles) and in open areas (white circles) after irrigation in a semiarid steppe in SE Spain. Soil moisture data represent means and standard errors (n = 10). Redrawn from Maestre et al. (2001).

such as lithology may be more important in determining forest floor properties (Sevink et al., 1989; Fons, 1995). Aphyllous species (crassulacean, many legumes, etc.), and species with no spontaneous leaf shedding (e.g. tussock grasses, palms) accumulate small amounts of litter; whereas species loosing a substantial amount of their foliage every year can create a relatively thick forest floor (Bochet et al., 1999; Peltier et al. 2001). Under Mediterranean conditions, conifers may accumulate thicker forest floor layers than hardwoods (Peltier et al., 2001).

Litter incorporation into the mineral soil and SOM stabilization can be greatly affected by soil fauna (Whitford, 2002). Soil fauna may comminute partly decomposed litter and favour SOM transport down the profile (Anderson, 1988; Bertrand and Lumaret, 1992; Romanyà et al., 2000a). It is also affected by plant species and plant cover type. For example, in New Mexico deserts, shrub cover may negatively affect the presence and activity of some mammals and promote the abundance of rabbits, but may not affect ants and termites (Krogh et al., 2002; Jackson et al., 2003). In Mediterranean environments, millipedes are more frequently found under *Quercus coccifera* than under *Q. ilex*, *Q. pubescens* or *Brachypodium ramosum* (Bertrand and Lumaret, 1992). However, Maestre and Cortina (2002) did not find any relationships between the spatial pattern of earthworm casts and the spatial pattern of grass species in a semi-arid steppe in SE Spain. Given the important effects of soil fauna on SOM dynamics, infiltration and nutrient turnover in drylands (Whitford, 2002), further studies on the interactions between plants, fauna and soils are needed.

SOM decomposition, as plant productivity, is negatively affected by water limitations (Meentemeyer, 1978), and SOM content in dryland soils is commonly low (Vallejo et al., 1998). SOM is usually higher underneath the canopy of isolated plants and vegetation patches than in open areas (see below). Organic matter accretion as shrubs increase in size has been described (Pugnaire et al., 1996a, Tirado, 2003). Fine root inputs and historical changes in the spatial pattern of vegetation distribution may, however, attenuate these differences (Puigdefábregas et al., 1999), although plant effects on SOM may last less than two decades (Burke et al., 1999; Romanyà et al., 2000b; Martínez-Mena et al., 2002). It is often difficult to evaluate species effect on SOM because the age of individuals under comparison may be very different. This is particularly true for resprouting species. For example, in a review on the effects of 20th century afforestations with *Pinus halepensis* (one of the most common tree species in the Mediterranean) on semiarid ecosystem properties, Maestre and Cortina (2004b) found that most studies on SOM dynamics reported lower SOM contents underneath pine canopies than underneath patches of relatively undisturbed shrubland. We don't know if differences arise from contrasted species properties or from time, and thus we don't know if pines will ever reach the SOM levels found under established shrubs. Contrasts in soil CO_2 efflux can be substantial under different types of vegetated patches within a given ecosystem (Maestre and Cortina, 2003; Fig.

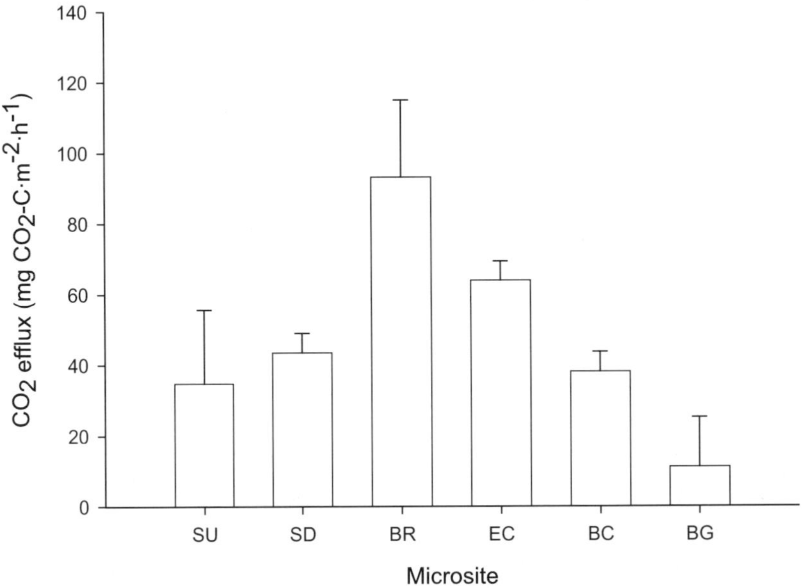

Figure 6. Soil respiration from different microsites in a *Stipa tenacissima* steppe in SE Spain. Measurements were taken on May 2001. Bars represent means and standard errors of ten replicates per microsite. SU: Upslope of *S. tenacissima* tussocks, SD: Downslope of *S. tenacissima* tussocks, BR: under the canopy of the sprouting perennial grass *Brachypodium retusum* EC: bare ground areas covered with earthworm casts, BC: bare ground areas covered with biological crusts, BG: bare ground areas covered with physical crusts. From original data in Maestre and Cortina (2003).

6), an indication that species may greatly differ in their overall effects on SOM content and quality. Plants can greatly modify nutrient accumulation and availability through mechanisms such as weathering, nitrogen fixation, runoff capture, concentration of animal feces, etc. (West, 1989; Kelly et al., 1998; Eviner and Chapin, 2003). Nitrogen fixing plants, legumes and actinorhizal plants, are widespread in drylands. This is remarkable, as soil properties that are common in drylands (such as low P availability, salinity and low water availability) may negatively affect N fixation rates (Reddell et al., 1991; Azcón and Altrash, 1997). Sprent (1987) suggested that deep rooting and adaptations to minimize water loss, such as the presence of phyllodes, may have contributed to their success. Experimental measures of N fixation rates range from less than 1 kg N ha^{-1} $year^{-1}$ to 390 kg N ha^{-1} $year^{-1}$ (Table 1).

We have reviewed published literature on changes in soil properties under vegetation patches in drylands. The data set includes 31 references, and more than 40 species and vegetation types covering a rainfall gradient from less than 100 mm to more than 1000 mm, with most studies in the range 200-250

Table 1. Some examples of N inputs through biological fixation in drylands.

Species	N fixation rate (kg ha^{-1} $year^{-1}$)
Prosopis glandulosa[1]	25-30
Prosopis glandulosa[2]	150
Prosopis glandulosa[7]	45-140
Prosopis glandulosa[8]	16
Prosopis glandulosa[11]	40-50
Casuarina equisetifolia[3,4]	40-60
Ceanothus velutinus[5]	108
Ceanothus greggi[6]	0.1
Adenocarpus decorticans[9]	0.5-1
Acacia caven[10]	0.5-18
Prosopis alba[10]	0.0-1.2
Prosopis chilensis[10]	0.2-1.7
Chamaecytisus proliferus[10]	8-193
Chamaecytisus proliferus[12]	390

[1]Rundel et al. (1982), [2]Felker (1979), [3]Dommergues (1963), [4]Gauthier et al. (1985), [5]Youngberg and Wollum (1976), [6]Kummerov et al. (1978), [7]Johnson and Mayeux (1990), [8]López-Villagra and Felker (1997), [9]Moro (1992), [10]Aronson et al. (2002), [11]Sharifi et al. (1982), [12]Unkovich et al. (2000).

mm. All communities show discontinous cover of woody vegetation and herbaceous tussocks, with most studies reporting cover values ranging from 20% to 50%. Lythology and soil types are diverse, from acidic (pH 4.5) to alkaline (pH 8.5). We used the index RII (Armas et al., 2004) to compare soil properties underneath the canopy and in intercanopy areas. RII is calculated as:

$$RII=(VARc-VARi)/(VARc+VARi)$$

where VARc and VARi are the values of a given soil property for the canopy and intercanopy areas, respectively. RII ranges from –1 to +1, with positive values indicating increases in the variable under study in canopy microsites.

For the 41 comparisons of soil organic carbon content (TOC) between soils underneath vegetated patches and soils in open areas (Fig. 7), average soil depth is 10 cm (range 1-40 cm). TOC ranged from 0.1 to 10%. Vegetated patches showed an average increase in TOC up to 450%, with most values falling between 50% and 100%. In no case TOC was higher in open areas than in vegetated areas.

Results from the 31 studies comparing total N (TKN) content are similar to those of TOC (Fig 8). TKN ranged from 0.02% to 1.7%. The relative

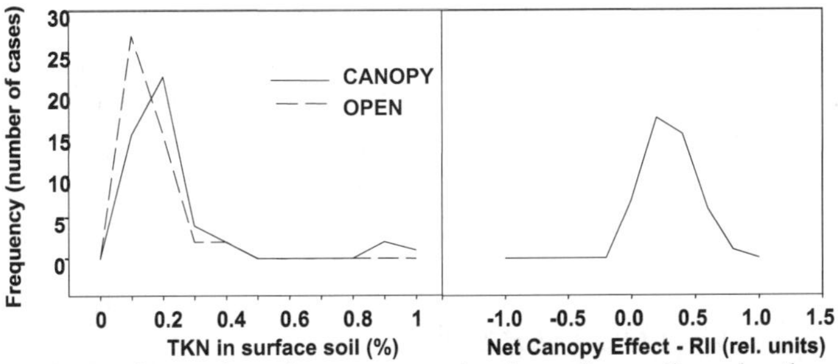

Figure 7. Left: Total organic carbon concentration in surface soils under the canopy of vegetation patches (CANOPY, solid line), and in open areas (OPEN, broken line) in several comparative studies conducted in drylands. Right: Net canopy effect (RII) on total soil carbon concentrations in surface soils. This effect was calculated as (TOCc-TOCi)/(TOCc+TOCi), where TOCc and TOCi are total carbon concentration in canopy and intercanopy areas, respectively.

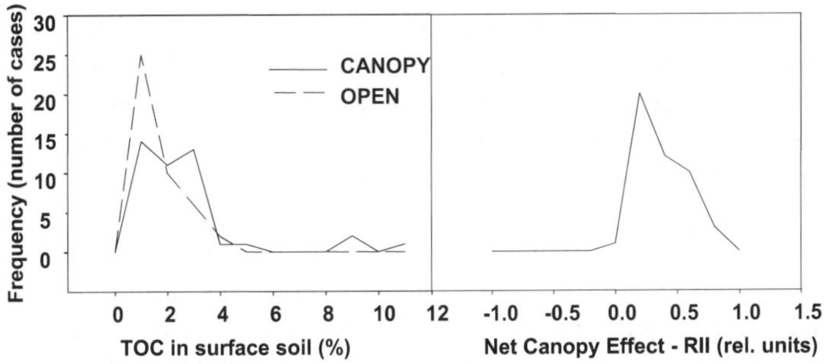

Figure 8. Left: Total organic nitrogen concentration in surface soils under the canopy of vegetation patches (CANOPY, solid line), and in open areas (OPEN, broken line) in several comparative studies in drylands. Right: Net canopy effect (RII) on total soil carbon concentration in surface soils. This effect was calculated as (TKNc-TKNi)/(TKNc+TKNi), where TKNc and TKNi are total nitrogen concentrations in canopy and intercanopy areas, respectively.

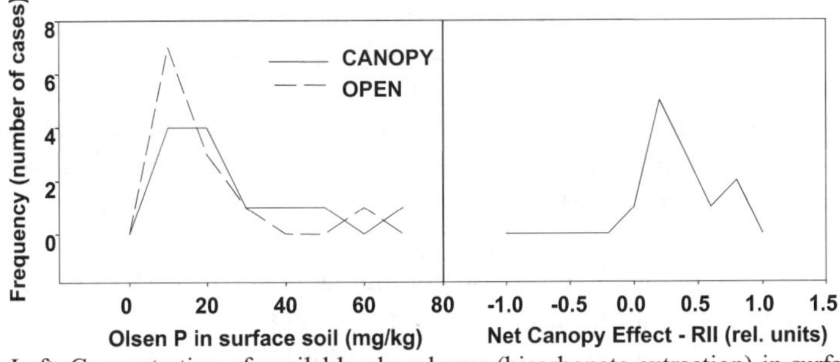

Figure 9. Left: Concentration of available phosphorus (bicarbonate extraction) in surface soils under the canopy of vegetation patches (CANOPY, solid line), and in open areas (OPEN, broken line) in several comparative studies in drylands. Right: Net canopy effect (RII) on total soil carbon concentration in surface soils. This effect was calculated as (PAVc-PAVi)/(PAVc+PAVi), where PAVc and PAVi are available phosphorus concentrations in canopy and intercanopy areas, respectively.

increses in TKN underneath vegetated patches were in the same range as for TOC. The number of studies showing higher TKN in open areas as compared to vegetated areas was very low, and of those, none showed a substantial decrease in TKN. Phosphorus availability (Olsen's bicarbonate extraction; n = 12; Fig. 9) ranged from 1-56 ppm in open areas to 2-65 ppm underneath vegetated patches. Most studies found a ca. 50% increase in P availability underneath vegetated patches.

BARE AREAS THAT ARE NOT

As mentioned before, precipitation in drylands is not high enough to maintain a continuous cover of vascular plants (Specht, 1988), leaving what are commonly, and wrongly, referred to as bare ground areas. Soil surface conditions in these areas is relevant for ecosystem functioning (Tongway and Ludwig, 1997), and its degradation may impair water fluxes to plant patches (Eldridge et al., 2000), and modify ecosystem-level processes like soil respiration (Maestre and Cortina, 2003). Bare areas are frequently covered by communities of cryptogams (mosses, lichens, cyanobacteria, liverworts and green algae), commonly referred to as biological crusts (West, 1990). Crusts are an important source of soil organic carbon (Beymer and Klopatek, 1991), fix atmospheric nitrogen (Rychert and Skujiņš, 1974), reduce wind and water erosion (Belnap, 1995), increase soil stability (Belnap and Gardner, 1993), and have an important effect on soil-water interactions (Eldridge et al., 2000). Biological crusts are preferentially established on fine textured slightly alkaline soils, with low content of surface rock fragments (Vitousek et al., 2002), In drylands, smooth and rugose crusts dominate (Belnap, 2001).

Depending on crust composition, net annual carbon input has been estimated between 0.4 and 37 g C m^{-2} $year^{-1}$ (Evans and Lange, 2003). In arid areas this may represent a significant input of organic matter. These figures correspond to relatively high instantaneous rates of C fixation, up to ca. 5 μmol CO_2 m^{-2} s^{-1} (García-Pichel and Belnap, 1996; Lange et al. 1997; Lange et al., 1998), as crusts remain inactive for long periods.

When biological crusts incorporate cyanobacteria and cyanolichens, they may fix substantial amounts of atmospheric nitrogen. Several studies have measured N fixation rates higher than 10 kg N ha^{-1} $year^{-1}$ (Rychert and Skujiņš, 1974; Belnap, 2002; Evans and Lange, 2003), although much lower rates are reported (Jeffries et al., 1992; Aranibar et al., 2003). Part of this N may be lost by denitrification and ammonia volatilization (Vitousek et al., 2002). These estimations are, however, subjected to great uncertainty, as substantial spatial and temporal variability adds to methodological limitations. Nevertheless, N inputs in arid ecosystems coming from biological crusts can be relevant at an ecosystem scale (Evans and Ehleringer, 1993; Billings et al., 2003).

Biological crusts affect water fluxes in various ways. Increased surface roughness may favour runoff reduction and water infiltration (Warren, 2001). However, the relatively flat morphology of biological crusts in arid and semiarid areas not subjected to freezing may reduce the magnitude of this effect. The net effect of biological crusts on infiltration rate depends on soil texture, and on the identity of the organisms dominating the crusts. In sandy soils, biological crusts may increase microporosity and reduce infiltration. In soils with higher fine particle content, biological crusts improve aggregation and create macropores, thus increasing infiltration rate. Moss-dominated crusts favor infiltration, and lichen- and cyanobacteria-dominated crusts reduce it (Maestre et al., 2002a). Crusts may also affect erosion and sediment transport by altering runoff, attaching soil particles together, and physically protecting the surface soil. Accordingly, disturbance of the biological crust may favour increased sediment yield (Belnap and Gillette, 1997).

Biological crusts and vascular plants interact in a number of ways. Vascular plants usually outcompete biological crusts, but they can also take advantage of the microenvironment created by the former. Environmental modifications promoted by plants, such as lower surface soil temperatures, reduced radiation, and decreased wind speed on the soil surface, promote changes in the composition, dominance and spatial pattern of the organisms forming these crusts (Eldridge, 1999; Maestre et al. 2002a, Maestre 2003). On the other hand, biological crusts can directly affect the establishment (Prasse and Bornkamm, 2000), survival (Eckert *et al.* 1986), nutrient status (Harper and Belnap, 2001), and water relations (DeFalco *et al.*, 2001) of vascular plants by altering soil surface topography, modifying water and nutrient fluxes, chelating metals, secreting growth promoting compounds, favouring mycorrhizal abundance or increasing pH (Belnap and Harper, 1995; Belnap et al., 2001; Li et al., 2002; Pendleton et al., 2003).

PLANT RESPONSE TO DISTURBANCES MODULATE PLANT EFFECTS ON SOILS

In environments that are prone to disturbances such as wildfire or grazing, resistance and resilience against disturbance are relevant traits affecting plant species effect on soils. As previously described, species that maintain a relatively humid microclimate under the canopy will affect soil processes directly. But they will also affect the probability of ignition and fire severity (Elvira and Lara, 1989; Wheelan, 1995), and thus attenuate the effects of wildfire, including changes in plant community composition. Persistance after disturbances that remove aboveground parts, such as wildfire, is increased by resprouting. This is a common trait in Mediterranean vegetation (Kummerov, 1981), that can be crucial to ensure fast soil protection after disturbance (Vallejo and Alloza, 1998). Resprouters are likely to affect soil properties in a given location more intensely than species that depend on the seed bank to reestablish, despite that some obligate seeders may show some degree of persistance at small spatial scales (Moreno and Oechel, 1994).

Plant architecture may also affect stability against disturbances. It is well known that trees may increase erosion by raindrop splash as compared to shrubs, due to the formation of big throughfall drops falling from more than 8-9 m above the surface soil (Viles, 1990). On the other hand, vertical continuity of the canopy may favour the combustion of all vegetation strata. Accordingly, self-pruning may influence fire effects on soils. Vertical structure can also be relevant in shrubs. *Ulex parviflorus* is a spiny leafless leguminous shrub that colonizes abandoned agricultural fields in the western Mediterranean. Senescent stems form early and, without abscission, remain as standing necromass, increasing the fuel load. When the canopy is not closed, the surface layer next to this species, devoid of forest floor, may be colonized by grasses such as *Brachypodium retusum*. *Cistus albidus* is a coexisting shrub in these areas; its leaves twist during drought and may fall as summer progresses. They usually form a relatively thick forest floor devoid of herbaceous layer, with little foliar biomass concentrating in the top of the branches. Vertical continuity of the fuel load is higher in *Ulex parviflorus*, and this may be the cause for higher combusting power underneath this species (Fig. 10).

Forest floor accumulation affects ecosystem resilience in several ways. Forest floor protects the surface soil from rainfall splash, limiting the formation of physical crusts (Thomas, 1988), and reducing the risk of erosion. This is particularly relevant in areas where plant cover has been previously removed by wildfire or clearing. For example, low severity fires may not completely destroy the forest floor (Gillon et al., 1999). Furthermore, patches of relatively thick forest floor may withstand further rains, even high intensity rains, and thus protect the underlying soil from erosion (García-Cano, 1998). Considering all factors involved in the comparison between *Ulex parviflorus* and *Cistus albidus*, soil protection after wildfire is lower in the former because it accumulates less forest floor, it shows higher combustion of surface litter, and because ashes are readily washed away after fire.

COMMUNITY DYNAMICS DRIVEN BY CHANGES IN SOIL PROPERTIES

Plant-plant interactions, resulting from the net output of positive and negative interactions, are crucial for ecosystem composition, structure and dynamics in drylands (Whitford, 2002). Most studies on negative plant-plant relations in these areas have focused on trophic interactions, resource depletion and competition, and allelopathy (Scholes and Archer, 1997; Kröpfl et al., 2002; Whitford, 2002). Other negative interactions have comparatively

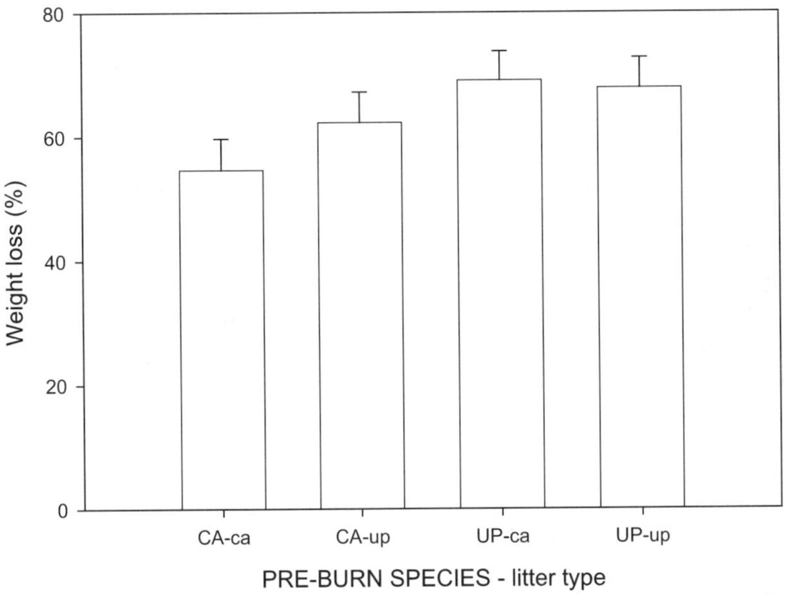

Figure 10. Weight loss through combustion of *Cistus albidus* (ca) and *Ulex parviflorus* (up) litter located underneath the canopy of individuals of *Cistus albidus* (CA) and *Ulex parviflorus* (UP) prior to an experimental fire. Bars represent means and standard errors of 30 individuals per species and litter type. Location was the only significant factor (F=88, p=0.01). Unpublished data from M.F. García-Cano.

received much less attention, probably due to the inherent experimental difficulty in differentiating trophic and non-trophic factors. Negative interactions can be the result of plant effects on soils, such as increased salinity, caliche formation, etc. (Cortina and Vallejo, 2004). On the other hand, positive interactions may result from increases in resource availability, amelioration of microclimate and soil conditions, increases in pollination and propagule dispersal rates, and defense against pathogens and herbivores (Callaway, 1995).

Positive and negative interactions are highly dynamic in drylands, and their balance may depend on site properties, climatic conditions, species identity and development stage (Pugnaire et al., 1996b; Maestre and Cortina, 2004c; Gómez-Aparicio et al., 2004). Bertness and Callaway (1995), and Callaway and Walker (1997) suggested that positive interactions should be more intense under high environmental stress or consumer pressure, and should depend on the size of the facilitator. Within individual shrub canopies, soil resources and microclimate show complex spatial patterns (Halvorson et al., 1994; Moro et al., 1997), and their interaction promote the emergence of different niches that increase in number and availability as shrubs increase in size (Pugnaire et al., 1996a). Increased habitat availability, together with the amelioration of the harsh climatic conditions, promote an increase in the strength of facilitative interactions with the increase in the size of the facilitator canopy (Callaway and Walker, 1997). Other plant traits –e.g. related to the capacity to concentrate resources or to deter predators- can be considered to estimate the 'nursing' power of a given species. For example, the capacity to build up *nebkhas* or mounds of sediments underneath plant canopies has been associated with canopy compactness (density) and openess (a trait related to the presence of branches at the soil level), as well as to the degree of mycorrhization (Carrillo-García et al., 1999). According to these authors, dense canopies, whether open or closed, should be prone to *nebkha* formation, whereas only open ones would allow the presence of understorey plants. These observations are likely to depend on slope. Evidences on the superior capacity of shrubs to alter soil properties are, however, not conclusive (Mazzarino et al., 1996; Schlesinger et al., 1996). The capacity to accumulate litter may also be relevant to establish nurse-protegé interactions, although evidence on the net effect of litter on seedling emergence and establishment is contrasting (Fowler, 1988; Owens et al., 1995; Milton, 1995). Functional matrices, as those proposed by Ervine and Chapin (2003) to characterize plant capacity to alter soils, could incorporate other relevant traits to describe nursing capacity.

It has been suggested that nurse-protegé interactions are more common in arid and semiarid communities than in other environments (Flores and Jurado, 2003). According to these authors, the Fabaceae and Mimosaceae (which may be capable of fixing atmospheric N) are among the most common nurse families (20% and 7% of the reported species, respectively), suggesting that N inputs may be a major driver of positive interactions in drylands. The capacity to fix N of a potential nurse plant is not, however, solid evidence of nutritional facilitation (Barnes and Archer, 1999). Defense against grazing and trampling may also be important, as many nurse species have thorns (*Acacia* spp., *Prosopis* spp., Cactaceae, with 11%, 4% and 5% of the nurse species, respectively), and unpalatable leaves.

As previously mentioned, nurse-protegé relations are commonly the combination of independent factors, both positive and negative. However, very few studies have performed manipulative field experiments to dissect the net effects of a given plant-plant interaction into their underlying positive and negative effects (Holzapfel and Mahall, 1999; Maestre et al., 2003a). This knowledge is necessary to understand community dynamics and to develop sound management programs. We evaluated the effects of the perennial tussock grass *Stipa tenacissima* on the native late-successional shrub *Pistacia lentiscus* in semiarid Mediterranean steppes. *Stipa tenacissima* (alpha grass) is a tussock grass distributed in the western Mediterranean basin from arid to dry subhumid conditions (100-500 mm; White, 1983). It is one of the dominant species in steppes, that have been strongly affected by human activities carried out during centuries, such as wood harvesting and fiber cropping (Barber et al., 1997). After cessation of human activities, shrub patches that were once part of these steppes hardly recover because of past management practices and inherent restrictions to plant growth. Shrub patches are, however, crucial for the composition, stability and function of semiarid steppes, despite their low contribution to the total plant cover (Maestre and Cortina, 2004a). In addition, the area covered by late-successional sprouting shrubs is the most influencing individual variable on perennial plant species richness in these steppes (Maestre, 2004).

Stipa tenacissima tussocks can retain runoff and sediments from upslope, a process that affects their own performance (Puigdefábregas et al., 1999). These tussocks also contribute to resource concentration, for example by favouring infiltration underneath the canopy (Cerdà, 1997). As a consequence of the changes in soil, microclimate and biological conditions, woody vegetation gets preferentially established close to alpha canopies (Maestre et al., 2001, 2003a), where they can withstand higher degrees of climatic stress (García-Fayos and Gasque, 2002; Maestre et al., 2003a). Previous studies had shown that *S. tenacissima* improved soil fertility (Puigdefábregas and Sánchez, 1996; Bochet et al. 1999; Aïdoud et al., 1999), but not nitrogen content (Bessah et al., 1999), reduced irradiation and soil temperature (Maestre et al., 2001), and received runoff water (Puigdefábregas et al., 1999) as compared to adjacent bare ground areas. We wanted to estimate the weight of each of these factors in increasing the survival and growth of introduced woody seedlings. We established a manipulative experiment in which we planted seedlings of a common shrub species, *Pistacia lentiscus*, upslope of alpha tussocks that were either undisturbed, herbicided or bended, or upslope of alpha tussocks where runoff had been excluded (Maestre et al., 2003a). We also planted *P. lentiscus* seedlings in undisturbed open areas. Finally, we carried out a laboratory experiment to test the effect of soil properties on *P. lentiscus* seedlings. In contrast to our expectations, runoff did not affect seedling survival. Seedlings planted in alpha soil showed a trend towards better growth and nutritional status, but the overall effect of soil type was not statistically significant. Finally, shadow was the most important factor affecting seedling establishment. So, in quantitative terms, the three factors ranked: shade>>soil fertility>runoff. No grazing on *Pistacia lentiscus* was detected.

We performed a similar experiment with *Pinus halepensis* in semiarid plantations of SE Spain. It is one of the most common tree species in the Mediterranean basin, indeed one of the few tree species that can thrive under semiarid conditions. *Pinus halepensis* forests have extended in the last decades due mainly to afforestation (Pausas et al., 2004). Under semiarid conditions, *P. halepensis* plantations show poor growth and cover, and spontaneous colonisation by sprouting shrubs is scarce (Maestre and Cortina, 2004b). Community composition (e.g. bird richness) has been related to the abundance and structure of shrubs in these forests (López and Moro, 1997). Pine canopies frequently show a relatively dense herbaceous understorey dominated by the perennial grass *Brachypodium retusum* (Bautista and Vallejo, 2002), whereas the establishment of woody seedlings in these microsites seems to be impeded (Maestre et al., 2003b). We established a series of field and glasshouse experiments to evaluate the relative importance of direct interactions (soil fertility, allelochemicals, shadow and water availability, and competition) and indirect interactions (through pine effects on the herbaceous layer) (Maestre et al., 2004). We found that the direct effects of pine on introduced seedlings, including competition, were rather small. Thus, pine death by girdling, and the resulting decrease in belowground competition, while maintaining a protective shadow, did not affect seedling performance. In contrast, suppression of the herbaceous layer greatly increased seedling survival and growth.

In a review of 31 studies on facilitation we found that improved soil properties, including increased nutrient availability, was the most common mechanism of facilitation that was mentioned (25 out of 31 studies). Of these, 8 studies attributed the positive interactions to higher nitrogen availability. Half of the studies mentioned shade as an important driver of positive interactions. Finally, only three studies attributed facilitation to the presence of litter. However, most of the studies reviewed were observational. On the other hand, for logistic reasons manipulative experiments commonly use simplified designs, focusing on integrated factors (such as soil fertility or shading) rather than specific processes. So our knowledge on the relative importance of the various drivers of facilitation in drylands is still very poor, and does not allow for sound generalizations.

IMPLICATIONS FOR ECOSYSTEM MANAGEMENT AND RESTORATION

Being a priority in land management in a wide variety of biomes, the restoration of degraded ecosystems is especially important in drylands, as they are being degraded and desertified at a fast rate throughout the globe (Reynolds, 2001, Abahussain et al., 2002; Reynolds and Stafford-Smith, 2002). Degraded ecosystems in drylands are usually characterised by a reduced plant cover and impoverished plant species diversity (Jauffret and Lavorel, 2003; Maestre, 2004; Cortina et al., in press). Despite the specific objectives of their restoration that may differ depending on the degree of degradation, and on climatic, biotic and socio-economic constraints, restoration programs often aim to increase plant cover by directly introducing plant individuals, primarily woody species (Whisenant, 1999; Young, 2000; Vallejo et al., 2000, Cortina et al., in press). This management action is crucial to stop further degradation, to combat desertification and to foster the recovery of the structure, composition and function of degraded ecosystems in these areas (Castillo et al., 1997; Reynolds, 2001; Cortina and Vallejo, 2004). However, if the target area is extensively degraded, restoration efforts could be initiated with actions focusing on the recovery of ecosystem structure by increasing the number of patches and reducing the downslope distance between them. This can be done by inserting brush piles parallel to land contours. Experiments conducted in Australia have shown the effectiveness of this technique in creating fertile patches and ultimately rehabilitating degraded landscapes (Ludwig and Tongway, 1996; Tongway and Ludwig, 1996). Such brush piles would reduce soil and nutrient losses, and would act as filters rather than barriers. They would also provide suitable microsites for enhancing the establishment, growth and survival of perennial plants in the short term. Once this intervention has reduced degradation, the next step to restore these systems should be the introduction of seedlings of native woody species.

If the target area holds some plant cover, restoration efforts can take advantage of facilitative interactions among plants. Positive interactions, and among them, those mediated by changes in soil properties, can be crucial to maintain the integrity of dryland ecosystems. Accordingly, they can be very helpful if not essential for reassembling pieces of degraded ecosystems. In the previous section we have seen that there are numerous evidences of facilitative interactions in drylands. These interactions can be used to promote the establishment of species of interest, and this has been increasingly recommended as a restoration technique (Maestre et al., 2001; Castro et al., 2002; Gómez-Aparicio et al., 2004). On the other hand, a thorough understanding of the mechanistic basis of positive interactions may allow the identification of the main drivers of facilitation, and thus their use in restoration. For example, shadow may be crucial in the first stages of seedling establishment. Nurse plants may not be always available, or their use may be restricted for logistic or economic reasons. In this case, the use of ecotechnological tools to reduce incoming radiation such as treeshelters, can be a suitable alternative to improve seedling performance (Cortina et al., in press). This technique is also convenient when consumer pressure is high. Amelioration of soil fertility can be easily achieved by using various types of soil amendments. Residues with high SOM and nutrient content, such as composted domestic refuses and sewage sludge are becoming increasingly available (Valdecantos et al., 2002; Fuentes et al., 2002a, 2002b; Valdecantos et al., in press). As the quality of these products improves, they will be increasingly used for the restoration of areas where soil fertility hampers succession. Nowadays, there is a vast array of techniques used in ecosystem restoration, deriving from mechanisms like those previously described (Table 2). Further knowledge on this type of interactions will help to develop new ecotechnology in this area, and to improve the success of restoration programmes.

Table 2. Some techniques used for the restoration of degraded semiarid ecosystems and the ecosystem component and main processes that are affected. From Tongway et al. (2004).

Technique	Ecosystem component	Process	References
Branches, mulch, etc.	Sinks	Sediment, runoff and seed capture	1,2
Perches	Birds rest	Propagule concentration, eventually water and nutrients	3,4
Organic amendments and fertilisation	Islands of fertility	Local soil improvement	5,6
Stones around introduced plants	Sinks	Resource retention including moisture	-
Stone pavements	Patches with low infiltration	Runoff	7,8,9
Treeshelters	Nurse plants	Protection against incomming radiation and herbivory	10,11
Field and nursery mycrorrhizae and rhizoflora inoculation	Exo and endosimbiotic microflora	Increase in resource availability, protection against pathogens and stress	12,13,14
Site preparation (microcatchments, terracing, etc.)	Sinks	Resource capture, mainly water. Increase in available soil	15,16
Cyanobacteria inoculation	Biological crusts	Soil protection, runoff generation	17,18
Nurse species plantation	Nurse plants	Processes associated with facilitation	19

[1]Ludwig and Tongway (1996), [2]Tongway and Ludwig (1996), [3]Bonet (2004), [4]Wunderlee (1997), [5]Cortina et al. (2001), [6]Valdecantos et al. (2002), [7]E. De Simón, pers. com., [8]Hillel (1991), [9]Lavee et al. (1997), [10]Vallejo et al. (2003), [11]Bellot et al. (2002), [12]Azcón and Barea (1997), [13]Caravaca et al. (2003), [14]Maestre et al. (2002b), [15]Whisenant et al. (1995), [16]Boeken and Shachak (1994), [17]Buttars et al. (1998), [18]Belnap (1993), [19]Vilagrosa et al. (1997).

CONCLUSIONS

Throughout the 20th century drylands have been the focus of an impressive amount of research. In some cases, findings have supported traditional knowledge and practices. But very often research has provided entirely new perspectives on dryland functioning. Still, it is somewhat discouraging that some of the questions that we are trying to answer, such as those on the rooting depth of some species, or on the effects of site preparation and the use of organic amendments, were already posed more than 2,000 years ago (e.g. Theophrastus, *Peri Phyton Historia*). This delay may partly result from the vast diversity of interactions that are involved. In this review, we have seen that interactions are strongly dependent on plant identity, or as a first approach, to plant functional types (Eviner and Chapin, 2003). We have also seen that plant-plant interactions may change depending on site properties, climatic conditions, or plant age. However, we know very little on the importance of such interactions on community dynamics and large-scale processes. In most cases, we just guess they are relevant. For example, we still don't know how important for succession is disturbance that may be followed by a short-term unfavourable climatic period. Must we asume that the interaction observed in a particular short-term study will prevail in the long-term, and will control community organization? When will facilitative interactions turn into competitive – or positive net balances turn into negative- as plants grow? It is clear that more research is needed in this area; particularly manipulative experiments, long-term studies and modelling.

This review has also shown that we can successfully estimate plant effects on background soil fertility. But as our knowledge on some areas of soil science such as soil biochemistry, soil microbiology and soil fauna increases, we realize that some plant-plant interactions that were previously attributed to other factors are actually occurring in the soil arena. This information is needed to evaluate the relative importance of soil processes on plant-plant interactions and community dynamics.

Drylands are threatened by a combination of natural factors and human activities (Reynolds and Stafford Smith, 2002). Further knowledge on the effects of plants on soils, and on the importance of such changes on long-term ecosystem dynamics will help to improve dryland management and sustainability.

ACKNOWLEDGEMENTS

We thank Dan Binkley and Oleg Menyailo for organizing the NATO Advanced Research Workshop *Trees and Soil Interactions. Implications to Global Climate Change*. This minireview has been possible thanks to funding from NATO, EC DGII (project CREOAK, QLRT-2001-01594), and CICYT-FEDER (projects FANCB, REN2001-0424-C02-02 / GLO, and BIOMON, REN2000-0181P4-03). Fernando T. Maestre is a Fulbright fellow, and greatly acknowledges support from the Dirección General de Universidades and Fondo Social Europeo. We thank our colleagues from Fundación CEAM and the Department of Ecology, Universitat d´Alacant for fruitful discussions on dryland ecology.

REFERENCES

Abahussain A A, Abdu A S, Al Zubari W K., El Deen N A. and Abdul-Raheem M 2002 Desertification in the Arab Region: analysis of current status and trends. J. Arid Environ. 51, 521-545.

Aerts R 1997 Climate, leaf litter chemistry and leaf litter decomposition in terrestrial ecosystems: a triangular relationship. Oikos 7, 439-449.

Aïdoud A 1989 Contribution à l'étude des écosystèmes steppiques pâturés des Hautes Plaines Algéro-Oranaises (Algérie). Ph. D. Thesis. Université des Sciences et Technologies, H. Boumedienne, Algeria. 248 pp.

Aïdoud A, Slimani H, Aidoud-Lounis F and Touffet J 1999 Changements édaphiques le long d'un gradient d'intensité de pâturage dans une steppe d'Algérie. Ecologia Mediterranea 25, 163-171.

Anderson J M 1988 Invertebrate-mediated transport processes in soils. Agr. Ecosyst. Environ. 24, 5-19.

Aranibar J N, Anderson I C, Ringrose S and Macko S A 2003 Importance of nitrogen fixation in soil crusts of southern African arid ecosystems: acetylene reduction and stable isotope studies. J. Arid Environ. 54, 345-358.

Armas C, Ordiales R and Pugnaire F I 2004 Measuring plant interactions: A new comparative index. Ecology 85, 2682-2686.

Arntz W E and Fahrbach E 1996 El Niño, experimento climático de la naturaleza. Fondo de Cultura Económica, México. 312 pp.

Aronson J, Ovalle C, Avendaño J, Longeri L and Del Pozo A 2002 Agroforestry tree selection in central Chile: biological nitrogen fixation and early plant growth in six dryland species. Agroforest. Syst. 56, 155-166.

Azcón R and Barea J M 1997 Mycorrhizal dependency of a representative plant species in Mediterranean shrublands (Lavandula spica L.) as a key factor to its use for revegetation strategies in desertification-threatened areas. Appl. Soil Ecol. 7, 83-92.

Azcón R and Elatrash F 1997 Influence of arbuscular mycorrhizae and phosphorus fertilization on growth, nodulation and N_2 fixation (N^{15}) in Medicago sativa at four salinity levels. Biol. Fert. Soils 24, 81-86.

Azcón-Aguilar C, Palenzuela J, Roldán A, Bautista S, Vallejo V R and Barea J M 2003 Analysis of the mycorrhizal potential in the rhizosphere of representative plant species from desertification-threatened Mediterranean shrublands. Appl. Soil Ecol. 22, 29-37.

Bais H P, Park S W, Weir T L, Callaway R M and Vivanco J M 2004 How plants communicate using the underground information superhighway. Trends Plant Sci. 9, 26-32.

Bais H P, Vepachedu R, Gilroy S, Callaway R M and Vivanco J M 2003 Allelopathy and exotic plant invasion: From molecules and genes to species interactions. Science 301, 1377-1380.

Barber A, Cabrera M R and Guardiola I 1997 Sobre la cultura de l´espart al territori valencià. Bancaixa, Valencia. 252 pp.

Barnes P W and Archer S 1999 Tree-shrub interactions in a subtropical savanna parkland: competition or facilitation?. J. Veg. Sci. 10, 525-536.

Bastin G, Ludwig J A, Eager R W, Chewings V H and Liedloff A C 2002 Indicators of landscape function: comparing patchiness metrics using remotely-sensed data from rangelands. Ecol. Indic. 1, 243-260.

Bautista S and Vallejo V R 2002 Spatial variation of post-fire plant recovery in Aleppo pine forests. *In* Fire and Ecological Processes. Eds. L Trabaud and L Prodon. pp. 13-24. Backhuys Publishers, Leiden.

Bellot J, Bonet A, Sánchez J R and Chirino E 2001 Likely effects of land use changes on the runoff and aquifer recharge in a semiarid landscape using a hydrological model. Landscape Urban Plan. 55, 41-53.

Bellot J, Ortíz de Urbina J M, Bonet A and Sánchez J R 2002 The effects of treeshelters on the growth of Quercus coccifera L. seedlings in a semiarid environment. Forestry 75, 89-106.

Bellot J, Maestre F T, Chirino E, Hernández N and Ortíz de Urbina J M 2004 Afforestation with Pinus halepensis reduces native shrub performance in a Mediterranean semiarid area. Acta Oecol. 25, 7-15.

Belnap J 1993 Recovery rates of cryptobiotic crusts: inoculant use and assesment methods. Great Basin Nat. 53, 89-95.

Belnap J 1995 Surface disturbances: their role in accelerating desertification. Environ. Monit. Assess. 37, 39-57.

Belnap J 2001 Comparative Structure of Physical and Biological Soil Crust. *In* Biological soil Crusts: Structure, Function and Management. Eds. J Belnap and O L Lange. pp. 177-191. Springer Verlag, Berlin.

Belnap J 2004 Nitrogen fixation in biological soil crusts from southeast Utah, USA. Biol. Fert. Soils. 35, 128-135.

Belnap J and Gardner J S 1993 Soil microstructure in soils of the colorado plateau the role of the cyanobacterium Microcoleus vaginatus. Great Basin Nat. 53, 40-47.

Belnap J and Gillette D A 1997 Disturbance of biological soil crusts: Impacts on potential wind erodibility of sandy desert soils in southeastern Utah. Land Degrad. Dev. 8, 355-362.

Belnap J and Harper K T 1995 Influence of cryptobiotic soil crusts on elemental content of tissue of two desert seed lants. Arid Soil Res. Rehab. 9, 107-115.

Belnap J, Prasse R and Harper K T 2001 Influence of biological soil crust on soil enviroments and vasculars plants. *In* Biological Soil Crusts: Structure, Function, and Management. Eds. J Belnap and O L Lange. pp. 281-300. Springer Verlag, Berlin.

Belsky A J, Mwonga S M, Amundson R G, Duxbury J M and Ali A R 1993 Comparative effects of isolated trees on their undercanopy environments in high- and low-rainfall savannas. J. Appl. Ecol. 30, 143-155.

Berg B, Albrektson A, Berg M P, Cortina J, Johansson M B, Gallardo A, Madeira M, Pausas J, Kratz W, Vallejo V R and Mc Claugherty C 1999 Amounts of litterfall in some pine forests in the northern hemisphere, especially Scots pine. Ann. For. Sci. 56, 625-639.

Berg B, Berg M, Cortina J, Escudero A, Gallardo A, Johansson M B, Madeira M and Virzo de Santo A 1993 Soil organic matter in some European coniferous forests. *In* Geography of Carbon Budget Processes in Terrestrial Ecosystems. Ed. A Breymeyer. Szymbark.

Bertness M D and Callaway R 1994 Positive interactions in communities. Trends Ecol. Evol. 9, 191-193.

Bertrand M and Lumaret J P 1992 The role of Diplopoda litter grazing activity on recycling processes in a Mediterranean climate. Vegetatio 99-100, 289-297.

Bessah R, Roze F and Nedjraoui D 1999 Activité cellulolytique in vitro des sols de deux steppes à alfa (Stipa tenacissima L.) d'Algérie. Ecologia Mediterranea 25, 185-192.

Beymer R J and Klopatek J M 1991 Potential contribution of carbon by microphytic crusts in pinion-juniper woodlands. Arid Soil Res. Rehab. 5, 187-198.

Billings S A, Schaeffer S M and Evans R D 2003 Nitrogen fixation by biological soil crusts and heterotrophic bacteria in an intact Mojave desert ecosystem with elevated CO_2 and added soil carbon. Soil Biol. Biochem. 35, 643-649.

Bochet E, Rubio J L and Poesen J 1999 Modified topsoil islands within patchy Mediterranean vegetation in SE Spain. Catena 38, 23-44.

Boeken B and Shachak M 1994 Desert plant communities in human-made patches-Implications for management. Ecol. Appl. 4, 702-716.

Bonet A 2004 Secondary succession of semi-arid Mediterranean old-fields in south-eastern Spain: insights for conservation and restoration of degraded lands. J. Arid Environ. 56, 213-233.

Breckle S W 2002 Walter's Vegetation of the Earth. Springer Verlag, Berlin.

Bromley J, Brouwer J, Barker A P, Gaze S R and Valentin C 1997 The role of surface water redistribution in an area of patterned vegetation in a semi-arid environment, south-west Niger. J. Hydrol. 198, 1-29.

Bruun H L, Fritzborger B, Rindel P O and Hansen U L 2001 Plant species richness in grasslands: the relative importance of contemporary environment and land-use history since the Iron Age. Ecography 24, 569-578.

Burch G J, Bath R K, Moore I D, and O'Loughlin E M 1987 Comparative hyddrological behaviour forested and cleared catchments in southeastern Australia. J. Hydrol. 90, 19-42.

Burguess S S O, Adams M A, Turner N C, and Ong C K 1998 The redistribution of soil water by tree root systems. Oecologia 115, 306-311.

Burke I, Lauenroth W K, Riggle R, Brannen P, Madigan B and Beard S 1999 Spatial variability of soil properties in the shortgrass steppe:the relative importance of topography, grazing, microsite, and plant species in controlling spatial patterns. Ecosystems 2, 422-438.

Buttars S M, St-Clair L L, Johansen J R, Sray J C, Payne M C, Webb B L, Terry R E, Pendleton B K and Warren S D 1998 Pelletized cyanobacterial soil amendments: laboratory testing for survival, escapability, and nitrogen fixation. Arid Soil Res. Rehab. 12, 165-178.

Butzer K W 1990 The realm of cultural-human ecology: Adaptation and change in historical perspective. *In* The Earth as Transformed by Human Action. B L Turner II, W C Clark, R W Kates, J F Richards, J T Mathews and W B Meyer. pp. 685-701. Cambridge University Press, Cambridge.

Caldwell M M, Dawson T E and Richards J H 1998 Hydraulic lift: consequences of water efflux from the roots of plants. Oecologia 113, 151-161.

Callaway R M 1995 Positive interactions among plants. The Botanical Review. 61, 306-349.

Callaway R M and Walker L R 1997 Competition and facilitation: a synthetic approach to interactions in plant communities. Ecology 78, 1958-1965.

Cannon W A 1911 The Root Habit of Desert Plants. Washington, Carnegie Institution. 96 pp.

Caravaca F, Barea J M, Palenzuela J, Figueroa D, Alguacil M M and Roldán A 2003 Establishment of shrub species in a degraded semiarid site after inoculation with native or allochthonous arbuscular mycorrhizal fungi. Appl. Soil Ecol. 22, 103-111.

Carrillo-Garcia Á, León de la Luz, Bashan Y and Bethlenfalvay GJ 1999 Nurse plants, mycorrhizae, and plant establishment in a disturbed area of the sonoran desert. Restor. Ecol. 7, 321-335.

Casals P, Romanyà J, Cortina J, Bottner P, Coûteaux M M and Vallejo V R 2000 CO_2 efflux from a Mediterranean semi-arid forest soil. I. Seasonality and effects of stoniness. Biogeochemistry 48, 261-281.

Castillo V M, Martínez-Mena M and Albaladejo J 1997 Runoff and soil loss response to vegetation removal in a semiarid environment. Soil Sci. Soc. Am. J. 61, 1116-1121.

Castro J, Zamora R, Hódar J A and Gómez J M 2002 Use of shrubs as nurse plants: a new technique for reforestation in Mediterranean mountains. Restor. Ecol. 10, 297-305.

Cerdà A 1997 The effect of patchy distribution of Stipa tenacissima L on runoff and erosion. J. Arid Environ. 36, 37-51.

Cortina J 1992 Efectes de les plantaciones de Pinus radiata sobre les propietats del sòl. Ph. D. Thesis. Univ. of Barcelona, Barcelona. 285 pp.

Cortina J, Bellot J, Vilagrosa A, Caturla R, Maestre F T, Rubio E, Martínez J M and Bonet A Restauración en semiárido. *In* In Avances en el Estudio de la Gestión del Monte Mediterráneo. Eds. V R Vallejo and J A Alloza. pp. 339-400.. Fundación CEAM-Bancaixa. In press.

Cortina J, Valdecantos A, Fuentes D, Casanova G, Vallejo V R, Díaz Bertrana J M, Llavador F and Ruano R 2001 El uso de biosólidos en el sector forestal valenciano. Foresta 3, 64-69.

Cortina J and Vallejo V R 1994 Effects of clearfelling on forest floor characteristics and litter decomposition in a radiata pine plantation. For. Ecol. Manag. 70, 299-310.

Cortina J and Vallejo V R 2004 Restoration Ecology. *In* The Science of Ecology for a Sustainable World. Eds. A Bodini and S Klotz. Encyclopaedia of Life Support Systems (EOLSS) developed under the auspices of the UNESCO, EOLSS Publishers, Oxford, (www.eolss.net).

DeFalco L A, Detling J K, Richard Tracy C and Warren S D 2001 Physiological variation among native and exotic winter annual plants associated with microbiotic crusts in the Mojave Desert. Plant Soil 234, 1-14.

De Luís M, García-Cano M F, Cortina J, Sánchez J R, González-Hidalgo J C and Raventós J 2001 Recent climatic trends favouring wildfires and erosion in the Region of Valencia (Spain) and their effect at an ecosystem level. For. Ecol. Manag. 147, 25-37.

De Luís M, González-Hidalgo J C, Raventós J, Sánchez J R and Cortina J 1997 Distribución espacial de la concentración y agresividad de la lluvia en el territorio de la Comunidad Valenciana. Cuaternario y Geomorfología 11, 33-44.

Dodd M B, Lauenroth W K and Welker J M 1998 Differential water resource use by herbaceous and woody plant life-forms in a shortgrass steppe community. Oecologia 117, 504-512.

Domingo F, Sánchez G, Moro M J, Brenner A J and Puigdefábregas J 1998 Measurement and modelling of rainfall interception by three semi-arid canopies. Agr. Forest Meteorol. 91, 275-292.

Dommergues Y R 1963 Evaluation du taux de fixation de 'azote dans un sol dunaire reboisé en filao (Casuarina equisetifolia). Agrochimica 7, 335-340.

Ducrey M 1998 Aspects écophysiologiques de la réponse et de l'adaptation des sapins méditerranéens aux extrêmes climatiques: gelées pruntanières et sécheresse estivale. Forêt Méditerranéenne 19, 105-116.

Duguay K J and Klironomos J N 2000 Direct and indirect effects of enhanced UV-B radiation on the decomposing and competitive abilities of saprobic fungi. Appl. Soil Ecol. 14, 157-164.

Dupouey J L, Dambrine E, Laffite J D and Moares C 2002 Irreversible impact of past land use on forest soils and biodiversity. Ecology 83, 2978-2984.

Eckert R E, Peterson E F, Meurisse M S and Stephens J L 1986 Effects of soil surface morphology on emergence and survival of seedlings in big sagebrush communities. J. Range Manage. 39, 414-420.

Ehleringer J R, Schwinning S and Gebauer R 1999 Water use in arid land ecosystems. *In* Physiological Plant Ecology. Eds. M C Press, J D Scholes and M G Barker. pp. 347-365. Blackwell Science, Oxford.

Eldridge D J 1999 Distribution and floristics of moss- and lichen-dominated soil crusts in a patterned Callitris glaucophylla woodland in eastern Australia. Acta Oecol. 20, 159-170.

Eldridge D J, Zaady E and Shachak M 2000 Infiltration through three contrasting biological soil crusts in patterned landscapes in the Negev, Israel. Catena. 40, 323-336.

Elvira L M and Hernando C 1989 Inflamabilidad y energía de las especies del sotobosque. INIA-MAPA Monografías nº 68, Madrid.

Evans R D and Ehleringer J R 1993 A break in the nitrogen cycle in aridlands? Evidence from δ^{15} N of soils. Oecologia 93, 314-317.

Evans R D and Lange O L 2004 Biological soil crust and ecosystem carbon and nitrogen dynamics. *In* Biological Soil Crusts: Structure, Function and Management. Eds. J Belnap and O L Lange. pp. 263-279. Springer Verlag, Berlin.

Eviner V T and Chapin III F S 2003 Functional matrix: A conceptual framework for predicting multiple plant effects on ecosystem processes. Annu. Rev. Ecol. Evol. S. 34, 455-485.

Felker P J 1979 Mesquite, an all purpose leguminose arid land tree. *In* New Agricultural Crops. Ed. G A Ritchie. pp. 89-132. West View Press, Washington.

Ferran A 1996 La Fertilitat de Sòls Forestals en la Regeneració Després del Foc de Diferents Ecosistemes Mediterranis. Ph. D. Thesis. Univ. de Barcelona, Barcelona.

Filella I and Peñuelas J 2003 Indications of hydraulic lift by Pinus halepensis and its effects on the water relations of neighbour shrubs. Biol. Plantarum 47, 209-214.

Flores J and Jurado E 2003 Are nurse-protegé interactions more common among plants from arid environments?. J. Veg. Sci. 14, 911-916.

Fons J 1995 Avaluació de la Fertilitat de Sòls Forestals Mediterranis. Ph. D. Thesis. Univ. of Barcelona. 138 pp.

Fowler N L 1988 What is a safe site?: neighbor, litter, germination date, and patch effects. Ecology 69, 947-961.

Fuentes D, Cortina J, Valdecantos A, Casanova G and Vallejo V R 2002a Seedling performancce in biosolid-amended degraded Mediterranean ecosystems. *In* Sustainable Use and Management of Soils in Arid and Semiarid Regions. Eds. A Faz, R Ortíz and A R Mermut. pp. 149-150. Quaderna Ed., Cartagena.

Fuentes D, Cortina J, Valdecantos A, Casanova G and Vallejo V R 2002b Sensitivity of Mediterranean woody species to high inputs of biosolids. *In* Sustainable Use and Management of Soils in Arid and Semiarid Regions. A Faz, R Ortíz and A R Mermut. pp. 151-153. Quaderna Ed., Cartagena.

Gallardo A and Merino J 1992 Leaf decomposition in two Mediterranean ecosystems of southwest Spain: Influence of substrate quality. Ecology 74, 152-161.

García Cano M F 1998 Avaluació de la pèrdua de nutrients i dels canvis en la fertilitat del sòl en un matollar de Ulex parviflorus afectat pel foc i la pluja torrencial. M. Sc. Thesis. Univerity of Alicante, Alicante.

García-Fayos P and Gasque M 2002 Consequences of a severe drought on spatial patterns of woody plants in a two-phase mosaic steppe of Stipa tenacissima L. J. Arid Environ. 52, 199-208.

García-Pichel F and Belnap J 1996 Microenvironments and microscale productivity of cyanobacterial desert crusts. J. Phycol. 32, 774-782.

Gauthier D, Diem H G, Dommergues Y R and Ganry F 1985 Assessment of N_2 fixation by Casuarina equisetifolia inoculated with Frankia ORS021001using ^{15}N methods. Soil Biol. Biochem. 17, 375-379.

Gillon D, Houssard C, Valette J C and Rigolot E 1999 Nitrogen and phosphorus cycling following prescribed burning in natural and managed Aleppo pine forests. Can. J. For. Res. 29, 1237-1247.

Gómez-Aparicio L, Zamora R, Gómez J M, Hódar J A, Castro J and Baraza E 2004 Applying plant facilitation to forest restoration: A meta-analysis of the use of shrubs as nurse plants. Ecol. Appl. 14, 1128-1138.

González-Hidalgo J C and Bellot J 1997 Soil moisture changes under shrub cover (Rosmarinus officinalis) and cleared shrub as response to precipitation in a semiarid environment: Stemflow effects. Arid Soil Res. Rehab. 11, 187-199.

González-Hidalgo J C, De Luis M, Raventós J, Cortina J and Sánchez J R 2003 Hydrological response of Mediterranean gorse shrubland under extreme simulation rainfall event. Zeitschrift für Geomorphologie. 48, 293-304.

Greene R S B, Valentin C and Esteves M 2001 Runoff and erosion processes. *In* Banded Vegetation Patterning in Arid and Semiarid Environments. Eds. D J Tongway, C Valentin and J Seghieri. pp. 52-76. Springer Verlag, New York.

Grove A T and Rackham O 2001 The Nature of Mediterranean Europe. Yale University Press, London. 384 pp.

Gutiérrez J and Squeo F A 2004 Importancia de los arbustos leñosos en los ecosistemas semiáridos de Chile. Ecosistemas 2004/1, http//www.aeet.org/ecosistemas/041 /investigacion2.htm.

Halvorson J J, Bolton H and Smith J L 1997 The pattern of soil variables related to Artemisia tridentata in a burned shrub-steppe site. Soil Sci. Soc. Am. J. 61, 287-294.

Harper K T and Belnap J 2001 The influence of biological soil crust on mineral uptake by associated vascular plants. J. Arid Environ. 47, 347-357.

Hart S C, Firestone M K and Paul E A 1992 Decomposition and nutrient dynamics of ponderosa pine needles in a Mediterranean-type climate. Can. J. For. Res. 22, 306-314.

Hillel D 1992 Out of the Earth. Aurum Press, London. pp 321.

Holzapfel C and Mahall B E 1999 Bidirectional facilitation and interference between shurbs and annuals in the Mojave Desert. Ecology 80, 1747-1761.

Huesca M, Cortina J and Bellot J 1998 Soil fertility as affected by gorse (Ulex parviflorus). *In* Forest Fire Research. Ed. D X Viegas. pp. 1643-1652. Coimbra.

Jackson E, Krogh S N and Whitford W G 2003 Desertification and biopedoturbation in the northern Chihuahuan Desert. J. Arid Environ. 53, 1-14.

Jackson R B and Caldwell M M 1992 Shading and the capture of localized soil nutrients: nutrient contents, carbohydrates, and root uptake kinetics of a perennial tussock grass. Oecologia 91, 457-462.

Jauffret S and Lavorel S 2003 Are plant functional types relevant to describe degradation in arid, southern Tunisian steppes?. J. Veg. Sci. 14, 399-408.

Jeffries D L, Klopatek J M, Link S O and Bolton H Jr 1992 Acetylene reduction by cryptogamic crusts from a blackbrush community as related to resaturation and dehydration. Soil Biol. Biochem. 24, 1101-1105.

Johnson H and Mayeux H S 1990 Prosopis glandulosa and the N balance of rangelands: extent and occurrence of nodulation. Oecologia 84, 176-185.

Kelly E F, Chadwick O A and Hilinski T E 1998 The effect of plants on mineral weathering. Biogeochemistry 42, 23-53.

Krogh S N, Zeisser M S, Jackson E and Whitford W G 2002 Presence/absence of a keystone specis as an indicator of rangeland health. J. Arid Environ. 50, 513-519.

Kröpfl A I, Cecchi G A, Villasuso N M and Distel R A 2002 The influence of Larrea divaricata on soil moisture and on water status and growth of Stipa tenuis in southern Argentina. J. Arid Environ. 52, 29-35.

Kummerov J 1981 Structure of roots and root systems. *In* Ecosystems of the World 11. Mediterranean-Type Shrublands. Eds. F Di Castri, G W Goodall and R L Specht. pp. 269-288. Elsevier, New York.

Kummerov J, Alexander J V, Neel J W and Fishbeck K 1978 Symbiotic nitrogen fixation in Ceanothus roots. Am. J. Bot. 65, 63-69.

Lange O L, Belnap J and Reichenberger H 1998 Photosynthesys of the cyanobacterial soil-crust lichen Collema tenax from arid lands in siuthern Utah, USA: role of water content on light and temperature responses of CO_2 exchange. Funct. Ecol. 12, 195-202.

Lange O L, Belnap J, Reichenberger H and Meyer A 1997 Photosynthesis of green algal soil crust lichens from arid lands in southern Utah, USA: role of water content, on light and temperature responses of CO_2 exchange. Flora 192, 1-15.

Lavee H, Poesen J and Yair A 1997 Evidence of high efficiency water-harvesting by ancient farmers in the Negev Desert, Israel. J. Arid Environ. 35, 341-348.

Le Houérou H N 1992 An overview of vegetation and land degradation in world arid lands. *In* Degradation and Restoration of Arid Lands. Ed. H E Dregne. pp. 127-163. ICASLS-Texas Tech. University, Lubbock.

Le Houérou H N 2001 Biogeography of the arid steppland north of the Sahara. J. Arid Environ. 48, 103-128.

Le Maitre D C, Van Wilgen B W, Gelderblom C M, Bailey C, Chapman R A and Nel J A 2002 Invasive alien trees and water resources in South Africa: case studies of the costs and benefits of management. For. Ecol. Manag. 160, 143-159.

Levitt J 1980 Water, Radiation, Salt, and other Stresses. *In* Responses of Plants to Environmental Stresses. Volume II. Academic Press, New York.

Li X R, Wang X P, Li T and Zhang J G 2002 Microbiotic soil crust and its effect on vegetation and habitat on artificially stabilized desert dunes in Tengger Desert, North China. Biol. Fert. Soils 35, 147-154.

Lloret F 1998 Fire, canopy cover and seedling dynamics in Mediterranean shrubland of northeastern Spain. J. Veg. Sci. 9, 417-430.

López G and Moro M J 1997 Birds of Aleppo pine plantations in south-east Spain in relation to vegetation composition and structure. J. Appl. Ecol. 34, 1257-1272.

López-Villagra G M L and Felker P 1997 Influence of understory removal, thinning and P fertilization on N_2 fixation in a mature mesquite (Prosopis glandulosa var. glandulosa) stand. J. Arid Environ. 36, 591-610.

Ludwig J A and Tongway D J 1996 Rehabilitation of semiarid landscapes in Australia. II. Restoring vegetation patches. Restor. Ecol. 4, 398-406.

Ludwig J A, Tongway D J and Marsden S G 1999 Stripes, strands or stipples: Modelling the influence of three landscape banding patterns on resource capture and productivity in semi-arid woodlands, Australia. Catena 37, 257-273.

Maestre F T 2003 Small-scale spatial patterns of two soil lichens in semi-arid Mediterranean steppe. The Lichenologist 35, 71-81.

Maestre F T 2004 On the importance of patch attributes, abiotic factors and past human impacts as determinants of plant species richness and diversity in Mediterranean semi-arid steppes. Divers. Distrib. 10, 21-29.

Maestre F T, Bautista S, Cortina J and Bellot J 2001 Potential for using facilitation by grasses to establish shrubs on a semiarid degraded steppe. Ecol. Appl. 11, 1641-1645.

Maestre F T, Huesca M T, Zaady E, Bautista S and Cortina J 2002a Infiltration, penetration resistance and microphytic crust composition in contrasted microsites within a Mediterranean semi-arid steppe. Soil Biol. Biochem. 34, 895-898.

Maestre F T, Bautista S, Cortina J, Honrubia M and Vallejo V R 2002b Microsite and mycorrhizal inoculum effects on the establishment of Quercus coccifera in a semi-arid degraded steppe. Ecol. Eng. 19, 289-295.

Maestre F T, Bautista S and Cortina J 2003a Positive, negative and net effects in grass-shrub interactions in Mediterranean semiarid grasslands. Ecology 84, 3186-3197.

Maestre F T and Cortina J 2002 Spatial pattern of soil properties and vegetation in a Mediterranean semi-arid steppe. Plant Soil 241, 279-291.

Maestre F T and Cortina J 2003 Small-scale spatial variation in soil CO_2 efflux in a Mediterranean semiarid steppe. Appl. Soil Ecol. 23, 199-209.

Maestre F T and Cortina J 2004a Insights on ecosystem composition and function in a sequence of degraded semiarid steppes. Restor. Ecol. 12, 494-502.

Maestre F T and Cortina J 2004b Are Pinus halepensis plantations useful as a restoration tool in degraded semiarid Mediterranean areas?. For. Ecol. Manag. 198, 303-317.

Maestre F T and Cortina J 2004c Do positive interactions increase with abiotic stress? A test from a semi-arid steppe. P. Roy. Soc. L. B. (Suppl.). 271, S331-S333.

Maestre F T, Cortina J, Bautista S and Bellot J 2003b Does Pinus halepensis facilitate the establishment of shrubs under semi-arid climate?. For. Ecol. Manag. 176, 147-160.

Maestre F T, Cortina J and Bautista S Mechanisms underlying the interaction between Pinus halepensis and the native late-successional shrub Pistacia lentiscus in a semiarid plantation. Ecography. 27, 776-786.

Martínez-Meza E and Whitford W G 1996 Stemflow, throughfall and channelization of stemflow by roots in three Chihuahuan desert shrubs. J. Arid Environ. 32, 271-287.

Martínez-Mena M, Alvarez-Rogel J, Castillo V and Albaladejo J 2002 Organic carbon and nitrogen losses influenced by vegetation removal in a semiarid Mediterranean soil. Biogeochemistry 61, 309-321.

Mauchamp A and Janeau J L 1993 Water funnelling by the crown of Flourensia cernua, a Chihuahuan Desert shrub. J. Appl. Ecol. 25, 299-306.

Mazzarino M J, Bertiller M B, Laos F and Coronato F R 1996 Spatial patterns of nitrogen availability, mineralizatión, and inmobilization in northern Patagonia, Argentina. Arid Soil Res. Rehab. 10, 295-309.

Meentemeyer V 1978 Macroclimate and lignin control of litter decomposition rates. Ecology 59, 465-472.

Milton S J 1995 Spatial and temporal patterns in the emergence and survival of seedlings in arid Karoo shrubland. J. Appl. Ecol. 32, 145-156.

Montaña C, Seghieri J and Cornet A 2001 Vegetation dynamics: Recruitment and regeneration in two-phase mosaics. *In* Banded Vegetation Patterning in Arid and Semiarid Environments. Eds. D Tongway, C Valentin and J Seghieri. pp. 132-145. Springer Verlag, New York.

Morehead D L, Westerfield M M and Zak J C 1998 Plant retards litter decay in a nutrient-limited soil: a case of exploitative competition?. Oecologia 113, 530-536.

Moreno J M and Oechel W C 1994 Fire intensity as a determinant factor of postfire plant recovery in southern California Chaparral. *In* The Role of Fire in Mediterranean-type Ecosystems. J M Moreno and W C Oechel. pp. 26-45. Springer Verlag, New York.

Moro M J 1992 Desfronde, descomposición y fijación de nitrógeno en una microcuenca con repoblación joven de coníferas y matorral de Cistus laurifolius y Adenocarpus decorticans en la Sierra de los Filabres (Almería). Ph.D. Thesis. U. Alicante.

Moro M J, Pugnaire F I, Haase P and Puigdefábregas J 1997 Effect of the canopy of Retama sphaerocarpa on its understorey in a semiarid environment. Funct. Ecol. 11, 425-431.

Moro M J, Pugnaire F I, Haase P and Puigdefábregas J 1997 Mechanisms of interaction between a leguminous shrub and its understorey in a semiarid environment. Ecography 20, 175-184.

Ne'eman G and Goubitz S 2000 Phenology of East Mediterranean vegetation *In* Life and Environment in the Mediterranean. Ed. L Trabaud. pp. 155-201. WIT Press, Southampton.

Northrup R R, Yu Z, Dahlgren R A and Vogt K A 1995 Polyphenol control of nitrogen release from pine litter. Nature 357, 227-229.

Noy Meir I 1973 Desert ecosystems: environment and producers. Annu. Rev. Ecol. Syst. 5, 195-214.

Ohte N, Koba K, Yoshikawa K, Sugimoto A, Matsuo N, Kabeya N and Wang L 2003 Water utilisation of natural and planted trees in the semiarid desert of Inner Mongolia, China. Ecol. Appl. 13, 337-351.

Olcina J 1994 Riesgos Climáticos en la Península Ibérica. Acción Divulgativa S.L., Madrid. 440 pp.

Owens M K, Wallace R B and Archer S R 1995 Landscape and microsite influences on shrub recruitment in a disturbed semi-arid Quercus Juniperus woodland. Oikos 74, 493-502.

Palenzuela J, Azcón-Aguilar C, Figueroa D, Caravaca F, Roldán A and Barea J M 2002 Effects of mycorrhizal inoculation of shrubs from Mediterranean ecosystems and composted residue application on transplant performance and mycorrhizal developments in a desertified soil. Biol. Fert. Soils 36, 170-175.

Pausas J G, Bladé C, Valdecantos A, Seva J P, Fuentes D, Alloza J A, Vilagrosa A, Bautista S, Cortina J and Vallejo V R 2004 Pines and oaks in the restoration of Mediterranean landscapes: New perspectives for an old practice-a review. Plant Ecol. 171, 209-220.

Peltier A, Ponge J F, Jordana R and Ariño A 2001 Humus forms in mediterranean scrublands with Aleppo Pine. Soil Sci. Soc. Am. J. 65, 884-896.

Pendleton R L, Pendleton B K, Howard G L and Warren S D 2003 Growth and nutrient content of herbaceous seedlings associated with biological soil crusts. Arid Land Res. Manag. 17, 271-281.

Piñol J, Lledó M J and Escarre A 1991 Hydrological balance of two Mediterranean forested catchments (Prades, NE Spain). Hydrolog. Sci. J. 36, 95-107.

Prasse R and Bornkamm R 2000 Effect of microbiotic soil surface crusts on emergence of vascular plants. Plant Ecol. 150, 65-75.

Pugnaire F I, Haase P, Puigdefábregas J, Cueto M, Clark S C and Incoll L D 1996a Facilitation and succession under the canopy of a leguminous shrub, Retama sphaerocarpa, in a semi-arid environment in south-east Spain. Oikos 76, 455-464.

Pugnaire F I, Haase P and Puigdefábregas J 1996b Facilitation between higher plant species in a semiarid environment. Ecology 77, 1420-1426.

Puigdefábregas J and Mendizábal T 1998 Perspectives on desertification: western Mediterranean. J. Arid Environ. 39, 209-224.

Puigdefábregas J and Sánchez G 1996 Geomorphological implications of vegetation patchiness on semi-arid slopes. *In* Advances in Hillslope Processes. Eds. M G Anderson and S M Brooks. pp. 1027-1060. John Wiley & Sons, London.

Puigdefábregas J, Solé A, Gutiérrez L, del Barrio G and Boer M 1999 Scales and processes of water and sediment redistribution in drylands: results from the Rambla Honda field site in Southeast Spain. Earth Sci. Reviews 48, 39-70.

Reddell P, Diem H G and Dommergues Y R 1991 Use of actynorhizal plants in arid and semiarid environments. Semiarid lands and deserts. *In* Soil resource and reclamation, Ed. J Skujins. pp. 469-485. Marcel Dekker, New York.

Reynolds J F 2001 Desertification, in: Encyclopaedia of Biodiversity. Ed. S Levin. pp. 61-78. Volume 2. Academic Press, New York.

Reynolds J F and Stafford Smith D M 2002a Do humans cause deserts?. *In* Global Desertification. Eds. J F Reynolds and D M Stafford Smith. pp. 1-21. Dahlem University Press, Berlin.

Reynolds J F and Stafford Smith D M. Eds. 2002b Global Desertification: Do humans cause deserts?, pp. 438. Dahlem University Press, Berlin.

Romanyà J, Casals P, Cortina J, Bottner P, Coûteaux M M and Vallejo V R 2000a CO_2 efflux from a Mediterranean semi-arid forest soil. II. Effects of soil fauna and surface stoniness. Biogeochemistry 48, 283-306.

Romanyà J., Cortina J, Falloon P, Coleman K and Smith P 2000b Modelling changes in soil organic matter after planting fast-growing Pinus radiata on Mediterranean agricultural soils. Eur. J. Soil Sci. 51, 627-641.

Rostagno C M 1989 Infiltration and sediment production as affected by soil surface conditions in a shrubland of Patagonia, Argentina. J. Range Manage. 42, 382-385.

Rundel P W, Nilson E T, Sharifi M R, Virginia R A, Jarrell W M, Kohol D H and Shearer G B 1982 Seasonal dynamics of nitrogen cycling for a Prosopis woodland in the Sonoran desert. Plant Soil 67, 343-353.

Rychert R C and Skujiņš J 1974 Nitrogen fixation by blue-green algae-lichen crusts in the Great Basin Desert. Soil Sci. Soc. Am. Pro. 38, 768-771.

Ryel R J, Caldwell M M, Leffler A J and Yoder C K 2003 Rapid soil moisture recharge to depth by roots in a stand of Artemisia tridentata. Ecology 84, 757-764.

Schlesinger W, Raikes J A, Hartley A E and Cross A F 1996 On the spatial pattern of soil nutrients in desert ecosystems. Ecology 77, 364-374.

Schlesinger W H and Pilmanis A M 1998 Plant-soil interactions in deserts. Biogeochemistry 42, 169-187.

Scholes R J and Archer S R 1997 Tree-grass interactions in savannas. Annu. Rev. Ecol. Syst. 28, 517-544.

Serrasolsas I and Vallejo V R 1999 Soil fertility after fire and clear-cutting. *In* Ecology of Mediterranean evergreen oak forests. Eds. F Rodà, C A Gracia, J Retana and J Bellot. pp. 601. Springer Verlag, Berlin.

Sevink J, Imeson A C and Verstraten J M 1989 Humus form development and hillslope runoff, and the effects of fire and management, under mediterranen forest in NE-Spain. Catena 16, 461-475.

Sharifi M R, Nilsen E T and Rundel P W 1982 Biomass and net primary productivity of Prosopis glandulosa (Fabaceae) in the Sonoran Desert of California. Am. J. Bot. 69, 760-767.

Soil Survey Staff 1990 Keys to Soil Taxonomy, SMSS Technical Monograph Nº 19. Virginia Polytech. Institute and State University, Blacksburg. 422 pp.

Specht R L 1988 Climate control of ecomorphological characters and species richness in mediterranean ecosystems of Australia. *In* Mediterranean-Type Ecosystems. Ed. R L Specht pp. 149-155. Kluwer Academic Publishers, Dordrecth.

Sprent J I 1987 The ecology of the nitrogen cycle. Cambridge Univ. Press, Cambridge.

Thomas D S G 1988 The biogeomorphology of arid and semi-arid environments. *In* Biogeomorphology. Ed. H A Viles. pp. 193-221. Blackwell, Oxford.

Tielborger K and Kadmon R 1997 Relationships between shrubs and annual communities in a sandy desert ecosystem: a three-year study. Plant Ecol. 130, 191-201.

Tirado R 2003 Interacciones positivas entre plantas: Mecanismos y consecuencias. Ph. D. Thesis. Universidad de Sevilla.

Tongway D J and Ludwig J A 1996 Rehabilitation of semiarid landscapes in Australia. I. Restoring productive soil patches. Restor. Ecol. 4, 388-397.

Tongway D J and Ludwig J A 1997 The nature of landscape dysfunction in rangelands *In* Landscape ecology: principles from Australia´s rangelands. Eds. J Ludwig, D Tongway, D Freudenberger, J Noble and K Hodgkinson. pp. 49-62. CSIRO Publishing, Collingwood.

Tongway D J, Valentin C and Seghieri C. Eds. 2001 Banded Vegetation Patterning in Arid and Semiarid Environments. pp. 251. Springer Verlag Ecological studies 149, New York.

Tongwa D, Cortina J and Maestre F T 2004 Heterogeneidad espacial y gestión de medios semiáridos. Ecosistemas 2004/1, www.aeet.org/ecosistemas/041/revision5.htm.

UNESCO 1977 World Map of Arid Regions. UNESCO, Paris.

Unkovich M, Pate J S, Lefroy E C and Arthur D J 2000 Nitrogen isotope fractionation in a fodder tree legume tagasaste (Chamaecytisus proliferus) and assessment of N_2 fixation inputs in deep sandy soils of Western Australia. Aust. J. Plant Physiol. 27, 921-929.

Valdecantos A and Cortina J Utilización de biosólidos para la restauración de ecosistemas mediterráneos. *In* Restauración de Ecosistemas Degradados. Fundación CEAM-Bancaixa. In press.

Valdecantos A, Cortina J, Fuentes D, Casanova G, Díaz-Beltrana J M, Llavador F and Vallejo V R 2002 Use of biosolids for reforestation in the Region of Valencia (E Spain). First results of a pilot project. Bioprocessing of Solid Waste and Sludge 1, 1-6.

Valentin C, d'Herbes J M and Poesen J 1999 Soil and water components of banded vegetation patterns. Catena 37, 1-24.

Valladares F 2003 Light heterogeneity and plants: from ecophysiology to species coexistence and biodiversity. Prog. Bot. 64, 439-471.

Valladares F and Pearcy R W 2002 Drought can be more critical in the shade than in the sun: a field study of carbon gain and photo-inhibition in a Californian shrub during a dry El Niño year. Plant Cell Environ. 25, 749-759.

Vallejo V R and Alloza J A 1998 The restoration of burned lands: the case of eastern Spain. *In* Large Forest Fires. Ed. J M Moreno pp. 91-108. Backhuys, Leiden, The Netherlands.

Vallejo V R, Cortina J, Ferran A, Fons J, Romanyà J and Serrasolsas I 1998 Sobre els trets distintius dels sòls mediterranis. Acta Bot. Barc. 45, 603-632.

Vallejo V R, Cortina J, Vilagrosa A, Seva J P and Alloza J A 2003 Problemas y perspectivas de la utilización de leñosas autóctonas en la restauración forestal. *In* Restauración de Ecosistemas en Ambientes Mediterráneos. Eds. J M Rey-Benayas, T Espigares and J M Nicolau. pp. 11-42. Publicaciones del la Universidad de Alcalá de Henares. Alcalá de Henares.

Vallejo V R, Serrasolses I, Cortina J, Seva J P, Valdecantos A, Vilagrosa A 2000 Restoration strategies and actions in Mediterranean degraded lands. *In* Desertification in Europe: mitigation strategies, land use planning. Eds. G Enne, Ch Zanolla and D Peter. pp. 221-233. Office for Official Publications of the European Communities, Luxembourg.

Van Wesemael B and Veer M A C 1992 Soil organic matter accumulation, litter decomposition and humus forms under Mediterranean-type forests in Southern Tuscany (Italy). J. Soil Sci. 43, 133-144.

Verhoef H A, Verspagen J M H and Zoomer H R 2000 Direct and indirect effects of ultraviolet-B radiation on soil biota, decomposition and nutrient fluxes in dune grassland soil systems. Biol. Fert. Soils 31, 366-371.

Vilagrosa A, Bellot J and Gil-Pelegrín E 2003 Cavitation, stomatal conductance, and leaf dieback in seedlings of two co-ocurring Mediterranean shrubs during an intense drought. J. Exp. Bot. 54, 2015-2024.

Vilagrosa A, Seva J P, Valdecantos A, Cortina J, Alloza J A, Serrasolsas I, Diego V, Abril M, Ferran A, Bellot J and Vallejo V R 1997 Plantaciones para la restauración forestal en la Comunidad Valenciana. *In* La Restauración de la Cubierta Vegetal en la Comunidad Valenciana. Ed. V R Vallejo. pp. 435-546. Centro de Estudios Ambientales del Mediterráneo and Generalitat Valenciana, Valencia, España.

Viles H A 1990 "The agency of organic beings": A selective review of recent work in biogeomorphology. *In* Vegetation and Erosion. Ed. J B Thornes. pp. 5-24. John Wiley and Sons Ltd, New York.

Vitousek P M, Cassman K, Cleveland C, Field C, Grimm N B, Howarth R W, Marinsanleandro P, Martinelli L, Ratetter E B and Sprent J I 2002 Towards an ecological understanding of biological nitrogen fixation. Biogeochemistry 57/58, 1-45.

Warren S D 2001 Biological soil crusts and water relations in North American deserts. *In* Biological Soil Crusts: Structure, Function and Management. Eds. J Belnap and O L Lange. pp. 327-337. Springer Verlag, Berlin.

West N E 1989 Spatial pattern-functional interactions in shrub-dominated plant communities. *In* The Biology and Utilization of Shrubs. Ed. C M McKell. pp. 283-305. Academic Press, San Diego.

West N E 1990 Structure and function of mycrophytic soil crusts in wildland ecosystems of arid to semi-arid regions. Adv. Ecol. Res. 20, 179-223.

Wheelan R J 1995 The Ecology of Fire. Cambridge University Press, Cambridge.

Whisenant S G 1999 Repairing damaged wildlands. Cambridge University Press, Cambridge.

Whisenant S G, Thurow T L and Maranz S J 1995 Initiating autogenic restoration on shallow semiarid sites. Restor. Ecol. 3, 61-67.

White F 1983 The vegetation of Africa. UNESCO, Paris.

Whitford W G 2002 Ecology of Desert Systems. Academic Press, London. 343 pp.

Whitford W G, Meentemeyer W, Seastedt T R, Cromack K Jr, Crossley D A Jr, Santos P F, Todd R L and Waide J B 1981 Exceptions to the AET model: deserts and clearcut forests. Ecology 62, 257-277.
Wu X B, Thurow T L and Whisenant S G 2000 Fragmentation and changes in hydrologic function of tiger bush landscapes, south-west Niger. J. Ecol. 88, 790-800.
Wunderlee J M 1997 The role of animal seed dispersal in accelerating native forest regeneration on degraded tropical lands. For. Ecol. Manag. 99, 223-235.
Young T P 2000 Restoration ecology and conservation biology. Biol. Conserv. 92, 73-83.
Youngberg C T and Wollum A G 1976 Nitrogen accretion in developing Ceanothus velutinus strand. Soil Sci. Soc. Am. J. 40, 109-112.

Chapter 7

THE RESPONSE OF BELOWGROUND CARBON ALLOCATION IN FORESTS TO GLOBAL CHANGE

Christian P. Giardina[1,4], Mark D. Coleman[2], Jessica E. Hancock[1,4], John S. King[4], Erik A. Lilleskov[1,4], Wendy M. Loya[4], Kurt S. Pregitzer[4], Michael G. Ryan[5,3] and Carl C. Trettin[6]

[1]USDA Forest Service, North Central Research Service, 410 MacInnes Drive, Houghton, MI 49931, USA, [2]USDA Forest Service, Southern Research Station, Savannah River, PO Box 700, New Ellenton, SC 29809, USA, [3]Graduate Degree Program in Ecology and Department of Forest, Rangeland and Watershed Stewardship, Colorado State University, Ft. Collins, CO 80523, USA, [4]The Ecosystem Science Center, School of Forest Resources and Environmental Science, Michigan Technological University, Houghton, MI 49931, USA, [5]USDA Forest Service, Rocky Mountain Research Service, 120 West Prospect, Fort Collins, CO 80523, USA, [6]USDA Forest Service, Southern Research Station, Charleston, SC, USA. Email: cgiardina@fs.fed.us

ABSTRACT

Belowground carbon allocation (BCA) in forests regulates soil organic matter formation and influences biotic and abiotic properties of soil such as bulk density, cation exchange capacity, and water holding capacity. On a global scale, the total quantity of carbon allocated belowground by terrestrial plants is enormous, exceeding by an order of magnitude the quantity of carbon emitted to the atmosphere through combustion of fossil fuels. Despite the importance of BCA to the functioning of plant and soil communities, as well as the global carbon budget, controls on BCA are relatively poorly understood. Consequently, our ability to predict how BCA will respond to changes in atmospheric greenhouse gases, climate, nutrient deposition, and plant community composition remains rudimentary. In this synthesis, we examine BCA from three perspectives: coarse-root standing stock, belowground net primary production (BNPP), and total belowground carbon allocation (TBCA). For each, we examine methodologies and methodological constraints, as well as constraints of terminology. We then examine available

D. Binkley and O. Menyailo (eds.),
Tree Species Effects on Soils: Implications for Global Change, 119–154.

data for any predictable variation in BCA due to changes in species composition, mean annual temperature, or elevated CO_2 in existing Free Air CO_2 Exposure (FACE) experiments. Finally, we discuss what we feel are important future directions for belowground carbon allocation research, with a focus on global change issues.

INTRODUCTION

Belowground carbon allocation (BCA) links the soil ecosystem and foodweb with the forest canopy, providing a flow of organic carbon (C) to the soil from the CO_2 fixed by photosynthesis from the air. From an evolutionary perspective, BCA represents the currency with which photosynthetic cyanobacterial endosymbionts in leaves (chloroplasts) acquire nutrients, water and structural support from their symbiotic partners belowground (plant roots, mycorrhizal fungi, and in some cases nitrogen-fixing bacteria). This flow of organic C between aboveground endosymbiont and belowground symbionts has a substantial impact on the global carbon cycle. BCA is the Earth's third largest biologically mediated C flux, after terrestrial photosynthesis (from which BCA is derived) and oceanic photosynthesis. Terrestrial plants allocate belowground some 60 Pg C out of the 120 Pg C fixed annually by terrestrial vegetation through photosynthesis, with most this gross carbon flux occurring in ecosystems with trees (Schimel 1995, Grace and Rayment 2000). By comparison, the annual flux of combusted fossil fuel C into the atmosphere is about 6 Pg C (Schimel 1995). At the stand scale, plants allocate large quantities of carbon belowground for the construction and maintenance of roots and mycorrhizae, such that BCA may represent the largest sink for gross primary production (Ryan et al. 1996, Janssens et al. 2001). In resource-limited environments typical of terrestrial ecosystems, high plant investment in BCA is necessary to secure the water and nutrients that drive terrestrial primary production.

Despite the magnitude of BCA, both globally and locally, BCA remains the least understood C flux in plant communities (Ryan et al. 1996, Clark et al. 2001a, Giardina and Ryan 2002, Giardina et al. 2004). In contrast to aboveground plant physiology, which is precisely captured in leaf-based physiological process models (Landsberg and Gower 1997), controls on belowground processes are poorly captured in process models. Efforts to validate belowground models are hindered by the complexity of above and belowground interactions with local and global changes in environmental factors (Figure 1). Further, the soil matrix complicates nearly all aspects of BCA. There are a wide range of approaches to characterizing belowground carbon cycling (Figure 2), but robust validation of these approaches remains problematic. As a result, conceptual and theoretical models describing BCA response to global change variables are highly uncertain (Giardina and Ryan 2000, Holland et al. 2000, Pendall et al. 2004), with ecosystem models often

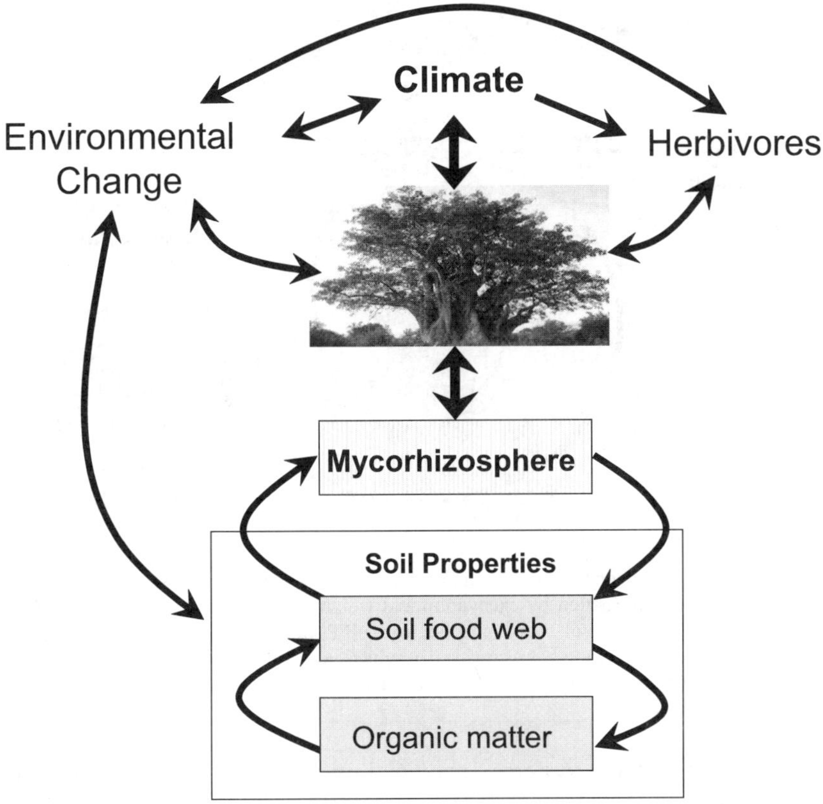

Figure 1. Diagram of the direct and indirect effects of environmental change (e.g., increased CO_2, O_3 and other greenhouse gases, change in fire regimes, elevated nutrient deposition rates, altered species succession, species invasion, etc.) on belowground carbon allocation through changes in canopy function, aboveground herbivore communities and soil properties (soil carbon and nutrient quality, soil food web including belowground herbivores).

relying on the assumption that the functioning and dynamics of aboveground tissues adequately describe those of belowground tissues (e.g., VEMAP et al 1994).

Forests are dynamic, with belowground process rates depending on factors such as tree species composition, nutrient and water supplies, and temperature. These factors influence BCA, and determine BCA response to global change. Species change is a dominant feature of global change (Figure 3, and other chapters in this volume), with composition varying over long and short periods. Dramatic species change can occur in response to climatic change in just centuries (Figure 3). Species change in response to exploitation (e.g., loss of white pine in the Great Lakes forests) or disease (e.g., loss of chestnut to blight) can occur in decades or less. Change can be even faster when short-term droughts are coupled with severe fires. Human management of ecosystems has altered species composition across plant life forms – annual

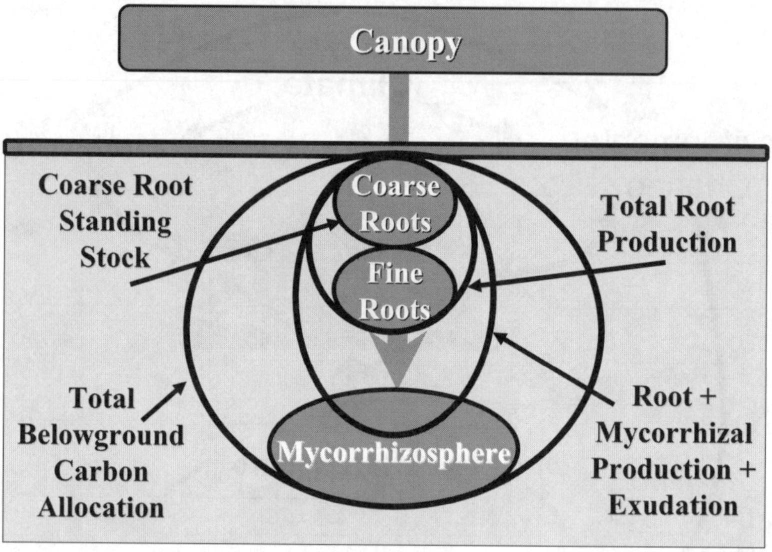

Figure 2. Various approaches to examining BCA in forests. Coarse root standing stock is a pool of carbon in soil measured by excavation and weighing at a single point in time. Total root production is comprised of coarse and fine root NPP, but typically excludes exudation or mycorrhizal production. Root + mycorrhizal production + exudation is all the C allocated belowground except for root respiration. Total belowground carbon allocation is all the C allocated belowground.

grasses in agricultural systems, perennial grasses and forbs in managed pastures, and long-lived trees in forest plantations.

Agricultural land use impacts soils, and most of these impacts are negative with regards to organic matter content (Paul and Clark 1996, Davidson et al. 2002). Forest management alters species composition (by planting, and use of fire and herbicides), nutrient supply (through fertilization or indirectly with harvesting and other silvicultural operations), and even water supply. These modifications typically increase aboveground process rates, but the response of BCA is less clear and probably variable. For example, intensive forest management usually increases aboveground net primary production (ANPP), but BCA may be reduced as a result of species (and genotype) change and improved tree nutrition. Afforestation in the 20th Century may have increased soil quality through increased organic matter content and reduced bulk density in many regions (see Six et al. 2002 for agricultural lands; Minkinnen et al. 1999 for a peatland case study), but we have little idea of the magnitude of changes in BCA that account for soil C changes (Bashkin and Binkley 1998, Binkley and Resh 1999, Paul et al. 2002, Giardina et al. 2004). Rising concentrations of gases in the atmosphere affect plants directly and indirectly. Increased concentrations of CO_2 in the atmosphere may stimulate productivity, including BCA (Pregitzer et al. 2000b, Zak et al. 2000a, King et al. 2001, Norby et al. 2002). Other gases

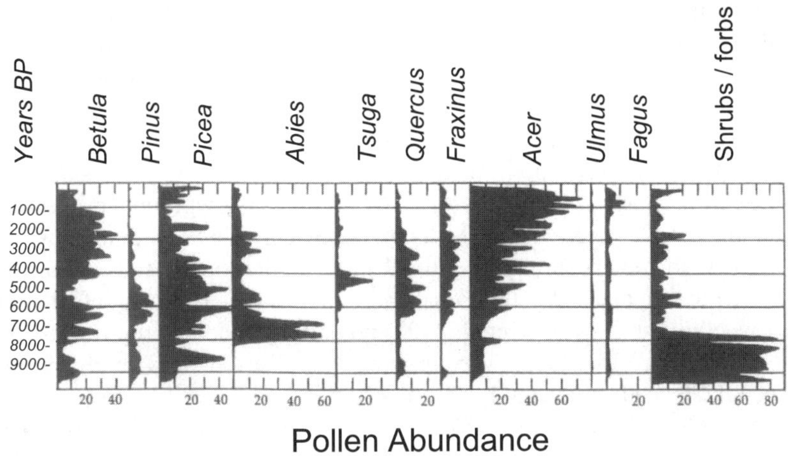

Figure 3. 9000 year record of change in vegetation (years before present) in Nova Scotia, as captured by a change in the quantity of pollen from different genera of common North American trees (Adapted from Livingston, 1968).

such as ozone (O_3) inhibit productivity (Reich 1983, Reich and Amundson 1985, King et al. 2001, Karnosky et al. 2003). Elevated CO_2 and other greenhouse gases also indirectly affect plants and ecosystems by changing global climate. Taken together, these direct and indirect effects on plants will likely impact BCA but above to belowground links (Figure 1) remain poorly quantified.

Rising greenhouse gases are likely to warm the biosphere, and micro to meso-scale studies often show strong temperature and moisture effects on plants and microbes (Uselman et al. 1999, Pregitzer et al. 2000a, Zak et al. 2000b, Pendall et al. 2004). The effects of increased greenhouse gases on BCA in forests remain uncertain, including direct alteration of canopy processes and indirect influences through warming and changing hydrology. Experiments on these individual processes have documented impacts on aboveground plant productivity (Reich 1983, Townsend et al. 1996, Holland et al. 1996, Karnosky et al. 2003), plant and mycorrhizal community composition (Karnosky et al. 2003, Lilleskov et al. 2001, Lilleskov et al. 2002), and soil heterotrophic organisms (Zak et al. 2000a, Zak et al. 2000b). However, the overall response of BCA to global change has been difficult to quantify King et al. (2001) because BCA integrates above and belowground changes, and the direct and indirect effects of global change factors on ecosystems can offset one another (Figure 1).

Few generalizations about controls on BCA have be made because methods and resulting estimates of BCA range widely (Ovington 1957, Raich and Nadelhoffer 1989, Albaugh et al. 1998, Reich and Bolstad 2001, Shaver and Jonasson 2001, Gower et al. 2001a, King et al. 2001, Davidson et al. 2002, Giardina and Ryan 2002), and responses to environmental variables are diverse (King et al. 1999, Pregitzer et al. 2000a, King et al. 2001, Giardina and Ryan 2002, Litton et al. *in review*). However, advances in belowground carbon science are occurring rapidly, particularly where stable isotopes permit investigators to track the flow of carbon through soil (Loya et al. 2003, Giardina et al. 2004, Matamala et al. 2004).

Two sets of findings point to an important change in our understanding of how global change factors control BCA. First , belowground processes in forests may be less responsive to temperature perturbations than previously believed, with root acclimation and substrate limitation on soil surface CO_2 efflux (Fitter et al. 1999, Giardina and Ryan 2000, Hogberg et al. 2001, Janssens et al. 2001, Melillo et al. 2002) potentially reducing the sensitivity of "soil respiration" to global warming (but see Burton and Pregitzer 2003, Burton et al. 2003). Secondly, plant canopies are tightly coupled to soil surface CO_2 efflux, with efflux being derived largely from recent photosynthesis (Horwath et al. 1994, Fitter et al. 1999, Janssens et al. 2001, Hogberg et al. 2001, Giardina et al. 2004). The degree of coupling was highlighted in a boreal forest by Hogberg et al. (2001), who reported up to a 40% reduction in soil CO_2 efflux within days of eliminating phloem transport of carbon to roots and mycorrhizae through girdling. Giardina et al. (2004) used ^{13}C isotopic methods to calculate that 90% of soil surface CO_2 efflux in a humid tropical forest was derived from current-year photosynthesis.

Further advances in belowground science have become possible with experiments exposing whole stands of trees to multiple global change variables (Karnosky et al. 2003). The free air CO_2 enrichment (FACE) experiment in Rhinelander, Wisconsin is especially important because three tree communities have been fumigated, singly and in combination, with gases that stimulate (CO_2) or reduce (O_3) plant primary production. The high cost of such replicated and multi-factorial ecosystem-scale experiments limits the number of interacting factors that can be examined. Consequently, the numerous interacting feedbacks originating both above and belowground will likely have to be examined through a combination of one and two-factor experiments, natural gradient studies, and modeling (Norby and Luo 2004).

A final feature of BCA complexity involves definitions of BCA. The terminology employed to describe carbon allocation within plants has been described as "varied, inconsistent, confusing, and often inadequate for understanding and integrating research results" (Dickson and Isebrands 1993). A similar lack of clarity continues to exist in BCA studies (Figure 2). BCA is often defined as fine or coarse root biomass standing stock (defined as partitioning by Dickson and Isebrands 1993), fine root production, total root production (coarse plus fine), total root production plus exudation and mycorrhizal production (which equals BNPP), and total belowground carbon allocation (which equals TBCA). Further, important components of BNPP are sometimes ignored in efforts to estimate whole stand or large scale patterns of NPP. While methodological ambiguities are not uncommon in ecological studies, the implications with respect to BCA are sizable because estimates are often scaled to entire regions or continents (e.g., Schimel et al. 1994, VEMAP 1994, Li et al. 2003), with uncertainties seriously impeding efforts to model climate change impacts on the global carbon cycle (Holland et al. 2000, Sarmiento 2000).

In this synthesis of global changes and the response of belowground production, we examine three BCA methods: coarse root standing stock, belowground net primary production (BNPP), and total belowground carbon allocation (TBCA). These three categories are methodologically and conceptually distinct, spanning the full spectrum of BCA studies (Figure 2). Coarse root standing stock is a pool determined through excavation and weighing, with quantification occurring at a single point in time. Changes over time are often inferred to be proportional to changes in aboveground biomass. BNPP is a flux arrived at by summing individually and periodically measured components, including period application of carefully determined allometries. Total belowground carbon allocation is a mass-balance approach that estimates a flux through periodic measurement of losses and changes in carbon storage. We examine available information on these three approaches to identify key features of each methodology, caveats, and data availability for examining BCA response to global change. Specifically, we examine whether aboveground measures can be used to predict BCA, and the likely magnitudes of BCA in relation to species, temperature and elevated CO_2. High rates of N deposition will likely impact BCA (Adams et al. 2004), but this was beyond the scope of our review. We finish the chapter with a list of the most pressing questions in the science of belowground carbon allocation.

COARSE ROOT STANDING STOCK

Quantifying root to shoot ratios has a long history in ecology (Ovington 1957, Cannel and Dewar 1994), with most studies measuring plant root standing stocks in non-woody plants or tree seedlings where roots serve primarily up-take and transport functions (McConnaughay and Coleman 1999, Giardina et al. 2001). The roots of older trees may be extensive for scavenging for resources, and very large to support massive aboveground structures. For example, Nepstad and colleagues (1994) showed that coarse roots can extend 7 m or more into soil in a seasonally dry tropical forest. Similarly, exploratory studies in riparian systems have shown that roots of trees including obligate phreatophytes (e.g., *Populus fremontii*) can extend many meters into soil to capture fluctuating groundwater (McElrone et al. 2004). As a result, accurately measuring the coarse-root standing stock of even young forests is challenging. Size is also complicated by variation in the horizontal distribution of coarse roots, with many studies sampling coarse roots between stumps to avoid digging up whole trees and their underlying tap roots. Because the largest mass of roots is located underneath the stump, this sampling bias renders between-tree coarse root mass estimates difficult to interpret. Variation in mass relating to species, fertility, or age also complicates efforts to estimate coarse-root standing stock, especially in mixed age / mixed species stands typically encountered in nature.

Variation in root to shoot relationships that relates to stand age (Figure 4; from Ovington 1957) can be dynamic, but process models typically use a single ratio to predict coarse-root standing stock, often set as a fixed proportion of aboveground biomass. For example, the function incorporated into a Canadian empirical model for pine relies on a single ratio of 0.22 (Li et al. 2003); this might be adequate for very broad assessments, but would miss important local detail if applied to individual stands. Recent papers demonstrating differences in root to shoot for conifers and hardwoods include Li et al. (2003) and Bolte et al. (2004).

The source of the discrepancy between model assumptions (e.g., Li et al. 2003) and the Ovington (1957) data presented in Figure 4 is unknown. However, if Ovington's Scots pine data are accurate, then process models may be under-predicting coarse-root standing stock in younger age classes of pine. Similar coarse-root standing stock data are available for hardwoods (Li et al 2003), but again age, species or site related patterns are poorly quantified. In general, uncertainty of coarse-root standing stock estimates has important implications for global C budgets. For example, of the estimated 60 Pg C allocated belowground by plants to roots and mycorrhizae, at least half occurs in wooded ecosystems (Grace and Rayment 2000). Based on limited

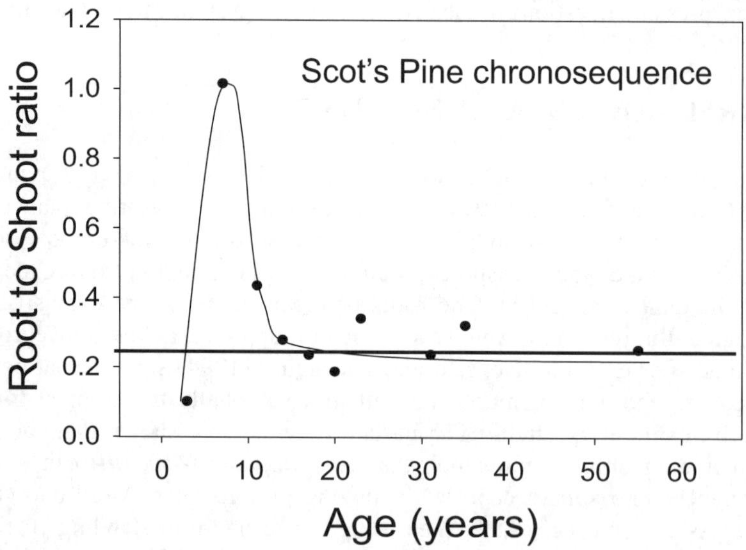

Figure 4. Data for Scots pine from Ovington (1957) and the root to shoot ratio (identified by the line at a root to shoot ratio of 0.22) used by Li et al. (2003) for estimating coarse root standing stocks for pine forests in Canada. The value of 1.0 for the 8-year-old stand of Ovington either indicates a major difference between the two studies, or it is an outlier and the studies support similar conclusions.

knowledge of how TBCA is partitioned belowground (Giardina and Ryan 2002, Giardina et al. 2004), approximately 10% of the C allocated belowground in wooded systems (3 Pg C) is allocated to coarse root production. If current coarse root allometries under-estimate coarse root to aboveground biomass ratios by an average error of 20%, then globally about 0.6 Pg C of coarse root NPP in forested ecosystems would be missed in current model estimates. To put this error in context, 0.6 Pg C is approximately 10% of annual global fossil fuel emissions.

While variation in coarse-root standing stock in relation to climate and species is poorly quantified, and errors in coarse root assumptions limit our ability to generalize about species or site differences, there appears to be some confidence that coarse root allometries within a species or climate zone are relatively insensitive to changes in fertility (King et al. 1999, Enquist et al. 2001, Giardina and Ryan 2002, Coleman et al. 2004, Coyle and Coleman 2005). Albaugh et al. (1998) harvested *Pinus taeda* trees from control, fertilized, irrigated and fertilized+irrigated stands after three years of treatment, and root to shoot allometry was constant despite a doubling of leaf area index and biomass in the fertilization and irrigation treatments (Figure 5). In contrast, Stape et al. (2004) observed a decrease in root to shoot ratios from 0.32 to 0.16 in *Eucalyptus* plantations with increasing moisture (Figure 5). Given that larger trees tend to have larger root systems, a key issue is whether the relationship between root and shoot biomass has a Y-intercept of 0 (as in the loblolly pine case in Figure 5), or not (as in the *Eucalyptus* case study in Figure 5). Litton et al. (2003) observed that root to shoot biomass ratio of young lodgepole pine trees increased with stand density but decreased with average stand basal area. In both cases, tree size varied with the treatment, such that ontogeny related effects (Figure 4) could not be ruled out.

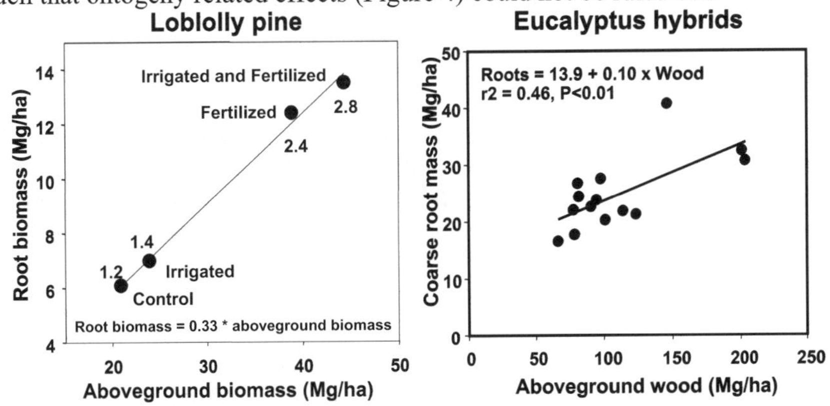

Figure 5. The root to shoot ratio was constant for loblolly pine (0.33) across treatments on a poor site ($R^2 = 0.98$, $P < 0.01$; from Albaugh et al. 1998). The same relationship for *Eucalyptus* in Brazil across 14 sites of varying productivity had a positive Y intercept, indicating that root mass increased as a proportion of aboveground mass from 16% on fertile sites to 32% on infertile sites (from Stape et al. 2004). Leaf area index (m^2/m^2) indicated next to loblolly points.

Given the importance of large roots for supporting large trees, King et al. (1999a) suggested that for a given site and species, root to shoot allometry should be less sensitive to changes in environmental conditions than root to shoot ratios of forbs (McConnaughay and Coleman 1999), or seedlings of woody plants (Gebauer et al 1996, King et al. 1999a, Giardina and Rhoades 2001). Specifically, these findings point to the important possibility that whole tree allometry may not change in response to anticipated changes in atmospheric CO_2 or global climate, though changes in moisture may alter allometry (Stape et al. 2004). This assertion is supported by King et al. (1999) where root to shoot allometry of *P. taeda* and *P. ponderosa* exposed to treatments of elevated temperature, CO_2 and nutrients, showed little effect of the treatments.

Even if changes in environmental conditions have little effect on root to shoot allometry, allometry may still change in response to alteration of site fertility, stand age, or species composition (King et al. 1996, King et al. 1999a, Bolte et al. 2004, Coleman et al. 2004, Coyle and Coleman, 2005). For example, increased tree growth due to elevated CO_2 or temperature may accelerate maturation and age-related changes coarse-root standing stock to aboveground allometry (Figure 4), and these changes could be misinterpreted as direct treatment effects on whole tree allometry rather than indirect effects of the treatments on allometry through accelerated ontogeny (see McConnaughay and Coleman 1999). We also note that changes in vegetation types may also occur; tree invasion of grasslands altered root architecture, BCA, and soil C storage (Jackson et al. 2002).

BELOWGROUND NET PRIMARY PRODUCTION

BNPP defined

Belowground net primary production (BNPP) has been defined as the mass of roots produced plus any root mortality occurring over a specified period of time. Increasingly, BNPP is defined as including all carbon allocated belowground by plants and not used for autotrophic respiration:

$$\text{BNPP} = \Delta B + D + H + E + M \qquad (1)$$

where ΔB is the change in root biomass, D is detritus generated, H is losses to herbivory, E is exudation from the roots, and M is C flowing to mycorrhizae. Change in biomass (ΔB) includes the increment in tap roots, structural roots and feeder-root tissue, measured over some increment of time (typically one year). Detritus (D) includes root mortality, root tissue loss, and mycorrhizal turnover during the year. Fine roots have received the most attention because the equivalent of their entire mass may be replaced (turnover) in one year or less (Eissenstat and Yanai 1997). Although the fraction of root tissue found in feeder roots at any time may be only 5% to 10% of belowground biomass, the rapid turnover rate makes this an important fraction of BNPP. Tap and coarse root mortality is typically low for healthy trees, but tree mortality is a normal component of forest development (and harvesting and fire!), and this would lead to significant mortality of tap roots and coarse roots. The loss of root cortical tissue during secondary thickening of feeder roots and sloughing of periderm tissue in large coarse roots should also be included as a component of BNPP, but it is unlikely to be a large fraction of total BNPP. The magnitude of insect herbivory on roots (*H*) remains poorly known, but may be large in some cases. The C allocated to mycorrhizae (*M*) has long been known to be a large component of BNPP, and probably remains the largest poorly quantified component of BNPP (Fogel and Hunt 1983, Eissenstat et al. 2000; Stevens et al. 2002; Wells et al. 2002b). Exudation (*E*) of organic compounds supporting rhizosphere organisms is difficult to quantify, yet *E* also may be a significant component of the BNPP budget and an important flux of labile carbon to soil (Uselman et al. 1999).

Methods of measuring BNPP

Each technique for measuring BNPP has advantages and disadvantages, and no perfect method is available to gauge the accuracy of other methods. Most effort has gone into assessing fine-root growth in part because of the importance of these tissues for nutrient uptake but also because they are the easiest component of BNPP to measure. Net production of fine roots has been studied using sequential coring, root in-growth cores or screens (Caldwell and Virginia 1991). Fine root biomass has also been estimated by coupling repeated soil coring with images from mini-rhizotrons (Hendrick and Pregitzer 1992).

With sequential coring, fine-root production and mortality are determined from changes in standing crops of live and dead fine-roots harvested from cores collected periodically throughout the year (e.g. Grier et al. 1981). The method assumes that incremental increases in live roots represent production and incremental increases in dead-roots represent mortality (Santantonio and Hermann 1985). The method also assumes that arbitrary size classes (e.g., $<$ 2.0 mm) accurately reflects the dynamic portion of the root system over the time steps of interest, that recovery of roots is unbiased, and that pools of live and dead roots are near steady-state, none of which may be necessarily true (Pregitzer 2002). Finally, sequential coring methods assume production and mortality do not occur simultaneously, and therefore the method can underestimate fine-root turnover and production (Publicover and Vogt 1993). The method can also overestimate root turnover if random variation in fine root estimates are mistaken for real gains and losses between sampling periods.

Viewing methods use rhizotrons, which involve using transparent viewing surfaces placed against the soil to allow measurement of the appearance, disappearance and lifespan of individual roots (Keyes and Grier 1981). Converting recorded images into estimates is labor intensive, but specialized software may be useful for image processing (Hendrick and Pregitzer 1992). Viewing methods simultaneously quantify fine-root production and loss, and rhizotron-based methods coupled with survival analysis techniques (Allison 1995; Wells and Eissenstat 2001) are leading to new insights into how environmental, developmental and phenological factors control fine-root turnover, especially when coupled with soil coring methods (Hendrick and Pregitzer 1996; Kern et al. 2004; Reuss et al. 2003; Wells et al. 2002a). Potential sources of error with rhizotron-based approaches to estimating BNPP include any effect of the observation window on root longevity (Withington et al. 2003), the difficulty of precisely measuring very small roots, especially in the surface few mm of soil (Vos and Groenwold 1987), disturbance effects of viewing windows installation on root growth (Coleman et al. 2000; Joslin and Wolfe 1999), and scaling from 2-dimensional area to the mass in a volume of soil.

Fine root lifespan can also be quantified with radiocarbon and stable carbon isotope depletion methods. These methods use bomb ^{14}C released during nuclear testing (Gaudinski et al. 2001) or a change in ^{13}C label from elevated CO_2 experiments (Matamala et al. 2003) to determine the mean residence time of root carbon. These isotope-based methods examine the isotopic composition of the total root pool at the end of some measurement interval. Notably, the survivorship of roots in soil is highly skewed, with a small portion living for a long period of time. However, short-lived roots that form and die within the measurement interval, perhaps the majority of roots in soil, will not be measured so that root longevity and turnover time may be overestimated. In fact, isotope based estimates appear to be many months to years longer than rhizotron-based approaches that track the birth and death of individual roots, although some of the discrepancy may derive from differences in size classes of roots of varying longevities. Roots that grow and die between measurement periods will not be sampled by isotope or rhizotron methods. With isotope methods, assumptions about the shape of the depletion curve and internal cycling of carbon also have strong effect on root lifespan estimates (Gaudinski et al 2001, Luo 2003). Reconciling discrepancies between rhizotron and isotope methods will almost certainly improve confidence in estimates of fine root NPP.

No validated estimates of mycorrhizal contribution to BNPP are available, in part because there are enormous challenges involved in trying to ascertain mycorrhizal fungal biomass, production and turnover. Three pools of mycorrhizal fungal biomass that differ sampling approach and quantification challenge are: reproductive sporocarps (mushrooms and spores), mycorrhizal roots and mycorrhizal mycelium in soil.

Saprotrophic sporocarps and their spores can be reliably distinguished from the sporocarps of mycorrhizal species, so sporocarp production is the most easily quantifiable component of mycorrhizal contribution to BNPP. Not all mycorrhizae fruit aboveground (e.g., truffles and truffle-like fungi), so quantification would require raking for hypogeous sporocarps of ectomycorrhizal fungi, and soil coring and extraction of the large spores of AM fungi. In Mediterranean climates, hypogeous sporocarps can be a large component of mycorrhizal sporocarp production. Sporocarp quantification also requires intensive sampling throughout the growing season, as fruiting can take place through the growing season

For mycorrhizal roots, visual estimates of fungal abundance (for AM fungi) and biochemical markers (for ectomycorrhizal fungi) are the primary approaches for estimating biomass. The arbuscular mycorrhizal (AM) component can be quantified by clearing and staining combined with some estimate of internal hyphal colonization. The ectomycorrhizal (EM) component can be quantified by using ergosterol, a sterol unique to fungi that has been used to quantify fungal biomass for basidiomycetes and ascomycetes, but appears to be absent in AM fungi (e.g., Grandmougin-Ferjani et al. 1999, Olsson et al 2003). Even when present, ergosterol concentrations in fungal tissue can vary several-fold, leaving large uncertainties in biomass conversions. Specific phospholipid fatty acids (PLFAs) are used as fungal biomarkers, but their concentrations are even more variable than that of ergosterol, making their use as biomass estimators untenable (Olsson et al. 2003). Production and turnover of mycorrhizal root tips can be estimated using minirhizotron systems, though we know of no such production estimates.

The quantification of mycorrhizal biomass and production in soils is probably the greatest challenge in estimating BNPP. Distinguishing between mycorrhizal and free-living heterotrophic fungi in soil is problematic, as dominant ectomycorrhizal and saprotrophic fungi are not taxonomically distinct, both being comprised primarily of Basidiomycetes and Ascomycetes. Recently natural abundance isotopes have been used in combination with ingrowth fine mesh bags with and without trenching, to estimate ectomycorrhizal fungal biomass production (e.g., Wallander et al 2001).

It is somewhat easier to distinguish AM fungi from saprotrophs, because the fungi that form arbuscular mycorrhizae are Glomeromycetes (formerly Glomales), which are taxonomically, morphologically and biochemically distinct from the Basidiomycetes and Ascomycetes (Smith and Read 1997). Given the uncertainties associated with biochemical markers described above, the best current method for biomass estimation is using microscopic methods (Bonfante-Fasolo 1986). Combined with in-growth fine mesh bags, some estimates of net production could be made.

Constraints of terminology on BNPP

Understanding the response of BNPP to global change is hindered by terminology. The estimates of BNPP in Figures 6, 7 and 8 were all termed BNPP, but none included all components of BNPP (Equation 1). Exudation is commonly excluded, or a "best guess" is used to constrain the magnitude of this component. Depending on how mycorrhizae are defined (heterotrophic or autotrophic), mycorrhizal contributions to BNPP are also poorly quantified.

These problems exist because it is extremely difficult to separate autotrophic from heterotrophic components of the total belowground C allocation (TBCA) at scales of stand and years. Attempts at quantifying BNPP have had to ignore key components to arrive at estimates, or have wrestled with the challenge of separating autotrophic and heterotrophic components. Recent advances have been made in the effort to separate heterotrophic from autotrophic components of soil respiration using trenching, stem girdling, or a components approach, but quantifying BNPP is still very difficult. First, girdling or more destructive approaches cannot separate root respiration (autotrophic) from exudation (heterotrophic) derived CO_2. Second, CO_2 from mycorrhizal respiration (autotrophic?) and mycorrhizal turnover (heterotrophic) cannot be separated (conceptually or physically) from root processes. Biologically, mycorrhizae are heterotrophic (and in some cases partially saprotrophic), but functionally they extend a plant's root system and therefore may be viewed as being autotrophic. For example, Gower et al. (2001a) identified mycorrhizae as a significant part of BNPP, and therefore autotrophic. Finally, even if soil respiration could be precisely divided into

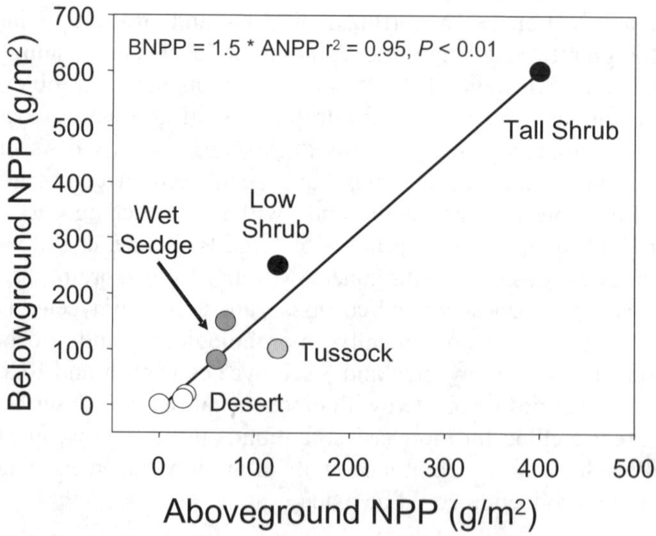

Figure 6. Data from review by Shaver and Jonasson (2001), showing the stability of BNPP to ANPP for Arctic ecosystems in North America.

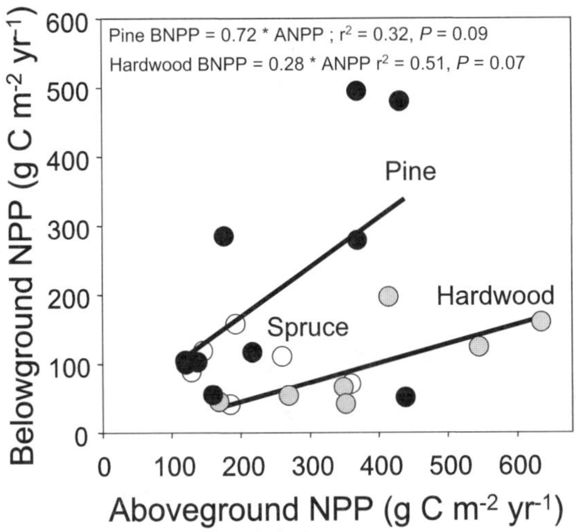

Figure 7. Data from review by Gower et al. (2001a), showing the stability of BNPP to ANPP for boreal hardwood and pine species across Russia and North America. There was no relationship between BNPP and ANPP for spruce species across sites, but the relationship appears negative.

heterotrophic and autotrophic respiration, the heterotrophic sources of soil respiration are themselves very complex and perhaps impossible to separate (Bond-Lamberty et al. 2004). For example, a significant but seasonally variable fraction of soil respiration is derived from aboveground litterfall carbon (Raich and Nadelhoffer 1989), but because leaves may be comminuted and transported within the soil by animals, heterotrophic decomposition of the leaf material may occur anywhere in the soil profile.

Other terminology issues complicate comparison among studies. Terms such as "root turnover" and "fine root" have been defined inconsistently. Root turnover is the rate at which roots are produced or lost during a specified period (based on mass or length) divided by the average standing crop during that period. Results are typically expressed in units of g g^{-1} d^{-1} or simply d^{-1}, which is the inverse of median root lifespan. The numerator may include production, mortality or the average of the two. The denominator may include maximum, minimum or average standing crop. For root systems at steady state, production and mortality should be equal, such that the choice of the parameter for the numerator is of little importance. However, steady-state conditions are rare within a season or through the development of a stand over years (Haynes and Gower 1995, Kern et al. 2004; Pregitzer et al. 2000b). Under non-steady-state conditions production and mortality differ and the choice of denominator or even numerator used for turnover calculations will influence estimates. For example, because turnover is the inverse of median lifespan, using lifespan emphasizes the importance of mortality rate. The use of survival or proportional hazard analysis provides powerful statistical tools for testing controls of turnover rate. Evaluating lifespan using root viewing or

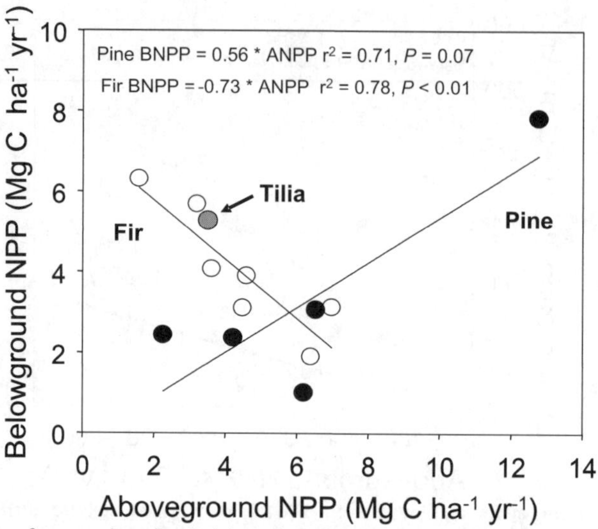

Figure 8. Data from review by Reich and Bolstad (2001) showing strong but opposing relationships between ANPP and BNPP for temperate pine and fir (true fir and Douglas-fir) species across sites. Only one point is reported for a hardwood point so no relationship is given.

isotopes methods are important techniques for determining lifespan that are free from choices of rate and standing crop.

Finally, definitions that include size classes can complicate comparisons. Fine roots are commonly distinguished based on their diameter, with definitions ranging from <1 mm to <5 mm. However, much of the perennial root system ranges between 1 and 5 mm, and ephemeral, small-diameter feeder roots increase in specific root length, nitrogen concentration, rate of root respiration, and risk of mortality from the proximal to distal end of the root system (Pregitzer, 2002, 2003). Accurately describing the range of individual root lifespan and primary function will require adopting terminology that more precisely describes the actual function of the branching root system and recognizes that root systems integrate a complex assembly of functionally distinct individuals (Pregitzer et al. 2000b, Pregitzer 2003).

Aboveground factors as predictors of BNPP

Can aboveground measures be used to predict belowground measures of BNPP? The answer depends in part on which components of BNPP are considered. Coarse root biomass often correlates highly with stem biomass (Figure 5; see above discussion; also Enquist 2002; Enquist and Niklas 2002). Our understanding of environmental controls on this ratio is also improving (Albaugh et al. 1998, Litton et al. 2003, Stape et al. 2004). These initial findings suggest that measures of aboveground stem biomass increment may adequately predict coarse root NPP (e.g. King et al. 1999a, Coleman et al.

2004). Yet, variation due to species and environmental factors may be large; some of this variation may be explained by accelerating development (Figure 4) due to treatments such as irrigation, mineral nutrients and elevated CO_2 (Gebauer et al. 1996, King et al. 1999a, McConnaughay and Coleman 1999, Coleman et al. 2004, Coleman et al. 2005), but the limited number of studies available to identify let alone quantify controls on above to belowground biomass relationships limits our ability to generalize across studies.

Relationships between aboveground measures and more complete measures of BNPP are more variable (Litton et al., *in review*). Shaver and Jonasson (2001) showed a strong correlation between BNPP and ANPP for arctic ecosystems (Figure 6). Gower et al. (2001a) also found strong correlations for boreal pine and hardwood forests (Figure 7), though the relationship was poor for boreal spruce. Gower et al. (2001a) suggested that total NPP (and BNPP by difference), could be predicted from commonly available forest inventory data. This assertion was supported by temperate forest data compiled by Reich and Bolstad (2001), showing a strong positive relationship for pine (Figure 8). Reich and Bolstad (2001) also found an inverse relationship between ANPP and BNPP for fir and Douglas-fir (Figure 8), in line with the inverse trend reported by Gower et al. (2001a) for spruce. Overall, a general relationship between ANPP and BNPP (Figure 9) is encouraging given variation in climatic conditions and soils. However, variation is substantial and meaningful generalizations about central tendencies, especially across species and biomes, will require more data.

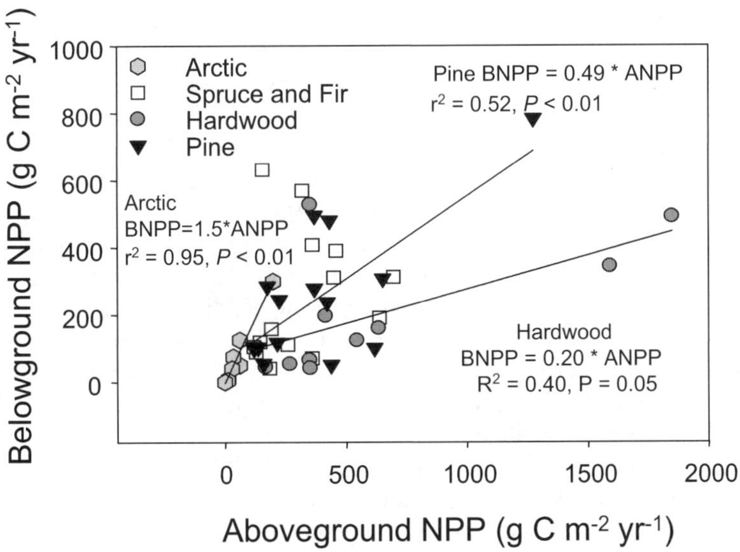

Figure 9. Global scale relationship between ANPP and BNPP with data from Figures 6 through Figure 8. The global relationships confirm regional patterns for pine and deciduous vegetation (hardwood trees plus larch, shrubs, forbs, and grasses), but fail to support patterns for spruce, fir, and Douglas-fir.

The use of aboveground factors to estimate BNPP requires that any measure integrates factors influencing both fine-root production and mortality. The response of root mortality to a range of environmental factors, along with the seasonal separation in fine-root production and mortality suggests that these processes might respond independently, and independence of two major components of BNPP would suggest that predicting the ephemeral root fractions from aboveground measures will remain difficult (Landsberg and Gower 1997; Pregitzer et al. 2000a, Pregitzer et al. 2000b).

Global change factors affecting BNPP

Global change factors that could influence BNPP include shifts in species composition with climatic change, elevated CO_2, temperature, moisture and the interactions among these factors. Data are limited, but some species have shown repeatable patterns, and experimental manipulations have yielded predicable changes in BNPP.

Species composition

Global change could affect species distribution by altering site temperature, precipitation and nitrogen deposition. Tree root systems acquire limiting resources from the soil, and changes in these resources will lead to changes in the belowground processes controlling species distribution (Norby and Jackson 2000). Both theory and paleoclimatic evidence indicate that global change will change species cover and distributions, but the interactions between species and sites with BNPP will likely be complex. Combining data from Figures 6, 7 and 8 into a single figure shows that arctic, pine and hardwood vegetation each fall on significant regression lines for aboveground-to-belowground NPP (Figure 9). In contrast, no relationship emerges for spruce and fir. There appears to be some variation among biome and species, but the two cannot be disentangled. Further, species segregate across landscapes in response to variation in site conditions, with both species and site altering BNPP. For example, arctic ecosystems appear to allocate substantially more carbon to BNPP relative to ANPP when compared with boreal and temperate pine ecosystems, but differences in species, hydrology, other site variables or methodology also could explain these patterns. In turn, pine species appear to allocate more carbon to BNPP relative to ANPP when compared with boreal and one temperate hardwood species, but again, the cause of the difference is difficult to ascertain. No apparent differences emerge across climate types for pine and hardwoods, and the variation is high. Forested wetlands tend to fall between the pine and hardwood trends (Figure 9), but these forests also exhibit considerable variation among species (Burke and Chambers, 2003) and site conditions (Finer and Laine, 1998). The absence of a clear trend for spruce, fir and Douglas-fir indicates that controls on variation in allocation patterns are still poorly understood within and across species.

Gill and Jackson (2000) considered available belowground data sets and found that fine root turnover generally increased with mean annual temperature. This pattern was consistent between various life forms and implies that warmer sites require greater production to maintain similar amounts of root biomass. In contrast to Figure 9, Gill and Jackson (2000) found no differences between temperate conifer and broadleaved tree species in either mean or temperature weighted root turnover, and that forest type in general explained little of the variation in turnover rates. Using similar data, Li et al. (2003) also found no differences for BNPP between hardwood and conifer forest types, despite differences in coarse-root standing stock.

Common garden studies present the most direct test of how species impact carbon allocation patterns in forests, and several studies have distinguished differences among evergreen conifers and deciduous hardwoods. In a planted species trial, Coleman et al. (2000) used minirhizotron methods, and estimated that fine-root production by *Pinus resinosa* was only 6% of that of *Populus* hybrid. Steele et al. (1997) used both sequential coring and minirhizotron techniques to show greater fine-root production for *Populus tremuloides* compared with *Pinus banksiana*, especially when adjusted for soil temperature.

These results agree with indirect nitrogen budget technique results, where evergreen conifers have lower annual fine-root biomass production than deciduous hardwoods across broad gradients in environment rather than in common gardens (Aber et al., 1985). However, most studies of conifers and hardwoods across wide ranges of site conditions have not found differences in fine root NPP (McClaugherty et al., 1982; Nadelhoffer and Raich, 1992). Interpreting discrepancies between comparisons of species at a single site versus across diverse sites is confounded by variation in site characteristics (and methodology). For example, pines might occupy nutrient poor sites, where high allocation to roots is required to meet water or nutritional needs, while hardwoods might occupy higher quality sites, where a greater allocation to ANPP is permitted. When pines and hardwoods are grown on the same site, as in a common garden, allocation patterns often change in response to altered soil conditions (Cannel and Dewar 1994, Giardina et al. 2003). Common garden studies are limited because patterns found between species at a single site may not represent the patterns that would be found across other sites (see Binkley and Menyailo, this volume).

Overall, new thinking is required to accurately predict climate change impacts on belowground productivity and allocation patterns in relationship to species and species interactions with climate. We suggest that greater attention to distinguishing root classes and characterizing site and stand characteristics will be particularly valuable.

Elevated CO_2: Free Air CO_2 Enrichment experiments.

Elevated CO_2 commonly increases BNPP in experiments with seedlings in growth chambers and in open-top chambers in the field (Berntson and Bazzaz 1996; Crookshanks et al. 1998; Godbold et al. 1997; King et al. 1996; Norby et al. 1992). The response of intact forest stands remains poorly known, but free air CO_2 enrichment (FACE) experiments provide some information. Several *Populus* species in the PopFACE experiment in Italy responded to three years of elevated CO_2 by increasing root production by 42 to 88% (Lukac et al. 2003). Minirhizotron observations of sweetgum forest stands during 6 years of treatment at the US Department of Energy Oak Ridge National Lab FACE facility increased root production, mortality and standing crop (Norby et al. 2002, 2004); the magnitude of these BNPP changes were large enough to account for the entire NPP response to elevated CO_2 (Figure 10). Further, elevated CO_2 shifted the partitioning of primary productivity from ANPP to BNPP, indicating that belowground resource demand increased with elevated CO_2. Tissue quality and nitrogen cycling were not reduced by elevated CO_2 relative to control plots, so the mechanism for the shift in allocation is not clear.

Similar trends have been reported for the loblolly pine FACE at Duke University in North Carolina, but the trends have not been significant. During one year of observation, fine root production increased 26%, fine root mortality increased 46%, and fine root standing crop increased 16 to 68% depending on the method of measurement (Matamala and Schlesinger 2000;

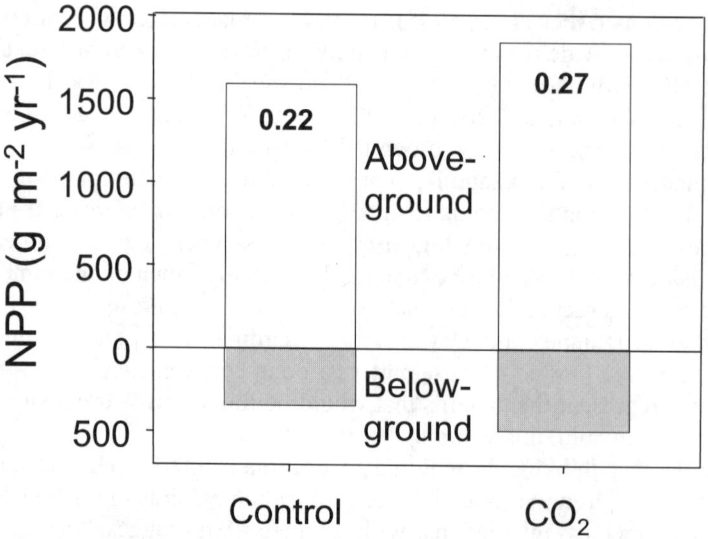

Figure 10. Data from Norby et al. 2002 showing the positive effects of elevated CO_2 on above and belowground NPP. Elevated CO_2 in this FACE experiment increased ANPP and BNPP, and also increased the BNPP to ANPP ratio (the ratio is identified in each bar).

Pritchard et al. 2001). For both US FACE studies, the effect of elevated CO_2 on fine-root turnover rate was limited, as observed in minirhizotrons and confirmed by carbon isotope depletion method in an inter-site comparison (Matamala et al. 2003).

Given numerous challenges associated with quantifying mycorrhizal fungal biomass, production and turnover, it is not surprising that there is limited information on the response of the mycorrhizal component of BNPP to elevated CO_2. Elevated CO_2 leads to an increase in mycorrhizal parameters (mostly measured as percent root colonization) of approximately 1.5 fold in field studies (Treseder, 2004). In the POPFACE study, root colonization after three years by arbuscular mycorrhizal and ectomycorrhizal fungi increased for two of three and one of three *Populus* species, respectively (Lukac et al 2003). The mycorrhizae of the hybrid *P. x euroamericana* did not respond to CO_2, despite strong responses in standing root biomass and fine root production. At the FACE site in Rhinelander, WI, mycorrhizal fungal sporocarp biomass production rates increased approximately 1.25 fold in elevated CO_2 when compared to ambient conditions, but increased 4.8 fold in elevated $CO_2 + O_3$ treatments compared to elevated O_3 (Lilleskov, unpublished), indicating a strong interaction of the effects of CO_2 and O_3 on this component of fungal production. Much more information will be needed to characterize the variety of mycorrhizal responses that alter BNPP.

Soil Temperature

Tree root growth commonly increases with soil temperature (Kaspar and Bland 1992; Lyr and Hoffmann 1967; Teskey and Hinckley 1981). Warm soil temperatures can increase fine-root production and decrease root longevity (Eissenstat and Yanai 1997; Hendrick and Pregitzer 1993; King et al. 1999b; Wan et al. 2004), with some studies showing that inter-annual variation in fine-root production relates strongly to inter-annual temperature fluctuations (Coleman et al. 2000; Tierney et al. 2003). Other studies have shown weak effects of temperature (Hendrick and Pregitzer 1997; Joslin et al. 2001). Seasonal changes in soil temperature are usually associated with seasonal changes in root growth, but this covariation confounds any temperature effect with normal seasonal phenology of plants (Pregitzer et al. 2000b). Other environmental factors such as drought, soil solution nutrient concentrations or freezing temperatures can also exert control over both production and mortality (Joslin et al. 2001; Tierney et al. 2003).

Soil Moisture – Flooding

Hydric soil conditions can cause root morphological and physiological adaptations to saturated conditions (McKevlin et al., 1998), but the influence of hydric conditions on BNPP remain largely unexplored. Trettin and

Jurgensen (2003) reviewed the state of knowledge for wetland forests, and found that BNPP in boreal bogs and fens was approximately 50% of ANPP, with that proportion declining in boreal swamps (30%) and temperate bottomland hardwoods (25%). This trend across sites may not be matched by responses to hydrologic regime within single sites. For example, studying BNPP in drained peatlands, Finer and Laine (1998) reported increased belowground allocation with increased temperature and aeration, and that the BNPP of the tree and shrub strata do not necessarily respond similarly to site conditions. Burke and Chambers (2003) found large differences among *Quercus* species responses to flooding in a southern bottomland forest, and that alternated aeration led to an increase in BNPP as trees adjusted to the variable soil conditions. Understanding BNPP dynamics in wetland soils is particularly important as they contain a disproportionate amount of the global terrestrial C (approximately 30%) and changes in the water cycle is a likely consequence of most global change scenarios (Trettin and Jurgensen, 2003).

Interactions among CO_2, forest type, and temperature

Interactions among global change factors are often the most intriguing and important responses of forests to multiple variables, highlighting the complex nature of environmental controls. In a study of two *Acer* species elevated CO_2 and temperature both increased root production and growth (Wan et al. 2004). In an earlier study, elevated CO_2 increased root biomass of *P. taeda* and *P. ponderosa*, with temperature interacting with CO_2 in *P. taeda* (King et al. 1996). Root growth may increase in response to combined elevated CO_2 and temperature, but there negative responses are also possible due to increased respiration, higher root N concentration, and altered soil microbial activity (Pendall et al. 2004). A multi-factored study of root exudation in *Robinia pseudoacacia* mesocosms found that elevated CO_2 did not influence exudation, whereas elevated temperature and additions of nitrogen stimulated exudation (Uselman et al. 1999). Overall, the lack of appropriate data from across species and for adult trees prevents any generalization about how forests will respond to multiple factors.

TOTAL BELOWGROUND CARBON ALLOCATION

Total belowground carbon allocation (TBCA) is defined as that carbon allocated belowground by plants to produce coarse and fine roots, root respiration, and root exudates and mycorrhizae (Figure 2). Belowground C allocation can be a large fraction of gross primary production (Ryan et al. 1994, 1997a; Giardina et al. 2003), sometimes exceeding aboveground net primary production (Law et al. 1999). Our understanding of the factors that control TBCA is poor, though increases in the numbers of experiments will help clarify the major role of TBCA in the C balance of terrestrial ecosystems, (Giardina et al. 2004).

Raich and Nadelhoffer (1989) originally proposed a mass-balance approach to quantify the total quantity of carbon allocated belowground by trees on an annual time step. This approach relies on mass balance to estimate TBCA with quantifiable uncertainty for all fluxes (unlike the BNPP methods described above). Plants send fixed C to roots. This C must either be respired by microbes or roots (measured as soil-surface CO_2 efflux or F_S) or stored in soil as organic matter, in the litter layer, or in living and dead roots. If C storage in soil, roots, or the litter layer does not change over the measurement period of interest, and leaching and erosion losses are negligible, then conservation of mass dictates (i.e., any soil carbon that is formed from TBCA will be offset by older carbon that is released through decomposition) that TBCA must equal F_S minus C inputs from aboveground litter (F_{AL}):

$$\text{TBCA} = F_S - F_{AL} \qquad (2)$$

The utility of TBCA estimates differs from that of soil respiration in several important ways. Soil surface CO_2 efflux ('soil respiration') is an integrator of the key components of the belowground C cycle, and consequently has been viewed as an index of belowground C cycling rates. From established information on soils, roots, and organisms inhabiting soils and the rhizosphere, soil surface CO_2 efflux (F_S) can be described by the following equation:

$$F_S = F_R + F_M + F_{AL} + F_{BL} + F_{SOC} \qquad (3)$$

where F_R is the flux of CO_2 from respiring roots, F_M is the flux of CO_2 from respiring mycorrhizae, F_{AL} and F_{BL} are fluxes of CO_2 from decomposing above and belowground litter (including root and mycorrhizal exudation and turnover), and F_{SOC} is the flux of CO_2 from decomposing organic C stored in mineral soil (microbial biomass, low-quality remains and by-products of litter decomposition). F_R represents CO_2 of autotrophic origin while F_M, F_{AL}, F_{BL} and F_{SOC} represent CO_2 released by heterotrophic organisms, though F_M has been described as autotrophic (Gower et al. 2001a).

Quantifying the individual components of soil surface CO_2 efflux is challenging because belowground C processes are intimately associated with the soil matrix. Sampling for individual components is often labor intensive (e.g., root excision to estimate F_R or trenching to estimate F_{SOC}), and estimates of the components of soil surface CO_2 efflux are often limited to a snapshot or a small area. Roots, mycorrhizae and soil are intimately connected, so these studies may not accurately represent belowground processes as they would occur in undisturbed soil (Högberg et al. 2001). Finally, the belowground (roots, microbes) and aboveground (leaf and branch litterfall) components of soil surface CO_2 efflux may not respond similarly to changes in the environment (e.g., Giardina et al. 2004). Where more than one variable is changing (e.g., temperature, moisture and nutrient supply), ecosystem responses to these multiple changes may be quite complex (Figure 1). Warming may increase decomposition rates, but associated increases in nutrient mineralization rates may alter plant allocation strategies, perhaps shifting C allocation away from roots to aboveground parts, lowering soil surface CO_2 efflux. Given the potential for offsetting effects, changes in soil surface CO_2 efflux are difficult to interpret, especially with respect to how component fluxes are altered.

The TBCA approach has the advantage of non-invasive, integrative over time and space, and bounded by mass balance. As conceived by Raich and Nadelhoffer (1989), the approach relies only on direct measures of soil surface CO_2 efflux and litterfall. Using Equation 2 and assuming that leaching losses of C are negligible, and that soil, forest floor and root C storage were in steady state, Raich and Nadelhoffer (1989) estimated TBCA for a wide variety of mature forests from published measurements of soil respiration and litterfall. They found that aboveground litter (F_{AL}) contributed 23% of soil surface CO_2 efflux at low efflux rates (400 g C m^{-2} yr^{-1}) to 31% of soil surface CO_2 efflux at high efflux rates (1500 g C m^{-2} yr^{-1}). By difference, the belowground sources of CO_2 (i.e., TBCA which equals [$F_R + F_M + F_{BL} + F_{SOC}$] when soil C is in steady state) contributed 69% to 77% of soil surface CO_2 efflux.

In an effort to address uncertainties resulting from steady state assumptions, Giardina and Ryan (2002) outlined a similar approach that accounts for changes in belowground and forest floor C storage:

$$\text{TBCA} = F_S - F_{AL} + \Delta\,[C_S + C_R + C_L] \qquad (4)$$

where C_S = carbon content of mineral soil, C_R = carbon content of root (coarse + fine) biomass, and C_L = carbon content of the litter layer. Increases in C storage will decrease soil respiration, while decreases in storage will increase soil respiration. This approach to estimating TBCA still requires that losses of C to leaching or erosion are negligible, but this will be true for most forests on level topography (Giardina and Ryan 2002).

An important finding of Giardina and Ryan (2002) was that litterfall was a poor predictor of TBCA across their treatments. More importantly, they found that changes in soil or forest floor carbon storage, while dynamic, contributed little to the TBCA budget in a young, fast growing plantation forest. They concluded that non-steady state conditions may not be a concern as long as both soil respiration and litterfall are measured; the failure of litterfall to predict TBCA did not relate to violation of steady state assumptions, but to the dynamic variation in the relationship between TBCA and litterfall.

The TBCA approach has limitations. Estimates of TBCA cannot be used to quantify BNPP, though TBCA can anchor BNPP estimates derived from other methods within a total belowground budget (Ryan et al. 1996, McDowell et al. 2001). Further, TBCA estimates as with all ecological measures rely on accurate estimates of soil respiration and litterfall, and in Equation 3, soil carbon, forest floor carbon and coarse roots. The latter measures are straight forward (Giardina and Ryan 2002), but soil respiration estimates can be sensitive (+/- 20%) to choice of equipment, frequency of measurement, and other factors.

The effects of species and temperature on TBCA

Here we summarize available TBCA data and examine relationships with mean annual temperature (MAT) and species. We also estimate TBCA for sites where soil CO_2 efflux and aboveground litterfall data are available to examine how elevated CO_2 affects TBCA. Finally, building on previously described relationship between TBCA and litterfall (Raich and Nadelhoffer 1989, Davidson et al. 2002), we examine large-scale patterns of TBCA to ANPP with the goal of understanding whether ANPP predicts TBCA.

TBCA is sensitive to changes in tree age (Giardina and Ryan 2002) and site fertility (Ryan et al. 1996). Tree age exerts a large influence on TBCA, with TBCA declining by as much as 30% from maximum rates at canopy closure (Smith and Resh 1999, Giardina and Ryan 2002). Reported responses of TBCA to fertilization also have been large. In plantations of *P. radiata* and *Eucalyptus saligna*, fertilization reduced TBCA by 28% and 12%, respectively. Using the TBCA approach in conjunction with stable C isotope measurements of SOC, Giardina et al. (2004) examined the belowground fate of TBCA in a *Eucalyptus* plantation, including the efficiency with which TBCA is retained in soil as new soil C, and the fraction returned to the atmosphere as soil surface CO_2 efflux (Figure 11). Increased nutrient supply shifted the allocation of carbon from fine roots and mycorrhizae to coarse roots and aboveground leaf and wood production, but did not alter the efficiency with which TBCA was converted into new soil carbon.

It would be difficult to extrapolate responses from stand age or fertilization studies to scenarios of global warming or species change. However, there are no experimental studies of TBCA response to changes in these variables. We addressed this information gap by assembling TBCA estimates for widely ranging forests, and examining how much variation across sites could be ascribed to mean annual temperature or species (Figure 12). TBCA at a site with an MAT of $20^{O}C$ was on average 1.8 times greater that TBCA at a site with an MAT of $10^{O}C$, yielding a Q_{10} of 1.8 for TBCA across sites. The relationship was robust ($R^2 = 0.47$; $P < 0.01$) considering the wide diversity of soil and vegetation types, methods and studies.

As with any natural gradient study, confounding factors may complicate interpretation of results. For example, temperature may co-vary with soil development, and as discussed above for BNPP, TBCA can change in response to differences in soil characteristics (Giardina et al. 2004). Efforts to

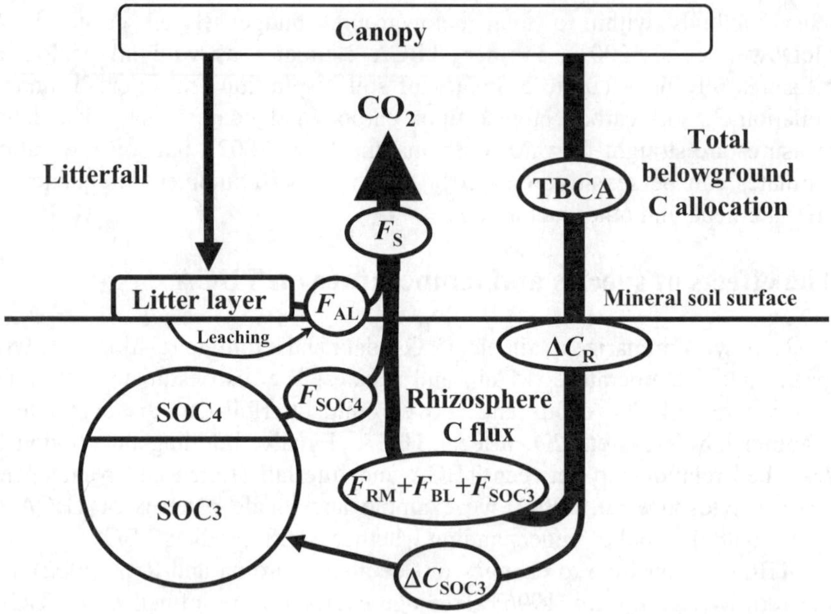

Figure 11. Conceptual model for how TBCA and stable isotopes can be used to examine the belowground fate of TBCA and the component sources of soil respiration (adapted from Giardina et al. 2004).

isolate the effects of species or climate on belowground processes may be compromised by legacy effects from earlier vegetation or by sampling periods that are too short to capture lags associated with yearly variations in above and belowground processes (Davidson et al. 2002). The lack of species effects on TBCA conflicts somewhat with data presented in Figure 9 but is consistent with several studies described above in the BNPP section. Further, across large scales, TBCA to ANPP ratios average about 1.5, indicating that TBCA generally represents a larger sink for GPP than does ANPP. While a TBCA to ANPP ratio of 1.5 represents a general central tendency of the data set, the variation in ANPP to MAT and TBCA to MAT relationships translates into some uncertainty about this tendency. When the 95% confidence intervals for both relationships are considered, a 95% confidence interval for a TBCA to ANPP ratio at $10^{o}C$ of 1.56 would include 1.33 to 1.82. Confidence would be lower at cooler or warmer ends of the relationship.

The fraction of TBCA that is BNPP is poorly quantified, but may be critical to correctly modeling ecosystem carbon cycling and the belowground carbon cycle. Because little data are available, it has been largely assumed that approximately 50% of TBCA is BNPP (Law et al. 1999, Giardina et al. 2003). It is noteworthy that comparing Figures 9 and 12, TBCA varied from 400 to 1500 g C m^{-2} yr^{-1}, while over a similar ANPP range, BNPP varied from 150 to 800 g C m^{-2} yr^{-1}, indicating that, despite high variance, BNPP is on average about 50% of TBCA.

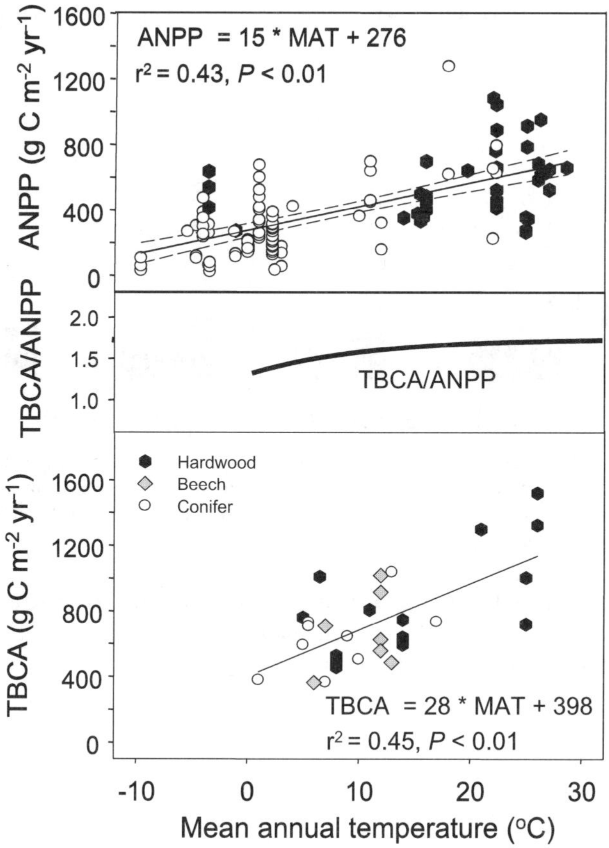

Figure 12. Global scale relationship between mean annual temperature (MAT) and ANPP (Top Panel), between MAT and TBCA (Bottom Panel) and the ratio of TBCA to ANPP derived from the equations describing the two relationships. Points are unfertilized forests across a global scale gradient in MAT. Data are from Ryan et al. (1996), Clark et al. (2001b), Gower et al. (2001b), McDowell et al. (2001), Reich and Bolstad (2001), Davidson et al. (2002), Giardina et al. (2003), and Litton et al. (2004). ANPP data were screened to include studies that reported at least wood and leaf NPP. If reported, branch NPP was included. We did not include forests growing on new or very young soils.

The effect of elevated CO_2 on TBCA

No data are available on how TBCA responds to changes in atmospheric CO_2. Several FACE studies have reported an increase in soil respiration under elevated CO_2, and these increases have been ascribed to increased fine root NPP (Norby et al. 2004), and increased exudation or litterfall (King et al. 2004). We modified the approach outlined by Giardina and Ryan (2002) to examine how FACE treatments altered TBCA at the Rhinelander and Oak Ridge FACE sites. We used these data to estimate TBCA as soil respired C *plus* coarse root increment C *minus* litterfall C. TBCA was 10 to 15% higher

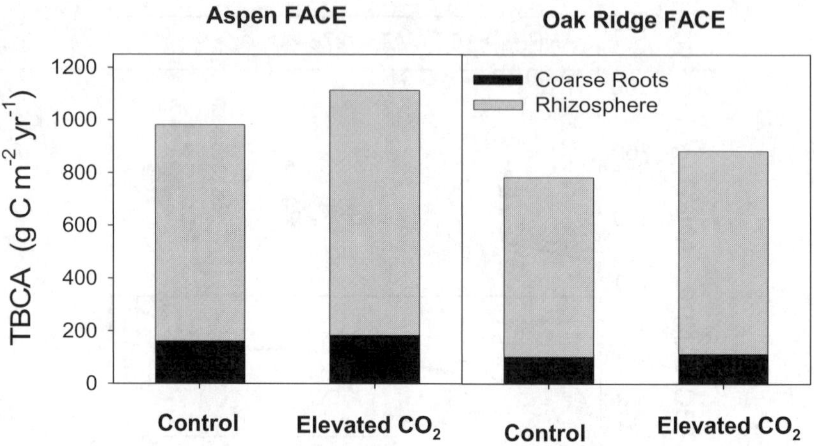

Figure 13. The effects of elevated CO_2 on mixed aspen-birch stands in Rhinelander Wisconsin (Left panel) and sweetgum (*Liquidambar styraciflua*) in Oak Ridge, Tennessee. Data for calculations of TBCA are from King et al. (2004), Norby et al. (2002), Norby et al. (2004), and unpublished data of the authors.

for both sites with elevated CO_2 (Figure 13). The increase in fine root NPP reported by Norby et al. (2002) for elevated CO_2 was similar to the increase in TBCA for the same plots, indicating that most of the increase in TBCA was allocated to fine root production.

CONCLUSIONS

A key frontier in global change science involves understanding the controls on belowground carbon allocation, processing and retention. Confidence in measurement techniques is constrained by our inability to directly measure the carbon flows of interest: how can we be sure that our data accurately represent a process of interest? Measurements of TBCA come closest to direct measurements, but this aggregated measure provides the least insight on the details of all the processes that comprise BCA. The perceived and actual sensitivity of the flux measurements to changes in the environment, and the sensitivity of these measures to artifacts, vary widely across methods. With these warnings in mind, we suggest three important generalizations:

- Changes in BCA will vary in concert with changes in aboveground productivity, because overall, BCA and ANPP vary in concert. While the fraction of GPP for each may change under global change, BCA and ANPP in general are closely linked.
- Greater integration of available data across biomes and species is needed to test what appear to be reasonable generalizations within a biome or species.

- The complete suite of BCA components needs to be measured for more forests, with explicitly defined populations (soil types, species, or gradients where both vary). These C budgets also need to be explicitly connected to experimental manipulations of resources and species within sites, to provide a gauge of the value of cross-site comparisons for predicting within-site responses.

Overall, BNPP studies have greatly advanced our understanding of how forest ecosystems function and will respond to global change. Studies that now combine isotopes of carbon with BNPP observations and mass balance approaches are building on these ground breaking BNPP studies. Future studies that combine TBCA, BNPP and isotope-based methods will lead to greater insights into how the belowground carbon cycle will respond to a changing world. A myriad of important questions remain unanswered about belowground carbon cycling (Table 1). It is our challenge to apply these new methods while continuing to develop new techniques for assessing belowground processes. We also need to prioritize these questions, as funding resources are limited and the potential combinations of conditions and factors are enormous. We feel that temperature and moisture gradients and manipulations of species and nutrients will serve as the basis to efficiently address the complex interactions of species, site, and global change factors.

ACKNOWLEDGMENTS

This work was supported by North Atlantic Treaty Organization; the US Department of Energy; the US Forest Service Northern and Southern Global Change Programs; and the North Central, Rocky Mountain, and Southern Research Stations of the US Forest Service.

Table 1. Our list of pressing questions in the science of belowground carbon cycling.

1. How does BCA vary by species?
2. Are stand characteristics such as species and stand age the ultimate drivers of BCA?
3. Alternatively, are moisture and temperature the ultimate drivers of BCA?
4. How will elevated CO_2, climate and nutrient deposition interact to impact BCA?
5. How will the impacts of CO_2, climate and nutrient deposition interact with species and site?
6. What is the efficiency with which BCA is converted into new soil carbon?
7. Does conversion efficiency vary by species, site or climate?
8. In a warmer world, will increases in BCA offset reductions in the conversion efficiency of BCA into soil organic matter, maintaining historic rates of formation of soil organic matter?

REFERENCES

Adams M A, Ineson P, Binkley D, Cadisch G, Tokuchi N, Scholes M, Hicks K and Chadwick M 2004 Soil functional responses to excess N inputs at global scales. Ambio In press.

Albaugh T J, Allen H L, Dougherty P M, Kress L W and King J S 1998 Leaf area and above- and belowground growth responses of loblolly pine to nutrient and water additions. For. Sci. 44, 317-328.

Allison P D 1995 Survival analysis using SAS: A practical guide. SAS Institute Inc., Cary, NC, USA. 292 pp.

Bashkin M A and Binkley D 1998 Changes in soil carbon following afforestation in Hawaii. Ecology 79, 828-833.

Berntson G M and Bazzaz F A 1996 The allometry of root production and loss in seedlings of *Acer rubrum* (*Aceraceae*) and *Betula papyrifera* (*Betulaceae*): Implications for root dynamics in elevated CO2. Am. J. Bot. 83, 608-616.

Bolte A, Rahmann T, Kuhr M, Pogoda P, Murach D and v Gadow K 2004 Relationships between tree dimension and coarse root biomass in mixed stands of European beech (Fagus sylvatica L.) and Norway spruce (Picea abies [L.] Karst.) Plant Soil 264, 1-11.

Bond-Lamberty, Chuankuan B W and Gower S T 2004 A global relationship between the heterotrophic and autotrophic components of soil respiration? Global Change Biol. 10, 1756-1766.

Bonfante-Fasolo P 1986 Anatomy and morphology of VA mycorrhizae. *In* VA Mycorrhiza. Eds. C Powell and D Bagyaraj. pp. 2-33. CRC Press, Boca Raton, FL.

Burke M K and Chambers J L 2003 Root dynamics in bottomland hardwood forests of the southeastern United States Coastal Plain. Plant Soil 250, 141-153.

Burton A J and Pregitzer K S 2003 Field measurements of root respiration indicate little to no seasonal temperature acclimation for sugar maple and red pine. Tree Physiol. 23, 273-280.

Burton A J, Pregitzer K S, Ruess R W, Hendrick R L and Allen M F 2002 Root respiration in North American forests: effects of nitrogen concentration and temperature across biomes. Oecologia 131, 559-568.

Binkley D and Resh S C 1999 Rapid changes in soils following *Eucalyptus* afforestation in Hawaii. Soil Sci. Soc. Am. J. 63, 222-225.

Caldwell MM, Virginia RA. 1991. Root systems. *In* Plant physiological ecology: field methods and instrumentation. R W Pearcy, J Ehleringer, H A Mooney and P W Rundel. pp. 367-398. Chapman and Hall, London.

Cannell M G R and Dewar R C 1994 Carbon allocation in trees: a review of concepts for modeling. Adv. Ecol. Res. 25, 59-104.

Clark D A, Brown S, Kicklighter D W, Chambers J Q, Thomlinson J R, Ni J and Holland E A. 2001a Net primary production in tropical forests: an evaluation and synthesis of existing field data. Ecol. Appl. 11, 371-384.

Clark D A, Brown S, Kicklighter D W, Chambers J Q, Thomlinson J R, Ni J and Holland E A 2001b NPP Tropical Forest: Consistent Worldwide Site Estimates. 1967-1999. Data set. Oak Ridge National Laboratory Distributed Active Archive Center, Oak Ridge, Tennessee, U.S.A. http://www.daac.ornl.gov.

Coleman M D, Dickson R E and Isebrands J G 2000 Contrasting fine-root production, survival and soil CO_2 efflux in pine and poplar plantations. Plant Soil 225, 129-139.

Coleman M D, Friend A L and Kern C C 2004 Carbon Allocation and Nitrogen Acquisition in a Developing Populus deltoides Plantation. Tree Physiol. In press.

Coyle D and Coleman M Forest production responses to fertilization and irrigation are not explained by shifts in allocation. For. Ecol. Manage. In press.

Crookshanks M, Taylor G and Broadmeadow M 1998 Elevated CO2 and tree root growth: contrasting responses in Fraxinus excelsior, Quercus petraea and Pinus sylvestris. New Phytol. 138, 241-250.

Davidson E A, Savage K, Bolstad P, Clark D A, Curtis P S, Ellsworth D S, Hanson P J, Law B E, Luo Y, Pregitzer K S, Randolph J C and Zak D 2002 Belowground carbon allocation in forests estimated from litterfall and IRGA-based soil respiration measurements. Agr. Forest Meteorol. 113, 39-51.

Dickson R E and Isebrands J G 1993 Carbon allocation terminology: should it be more rational? Bulletin of Ecological Society of America 74, 175-177.

Eissenstat D M, Wells C E, Yanai R D and Whitbeck J L 2000 Building roots in a changing environment: implications for root longevity. New Phytol. 147, 33-42.

Eissenstat D M and Yanai R D 1997 The ecology of root lifespan. Advances in Ecological Research. 27, 1-60.

Enquist B J 2002 Universal scaling in tree and vascular plant allometry: Toward a general quantitative theory linking plant form and function from cells to ecosystems. Tree Physiology. 22, 1045-1064.

Enquist B J and Niklas K J 2002 Global allocation rules for patterns of biomass partitioning in seed plants. Science 295, 1517-1520.

Finer L and Laine J 1998 Root dynamics at drained peatland sites of different fertility in southern Finland. Plant Soil 201, 27-36.

Fitter A H, Self G K, Brown T K, Bogie D S, Graves J D, Benham D and Ineson P 1999 Root production and turnover in an upland grassland subjected to artificial soil warming respond to radiation flux and nutrients, not temperature. Oecologia 120, 575-581.

Fogel R and Hunt G 1983 Contribution of mycorrhizae and soil fungi to nutrient cycling in a Douglas-fir ecosystem. Can. J. For. Res. 13, 219-232.

Gaudinski J B, Trumbore S E, Davidson E A, Cook A C, Markewitz D and Richter D D 2001 The age of fine-root carbon in three forests of the eastern United States measured by radiocarbon. Oecologia 129,420-429.

Gebauer R L E, Reynolds J F and Strain B R 1996 Allometric relations and growth in *Pinus taeda*: the effect of elevated CO_2 and changing N availability. New Phytol. 134, 85-93.

Giardina C P and Rhoades C C 2001 Clear cutting and burning affect nitrogen supply, phosphorus fractions and seedling growth in soils from a Wyoming lodgepole pine forest. For. Ecol. Manag. 140, 19-28.

Giardina C P and Ryan M G 2000a Evidence that decomposition rates of organic carbon in mineral soil do not vary with temperature. Nature 404, 858-861.

Giardina C P and Ryan M G 2002 Soil surface CO_2 efflux, litterfall, and total belowground carbon allocation in a fast growing *Eucalyptus* plantation. Ecosystems 5, 487-499.

Giardina C P, Ryan M G, Binkley D, and Fownes J H 2003 Primary production and carbon allocation in relation to nutrient supply in an experimental tropical forest. Global Change Biol. 9, 1438-1450.

Giardina C P, Binkley D, Ryan M G and Fownes J H 2004 Belowground carbon cycling in a humid tropical forest decreases with fertilization. Oecologia 139, 545-550.

Gill R A and Jackson R B 2000 Global patterns of root turnover for terrestrial ecosystems. New Phytol. 147, 13-31.

Godbold D L, Berntson G M and Bazzaz F A 1997 Growth and mycorrhizal colonization of three North American tree species under elevated atmospheric CO2. New Phytol. 137, 433-440.

Gower, S.T., O. Krankina, R.J. Olson, M. Apps, S. Linder and C. Wang 2001a. Net primary production and carbon allocation patterns of boreal forest ecosystems. Ecol. Appl. 11:1395-1411.

Gower S T, Krankina O, Olson R J, Apps M, Linder S and Wang C 2001b NPP Boreal Forest: Consistent Worldwide Site Estimates. 1977-1994. Data set. Oak Ridge National Laboratory Distributed Active Archive Center, Oak Ridge, Tennessee, U.S.A. http://www.daac.ornl.gov.

Grace J and Rayment M 2000 Respiration in the balance. Nature 404, 819-820.

Grandmougin-Ferjani A, Dalpe Y, Hartmann M A, Laruelle F and Sancholle M 1999 Sterol distribution in arbuscular mycorrhizal fungi. Phytochemistry 50, 1027-1031.

Grier C C, Vogt K A, Keyes M R and Edmonds R L 1981 Biomass distribution and above- and below-ground production in young and mature Abies amabilis zone ecosystems of the Washingtion Cascades. Can. J. For. Res. 11, 155-167.

Haynes B E and Gower S T 1995 Belowground carbon allocation in unfertilized and fertilized red pine plantations in northern Wisconsin. Tree Physiol.153, 17-325.

Hendrick R L and Pregitzer K S 1992 The demography of fine roots in a northern hardwood forest. Ecology 73, 1094-1104.

Hendrick, R.L. and K.S. Pregitzer 1993. Patterns of fine root mortality in two sugar maple forests. Nature 361, 59-61.

Hendrick R L and Pregitzer K S 1996 Temporal and depth-related patterns of fine root dynamics in northern hardwood forests. J. Ecol. 84, 167-176.

Hendrick R L and Pregitzer K S 1997 The relationship between fine root demography and the soil environment in northern hardwood forests. Ecoscience 4, 99-105.

Högberg P, Nordgren A, Buchmann N, Taylor A F S, Ekblad A, Högberg M, Nyberg G, Ottosson-Löfvenius M, and Read D J 2001 Large-scale forest girdling shows that current photosynthesis drives soil respiration. Nature 411, 789-792.

Holland E A, Neff J C, Townsend A R and McKeown B 2000 Uncertainties in the temperature sensitivity of decomposition in tropical and subtropical ecosystems: Implications for models. Global Biogeochem. Cy. 14, 1137-1151.

Holland E A, Braswell B H, Lamarque J F, Townsend A, Sulzman J, Muller J F, Dentener F, Brasseur G, Levy H, Penner J E, Roelofs G J 1997 Variation in the predicted spatial distribution of atmospheric nitrogen deposition and their impact on terrestrial carbon uptake. J. Geophys. Res. 102, 15849-15866.

Horwath W R, Pregitzer K S and E A Paul 1994 ^{14}C allocation in tree soil systems. Tree Physiol. 14, 1163-1176.

Jackson R B, Banner J L, Jobbágy E G, Pockman W T and Wall D H 2002 Ecosystem carbon loss with woody plant invasion of grasslands. Nature 418, 623-626

Janssens I A, Lankreijer H, Matteucci G, Kowalski A S, Buchmann N, Epron D, Pilegaard K, Kutsch W, Longdoz B, Grunwald T, Montagnani L, Dore S, Rebmann C, Moors E J, Grelle A, Rannik U, Morgenstern K, Oltchev S, Clement R, Gudmundsson J, Minerbi S, Berbigier P, Ibrom A, Moncrieff J, Aubinet M, Bernhofer C, Jensen NO, Vesala T, Granier A, Schulze E D, Lindroth A, Dolman A J, Jarvis P G, Ceulemans R, Valentini R. 2001 Productivity overshadows temperature in determining soil and ecosystem respiration across European forests. Global Change Biol. 7, 269-278.

Johnson M G, Tingey D T, Phillips D L and Storm M J 2001 Advancing fine root research with minirhizotrons. Environ. Exp. Bot. 45, 263-289.

Joslin J D and Wolfe M H 1999 Disturbances during minirhizotron installation can affect observation data. Soil Sci. Soc. Am. J. 63, 218-221.

Joslin J D, Wolfe M H and Hanson P J 2001 Factors controlling the timing of root elongation intensity in a mature upland oak stand. Plant Soil 228, 201-212.

Karnosky D F, Zak D R, Pregitzer K S, Awmack C S, Bockheim J G, Dickson R E, Hendrey G R, Host G E, King J S, Kopper B J, Kruger E L, Kubiske M E, Lindroth R L, Mattson W J, Mcdonald E P, Noormets A, Oksanen E, Parsons W F J, Percy K E, Podila G K, Riemenschneider D E, Sharma P, Thakur R, Sober A, Sober J, Jones W S, Anttonen S, Vapaavuori E, Mankovska B, Heilman W, Isebrands J G. 2003. Tropospheric O_3 moderates responses of temperate hardwood forests to elevated CO_2: a synthesis of molecular to ecosystem results from the Aspen FACE project. Funct. Ecol. 17, 289-304.

Kaspar TC and Bland W L 1992 Soil-Temperature and Root-Growth. Soil Sci. 154, 290-299.

Kern C C, Friend A L, Johnson J M F and Coleman M D 2004 Fine-root dynamics in a developing *Populus deltoides* plantation. Tree Physiol. 24, 651-660.

Keyes M R and Grier C C 1981 Above- and below-ground net production in 40-year-old Douglas-fir stands on low and high productivity sites. Can. J. For. Res. 11, 599-605.

King J S, Albaugh T J, Allen H L and Kress LW 1999a Stand-level allometry in *Pinus taeda* as affected by irrigation and fertilization. Tree Physiol. 19, 769-778.

King J S, Pregitzer K S and Zak D R 1999b Clonal variation in above- and below-ground growth responses of Populus tremuloides Michaux: Influence of soil warming and nutrient availability. Plant Soil 217, 119-130.

King J S, Pregitzer K S, Zak D R, Sober J, Isebrands J G, Dickson R E, Hendrey G R, Karnosky D F. 2001 Fine-root biomass and fluxes of soil carbon in young stands of paper birch and trembling aspen as affected by elevated atmospheric CO_2 and tropospheric O_3. Oecologia 128, 237-250.

King J S, Hanson P J, Bernhardt E, DeAngelis P, Norby R J and Pregitzer K S 2004 A multiyear synthesis of soil respiration responses to elevated atmospheric CO2 from four forest FACE experiments. Global Change Biol. 10, 1027-1042.

King J S, Thomas R B and Strain B R 1996 Growth and carbon accumulation in root systems of Pinus taeda and Pinus ponderosa seedlings as affected by varying CO2, temperature and nitrogen. Tree Physiol. 16, 635-642.

Landsberg J J and Gower S T 1997 Applications of physiological ecology to forest management. Academic Press, New York. 354 pp.

Law B E, Ryan M G and Anthoni P M 1999 Seasonal and annual respiration of a ponderosa pine ecosystem. Global Change Biol. 5, 169-182.

Li Z, Kurz W A, Apps M J and Beukema S J 2003 Belowground biomass dynamics in the Carbon Budget Model of the Canadian Forest Sector: recent improvements and implications for the estimation of NPP and NEP. Can. J. For. Res. 33, 126-136.

Lilleskov E A, Fahey T J and Lovett G M 2001 Ectomycorrhizal fungal aboveground community change over an atmospheric nitrogen deposition gradient. Ecol. Appl. 11, 397-410.

Lilleskov E A, Fahey T J, Horton T R, Lovett G M 2002 Belowground ectomycorrhizal fungal community change over a nitrogen deposition gradient in Alaska. Ecology 83(1), 104-115.

Litton C M, Ryan M G and Raich J Carbon allocation in forest ecosystems. Oecologia In Review.

Litton C M, Ryan M G, Tinker D B and DH Knight 2003 Below- and aboveground biomass in young post-fire lodgepole pine forests of contrasting tree density. Can. J. For. Res. 33, 351-363.

Litton C M, Ryan M G and Knight D H 2004 Effects of tree density and stand age on carbon allocation patterns in a postfire lodgepole pine ecosystem. Ecol. Appl. 14, 460-475.

Livingston D A 1968 Some inter-stadial and post-glacial pollen diagrams from eastern Canada. Ecol. Monogr. 38, 87-125.

Loya W, Pregitzer K, Karberg N, King J and Giardina C 2003 Reduction of soil carbon formation by tropospheric ozone under elevated carbon dioxide. Nature 425, 705-707.

Lukac M, Calfapietra C and Godbold D L 2003 Production, turnover and mycorrhizal colonization of root systems of three Populus species grown under elevated CO2 (POPFACE). Global Change Biol. 9, 838-848.

Luo Y Q 2003 Uncertainties in interpretation of isotope signals for estimation of fine root longevity: theoretical considerations. Global Change Biol. 9, 1118-1129.

Lyr H and Hoffmann G 1967 Growth rates and growth periodicity of roots. International Review of Forestry Research 2, 181-236.

Matamala R, Gonzalez-Meler M A, Jastrow J D, Norby R J and Schlesinger W H 2003 Impacts of fine root turnover on forest NPP and soil C sequestration potential. Science 302, 1385-1387.

Matamala R and W H Schlesinger 2000 Effects of elevated atmospheric CO2 on fine root production and activity in an intact temperate forest ecosystem. Global Change Biol. 6, 967-979.

McClaugherty C A, Aber J D and Melillo J M. The role of fine roots in the organic matter and nitrogen budgets of forested ecosystems. Ecology 63, 1481-1490.

McConnaughay K D M and Coleman J S 1999 Biomass allocation in plants: Ontogeny or optimality? A test along three resource gradients. Ecology 80, 2581-2593.

McDowell N G, Balster N J and Marshall J D 2001 Belowground carbon allocation of Rocky Mountain Douglas-fir. Can. J. For. Res. 31, 1425-1436.

McElrone A J, Pockman W T, Martinez-Vilalta J and Jackson R B 2004 Variation in xylem structure and function in stems and roots of trees to 20 m depth. New Phytol. 163, 507-517.

McKevlin M R, Hook D H and Rozelle A A 1998 Adaptations of plants to flooding and soil waterlogging. *In* Southern Forested Wetlands – Ecology and Management. Eds. M G Messina and W H Conner. pp. 173-204. Lewis Press. Boca Raton, FL.

Minkkinen K and Laine J 1998 Long term effect of forest drainage on peat carbon stores of pine mires in Finland. Can. J. For. Res. 28, 1267-1275.

Melillo J M, Steudler P A, Aber J D, Newkirk K, Lux H, Bowles F P, Catricala C, Magill A, Ahrens T, Morrisseau S 2002 Soil warming and carbon-cycle feedbacks to the climate system. Science 298, 2173-2176.

Nadelhoffer K J and Raich J W 1992 Fine root production estimates and belowground carbon allocation in forest ecosystems. Ecology 73, 1139-1147.

Nepstad D C, de Carvalho C R, Davidson E A, Jipp P H, Lefebvre P A, de Negreiros G H, da Silva E D, Stone T A, Trumbore S E and Vieira S 1994 The role of deep roots in the hydrological and carbon cycles of Amazonian forests and pastures. Nature 372(6507), 666-669.

Norby R J and Jackson R B 2000 Root dynamics and global change: seeking an ecosystem perspective. New Phytol. 147, 3-12.

Norby R J and Luo Y 2004 Evaluating ecosystem responses to rising atmospheric CO_2 and global warming in a multi-factor world. New Phytol. 162, 281-293.

Norby R J, Gunderson C A, Wullschleger S D, O'Neill E G and McCracken M K 1992 Productivity and compensatory responses of yellow-poplar trees in elevated CO2. Nature 357, 322-324.

Norby R J, Ledford J, Reilly C D, Miller N E and O'Neill E G 2004 Fine-root production dominates response of a deciduous forest to atmospheric CO2 enrichment. P. Natl. Acad. Sci. USA. 101, 9689-9693.

Norby R J, Hanson P J, O'Neill E G, Tschaplinski T J, Weltzin J F, Hansen R A, Cheng W X, Wullschleger S D, Gunderson C A, Edwards N T, Johnson D W 2002 Net primary production of a CO2-enriched deciduous forest and the implications for carbon storage. Ecol. Appl. 12, 1261-1266.

Olsson P A, Larsson L, Bago B, Wallander H and van Aarle I M 2003 Ergosterol and fatty acids for biomass estimation of mycorrhizal fungi. New Phytol. 159, 7-10.

Ovington J D 1957 Dry-matter production by *Pinus sylvestris* L. Ann. Bot. (N.S.) 82, 288-314.

Paul E and Clark F 1996 Soil Microbiology and Biochemistry. Academic Press, New York.

Paul K I, Polglase P J, Nyakuengama J G, Khanna P K 2002 Change in soil carbon following afforestation. For. Ecol. Manag. 168, 241-257.

Pendall E, Bridgham S, Hanson P J, Hungate B, Kicklighter D W, Johnson D W, Law B E, Luo Y Q, Megonigal J P, Olsrud M, Ryan M G and Wan S Q 2004 Below-ground process responses to elevated CO2 and temperature: a discussion of observations, measurement methods, and models. New Phytol. 162, 311-322.

Pregitzer K S 2002 Fine roots of trees – a new perspective. New Phytol. 154, 267-273.

Pregitzer K S 2003 Woody plants, carbon allocation and fine roots. New Phytol. 158, 421-424.

Pregitzer K S, King J S, Burton A J and S E Brown 2000a Response of tree fine roots to temperature. New Phytol. 147, 105-115.

Pregitzer K, Zak D, Curtis P, Kubiske M, Teeri J and Vogel C 1995 Atmospheric CO_2, soil nitrogen, and turnover of fine roots. New Phytol. 129, 579-585.

Pregitzer K S, Zak D R, Maziasz J, DeForest J, Curtis P S and Lussenhop J 2000b Interactive effects of atmospheric CO_2 and soil-N availability on fine roots of *Populus tremuloides*. Ecol. Appl. 10, 18-33.

Pritchard S G, Rogers H H, Davis M A, Van Santen E, Prior S A and Schlesinger W H 2001 The influence of elevated atmospheric CO2 on fine root dynamics in an intact temperate forest. Global Change Biol. 7, 829-837.

Publicover D A and Vogt K A 1993 A comparison of methods for estimating forest fine root production with respect to sources of error. Can. J. For. Res. 23, 1179-1186.

Raich J W and Nadelhoffer K J 1989 Belowground carbon allocation in forest ecosystems: Global trends. Ecology 70, 1346-1354.

Reich P B 1983 Effects of low concentrations of O_3 on net photosynthesis, dark respiration, and chlorophyll contents in aging hybrid poplar leaves. Plant Physiol 73, 291-296.

Reich P B and Amundson R G 1985 Ambient levels of O_3 reduce net photosynthesis in tree and crop species. Science 230, 566-570.

Reich P and Bolstad P 2001 Productivity of evergreen and deciduous temperate forests. *In* Terrestrial Global Productivity. J Roy, B Saugier and H A Mooney. pp. 245-283 Academic Press, San Diego, USA.

Reuss R W, Hendrick R L, Burton A J, Pregitzer K S, Bjartmar Sveinbjornsson, Allen M F and Maurer G E 2003 Coupling fine root dynamics with ecosystem carbon cycling in black spruce forests of interior Alaska. Ecol. Monogr. 73, 643-662.

Ryan M G, Linder S, Vose J M and Hubbard R M 1994 Dark respiration in pines. *In* Pine Ecosystems. Eds. H L Gholz, S Linder, R E McMurtrie. pp. 50-63. Ecological Bulletins 43, Uppsala.

Ryan M G, Hubbard R M, Pongracic S, Raison R J and McMurtrie R E 1996 Autotrophic respiration in *Pinus radiata* in relation to nutrient status. Tree Physiol. 16, 333-343.

Ryan M G, Lavigne M B, Gower S T 1997 Annual carbon cost of autotrophic respiration in boreal forest ecosystems in relation to species and climate. J. Geophys. Res. 102(D24), 28871-28884.

Rygiewicz P T, Johnson M G, Ganio L M, Tingey D T and Storm M J 1997 Lifetime and temporal occurrence of ectomycorrhizae on ponderosa pine (Pinus ponderosa Laws) seedlings grown under varied atmospheric CO2 and nitrogen levels. Plant Soil 189, 275-287.

Santantonio D and Hermann R K 1985 Standing crop, production and turnover of fine roots on dry, moderate, and wet sites of mature Douglas-fir in western Oregon. Ann. Sci. Forest. 42, 113-142.

Sarmiento J 2000 That sinking feeling. Nature 408, 155-156.

Shaver G R and Jonasson S 2001 Productivity of Arctic Ecosystems. *In* Terrestrial Global Productivity. Eds. J Roy, B Saugier and H A Mooney. pp. 189-210. Academic Press, San Diego, USA.

Schimel D S, Braswell B H, Holland E A, McKeown R, Ojima D S, Painter T H, Parton W J and Townsend A R 1994 Climatic, edaphic, and biotic controls over storage and turnover of carbon in soils. Global Biogeochem. Cy. 8, 279-293.

Six J, Callewaert P, Lenders S, Degryze S, Morris S J, Gregorich E G, Paul E A and Paustian K 2002 Measuring and understanding carbon storage in afforested soils by physical fractionation. Soil Sci. Soc. Am. J. 66, 1981-1987.

Smith S E and Read D J 1997 Mycorrhizal Symbiosis. London, UK: Academic Press.

Smith F W Resh S C 1999 Age-related changes in production and below-ground carbon allocation in *Pinus contorta* forests. For. Sci. 45:333-341.

Stape J L, Binkley D and Ryan M G 2004 Eucalyptus production and the supply, use and the efficiency of use of water, light and nitrogen across a geographic gradient in Brazil. For. Ecol. Manag. 193, 17-31.

Steele S J, Gower S T, Vogel J G, and Norman J M 1997 Root mass, net primary production and turnover in aspen, jack pine and black spruce forests in Saskatchewan and Manitoba, Canada. Tree Physiol. 17, 577-587.

Stevens G N, Jones R H and Mitchell R J 2002 Rapid fine root disappearance in a pine woodland: a substantial carbon flux. Can. J. For. Res. 32, 2225-2230.

Teskey R O and Hinckley T M 1981 Influence of temperature and water potential on root growth of white oak. Physiol. Plantarum 52, 363-369.

Tierney G L, Fahey T J, Groffman P M, Hardy J P, Fitzhugh R D, Driscoll C T and Yavitt J B 2003 Environmental control of fine root dynamics in a northern hardwood forest. Global Change Biol. 9, 670-679.

Townsend A R, Braswell B H, Holland E A and Penner J E 1996 Spatial and temporal patterns in terrestrial carbon storage due to deposition of fossil fuel nitrogen. Ecol. Appl. 6, 806-814.

Treseder K K 2004 A meta-analysis of mycorrhizal responses to nitrogen, phosphorus, and atmospheric CO2 in field studies. New Phytol. 164(2), 347-355.

Trettin C C and Jurgensen M F 2003 Carbon cycling in wetland forest soils. *In* The Potential of U.S. Forest Soils to Sequester Carbon and Mitigate the Greenhouse Effect. Eds. J Kimble, R Birdsie and R Lal. pp. 311-328. CRC Press. Boca Raton, FL.

Uselman S M, Qualls R G and Thomas R B 1999 Effects of increased atmospheric CO2, temperature, and soil N availability on root exudation of dissolved organic carbon by a N-fixing tree (*Robinia pseudoacacia* L.). Plant Soil 222, 191-202.

VEMAP members 1995 Vegetation/ecosystem modeling and analysis project: Comparing biogeography and biogeochemistry models in a continental-scale study of terrestrial ecosystem response to climate change and CO_2 doubling. Global Biogeochem. Cy. 9, 407-437.

Vos J and J Groenwold 1987 The relation between root growth along observation tubes and in bulk soil. *In* Minirhizotron observation tubes: Methods and applications for measuring rhizosphere dynamics. Ed. H M Taylor. pp. 39-49. American Society of Agronomy, Inc. Madison, WI.

Wan S Q, Norby R J, Pregitzer K S, Ledford J and O'Neill E G 2004 CO2 enrichment and warming of the atmosphere enhance both productivity and mortality of maple tree fine roots. New Phytol. 162, 437-446.

Wells C E and Eissenstat D M 2001 Marked differences in survivorship among apple roots of different diameters. Ecology 82, 882-892.

Wells C E, Glenn D M and Eissenstat D M 2002a Changes in the risk of fine-root mortality with age: A case study in peach, Prunus persica (Rosaceae). Am. J. Bot. 89,79-87.

Wells C E, Glenn D M and Eissenstat D M 2002b Soil insects alter fine root demography in peach (*Prunus persica*). Plant Cell Environ. 25,431-439.

Withington J M, Elkin A D, Bulaj B, Olesinski J, Tracy K N, Bouma T J, Oleksyn J, Anderson L J, Modrzynski J, Reich P B and Eissenstat D M 2003 The impact of material used for minirhizotron tubes for root research. New Phytol. 160, 533-544.

Wallander H, Nilsson L O, Hagerberg D and Baath E 2001 Estimation of the biomass and seasonal growth of external mycelium of ectomycorrhizal fungi in the field. New Phytol. 151, 753-760

Zak D R, Pregitzer K S, King J S and Holmes W E 2000a Elevated atmospheric CO_2 and the composition and function of soil microbial communities. Ecol. Appl. 10, 47-59.

Zak D R, Pregitzer K S, Curtis P S and Holmes W E 2000b Atmospheric CO_2, fine roots and the response of soil microorganisms: a review and hypothesis. New Phytol. 147, 201-222.

Chapter 8

HOW NITROGEN-FIXING TREES CHANGE SOIL CARBON

Dan Binkley

Department of Forest, Rangeland and Watershed Stewardship, Graduate Degree Program in Ecology, and Natural Resource Ecology Laboratory, Colorado State University, Fort Collins, Colorado 80523 USA

Keywords: soil nitrogen, carbon accretion, soil organic matter

ABSTRACT

Nitrogen-fixing tree species have larger effects on forest soils than other species, and these effects include consistent increases in soil organic matter and carbon (C). Across 19 case studies, an increase of 1 g N was associated with an increase of 12 to 15 g C. Few studies have examined the mechanisms behind the accumulation of soil C; documented processes include reduced decomposition of older, stabilized soil C, increased rates of formation of stabilized soil C, and higher rates of input of C in detritus. The influences of N-fixing trees on stabilized soil C may not derive directly from the increased supply of N, as fertilization with inorganic N did not alter soil C. More experimentation is needed on the influence of N-fixing trees on the soil biotic community, as the soil microbes and invertebrates may hold the key to the influence of N-fixation on soil C.

INTRODUCTION

Nitrogen-fixing trees change soils more rapidly than other species. The changes in soil nitrogen (N) begin with the fixation of N by symbiotic bacteria in root nodules and the incorporation of this N into tree tissues. With the senescence of leaves, roots, and woody material, the fixed N enters the detrital

D. Binkley and O. Menyailo (eds.),
Tree Species Effects on Soils: Implications for Global Change, 155–164.

part of the ecosystem nutrient cycle. Forests with major components of N-fixing trees generally add two-to-three times as much N to the soil in litterfall annually compared with non-N-fixing forests (Binkley and Giardina 1998). In forests without N-fixing species, soil N supply commonly differs by up to 2-fold beneath different species on the same sites; the presence of N-fixing trees generally increases the N content of litterfall by 4-10 times (Binkley and Giardina 1998).

These massive changes in soil N cycling lead to large changes in many other soil properties. Rates of cycling of P in litterfall seem to always be higher under N-fixing species, whereas the effects on the pools of labile P in the soil sometimes increase or decrease (Compton and Cole 1998, Binkley et al. 2000, Kaye et al. 2000). Nitrogen-fixing trees have also led to overall declines in pools of exchangeable cations (Binkley and Sollins 1990, Rhoades and Binkley 1996), no change in cations (Binkley and Sollins 1990), increases in cations (Johnson 1995), shifts in the depth profile of a constant quantity of cations (Binkley et al. 1992), and even a reduction in base saturation despite no change in cation quantity, owing to an increase in cation exchange capacity (Homann et al. 1992).

The effects of N-fixation are driven in part by changes in the soil biotic community, and the community in turn drives changes in soil biogeochemistry. Fewer case studies have examined these biotic effects of N-fixing trees; some documented changes have included increased biomass of bacteria, reduced biomass of fungi, and increased populations of soil invertebrates (e.g. Zou 1993).

The most universal effect of N-fixing trees on biogeochemistry other than nitrogen appears to be on soil carbon (C). This paper explores the magnitude of C accretion under N-fixing trees, how this relates to the rate of N accretion, and how the rates of C accretion derive from changes in rates of both decomposition and C stabilization.

MAGNITUDE OF C AND N ACCRETION

Nineteen case studies were available in the literature with information the pool sizes of N and C under an N-fixing species and a paired non-N-fixing species (Table 1). The N-fixing species in temperate forests were trees and shrubs of *Alnus* and *Ceanothus*, and in the tropics *Leucaena*, *Falcataria* , and *Casuarina*. The rates of C accretion averaged 87 g m^{-2} yr^{-1} with a range of 25 to 180 g C m^{-2} yr^{-1} (Table 1, Figure 1). Some of this variation in C accretion was an artifact of differences in the depths of soil sampled among the studies, but in all cases the contrast of C and N are from the same horizons. Rates of N accretion averaged 7.3 g N m^{-2} yr^{-1} across all studies, with a range from 0.8 to 15.3 g N m^{-2} yr^{-1}. Overall, these soils accumulated about 12 to 15 g of C for every g of N accumulated (Figure 1).

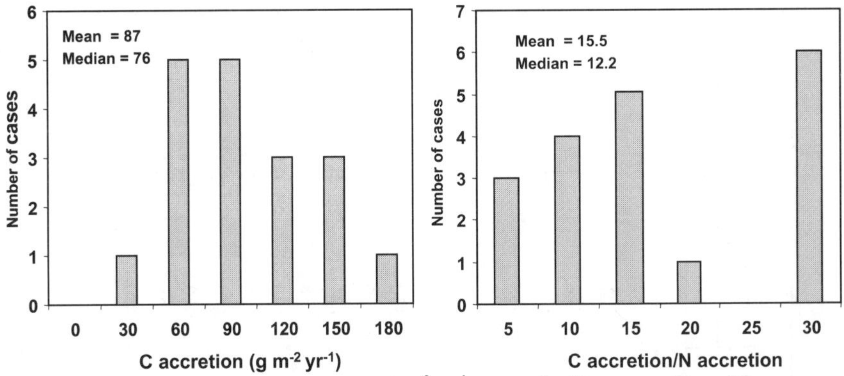

Figure 1. Accretion of soil C averaged 87 g m^{-2} yr^{-1} across the 19 case studies, with an average ratio of C accretion to N accretion of 15.5.

What patterns of soil C accretion were consistent across the case studies? The rate of C accretion in the soil related only weakly to the rate of N accretion (Figure 2); this means that the new material accumulated under the N-fixing trees had only moderate consistency in C:N. The C accretion did not relate to the initial soil C content, giving no evidence that antecedent soil C limits or influences the quantity of new C that may accumulate. However, sites with the highest amounts of soil C tended to accumulate relatively little C relative to N accretion (Figure 2); these sites may accumulate more N without a concomitant increase in already high quantities of C. The accretion of C relative to N also did not relate to the initial C:N of the soil, although the C:N of new material was significantly lower than the C:N of the initial soil.

MECHANISMS DRIVING C ACCRETION

The accumulation of N beneath N-fixing trees is clearly associated with accumulation of soil C, but the statistical relationships show high variation. The variation probably depends in part on the fundamental difference in N and C biogeochemistry; most of the fixed N added to the soil remains in the soil, whereas most of the C added to the soil is oxidized and lost as CO_2. At a process level, the relationship between N and C accretion needs to consider the effects of N on rates of C inputs, on C transformation to stabilized forms (humus), and on C losses.

The pools of soil C can be illustrated simply as comprised of recent detritus (rapid turnover) and stabilized (slow turnover) pools (Figure 3). Nitrogen fixation could increase these pools in two fundamental ways. The pool sizes would increase if the rate of C input increased, without any change in the rates of turnover. Inputs of C would increase if N-fixing species increased the rates of litterfall or fine root and mycorrhizal production.

Table 1. Soil N and C pool sizes from case studies of N-fixing species. Some values were approximated from graphs.

N-fixing species	Comparison species	Number of years	Soil depth sampled	N- fixer C (g/m^2)	Non-N-fixer C (g/m^2)	N-fixer N (g/m^2)	Non-N-fixer N (g/m^2)	N accretion (g m^{-2} yr^{-1})	C accretion (g m^{-2} yr^{-1})
Alnus rubra[1]	*Pseudotsuga menziesii*	23	O + 0 to 50 cm	6320	3720	322	173	6.5	113.0
Alnus rubra[2]	*Pseudotsuga menziesii*	55	O + 0 to 90 cm	26950	24750	1454	1052	7.3	40.0
Alnus rubra[3]	*Pseudotsuga menziesii*	19	O + 0 to 45 cm	14700	12400	787	621	8.7	121.1
Alnus rubra[2]	*Pseudotsuga menziesii*	55	O + 0 to 90 cm	11565	8580	503	223	5.1	54.3
Alnus rubra[1]	*Pseudotsuga menziesii*	23	O + 0 to 50 cm	8880	7670	634	537	4.2	52.6
Alnus rubra4	*Pseudotsuga menziesii*	50	O + 0 to 45 cm	16405	12755	976	584	7.8	73.0
Alnus rubra[5]	*Pseudotsuga menziesii*	40	O + 0 to 20 cm	8415	5415	392	282	2.8	75.0
Alnus sinuata[6]	*Pseudotsuga menziesii*	23	O + 0 to 50 cm	5170	3720	231	173	2.5	63.0
Casuarina equisetifolia[7]	*Eucalyptus robusta*	16	0-40 cm	9000	7000	620	550	4.4	125.0
Casuarina equisetifolia[7]	*Eucalyptus robusta*	7	0-40 cm	3400	3100	400	330	10.0	42.9
Casuarina equisetifolia[8]	*No vegetation*	34	O + 0-100 cm	7800	6700	430	220	7.5	32.4
Ceanothus velutinus[9]	*Pinus jeffreyi*	115	0-30 cm	7009	4111	215	121	0.8	25.2
Ceanothus velutinus[10]	*Pseudotsuga menziesii*	12	0-30	4990	3560	188	138	4.2	119.2
Ceanothus velutinus[10]	*Pseudotsuga menziesii*	12	0-30	5640	3470	232	149	6.9	180.8
Facaltaria mollucana[7]	*Eucalyptus saligna*	15	0-40 cm	15860	14700	1040	900	9.3	77.3
Facaltaria mollucana[7]	*Eucalyptus saligna*	15	0-50 cm	14800	12700	920	690	15.3	140.0
Leucaena leucocephala[7]	*Eucalyptus robusta*	16	0-40 cm	9600	7000	770	550	13.8	162.5
Leucaena leucocephala[7]	*Eucalyptus robusta*	7	0-40 cm	3600	3100	430	330	14.3	71.4
Acacia mearnsii[11]	*Eucalyptus globulus*	10	0-30 cm	7930	6630	430	330	10.0	130.1

[1]Binkley 1983
[2]Binkley et al. 1992
[3]Rothe et al. 2002
[4]Johnson and Lindberg 1992
[5]Bormann and DeBell 1981
[6]Binkley et al. 1984
[7]Resh et al. 2002
[8]Mailly and Margolis 1992
[9]Johnson 1995
[10]Binkley et al. 1982
[11]Pares 2002

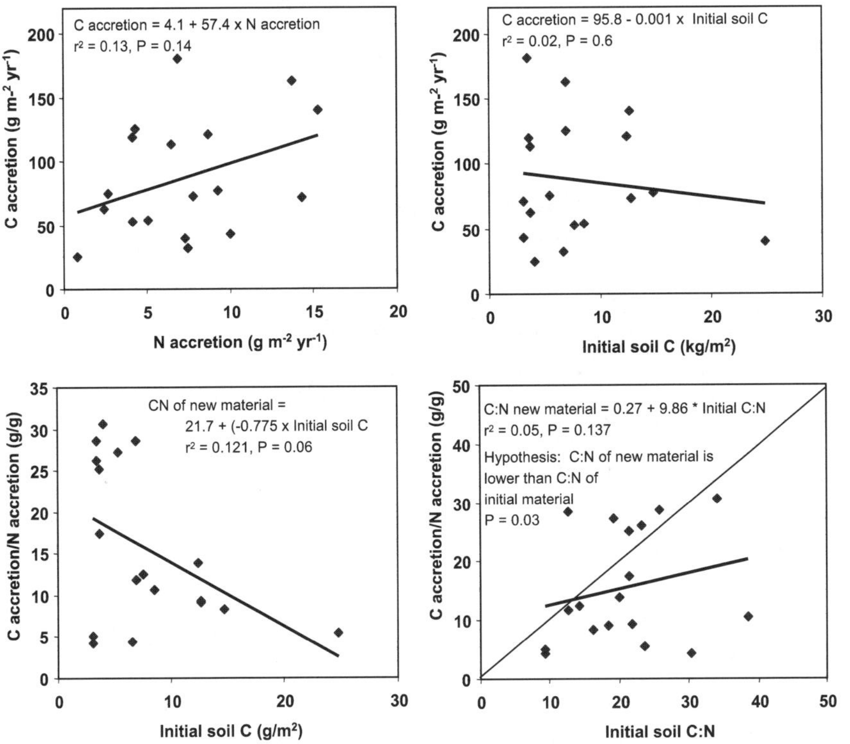

Figure 2. The accretion of soil C related weakly with the rate of N accretion (top left), and did not relate to the initial soil content of C (top right). The rate of C accretion to N accretion did decline with increasing initial soil C (bottom left); this relationship could also be interpreted as a wide range of C:N accretion at low initial soil C, but only low C:N for soils with high C. The accretion of C:N did not relate to the initial C:N of the soil, but the C:N of new material did average less than that of initial soil C (bottom right).

Alternatively, decreases in rates of decomposition would also lead to C accumulation. Decomposition rates might decrease if more of the recent detritus was stabilized as humus under the N-fixing trees. This may sound unlikely, as litter from species with high N contents clearly show more rapid initial decomposition than litter from low-N species; however, the stabilization of longer-term C may be higher for material with high N contents (cf. Berg 2000).

A few studies have used ^{13}C isotope methods to investigate the accumulation of C derived from N-fixing trees, and the rates of loss of older soil C. Kaye et al. (2000) used a replacement series experiment with *Eucalyptus saligna* and N-fixing *Falcataria moluccana* (= *Albizia falcataria*) to investigate whether N fixation increased the pool of C derived from the current generation of trees (= increase in Process 1 and 3 in Figure 3), or greater retention of older C from previous land use (= a increase in Process 2b in Figure 3). This soil had been cropped with sugarcane for more than 50

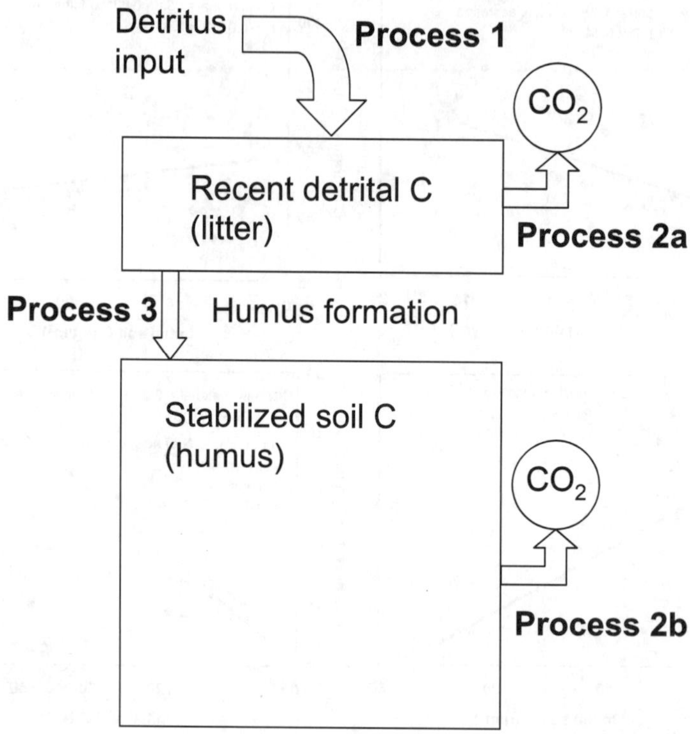

Figure 3. Diagram of soil C flows. Nitrogen-fixing species could increase soil C pools by increasing detritus inputs (1), by decreasing rates of decomposition (2a, 2b), or by increasing humus formation (3) while decreasing decomposition of recent detritus (2a) The flows are constraint by the requirement that 1 = (2a + 2b + 3), plus any change in mass of the recent detrital C pool.

years prior to afforestation, and the C-4 photosynthetic pathway of the cane had labelled the soil with a ratio of ^{13}C:^{12}C that differed substantially from that of the C-3 trees. The accretion of C increased linearly with the proportion of *Falcataria* planted 17 years earlier (total tree density was held constant in the plantation design). About 1/3 of the apparent accretion of soil C resulted from a higher retention of older soil C, and 2/3 from C derived from the current generation of trees (Figure 4).

Which processes of Figure 3 accounted for the pattern in Figure 4? The greater retention of older soil C indicated an important decrease in Process 2b of Figure 3. Resh et al. (2002) incubated soils from some of these stands, as well as other N-fixing stands, and concluded that recent C derived from the current generation of trees decayed as fast or faster for N-fixer than for non-N-fixers, indicating that Process 2a either played no role in C accretion, or actually worked against C accretion. Binkley and Ryan (1998) estimated net primary production in 3 sets of pure stands of *Eucalyptus* and *Falcataria*

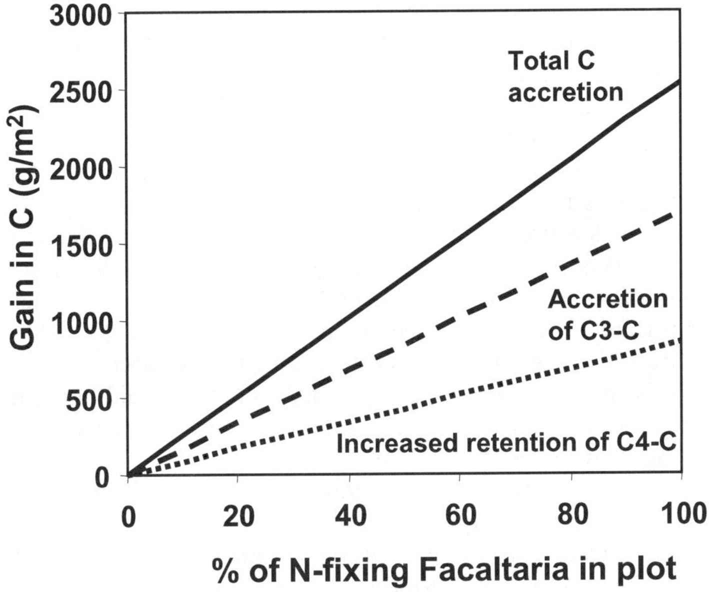

Figure 4. About 1/3 of the greater C found under N-fixing *Falcataria* after 17 years resulted from greater retention of older C derived from sugarcane under the prior land use (C4-C). The other 2/3 resulted from increased accumulation of C derived from the current generation of trees. The trend in greater retention of older C4-C had a P of 0.06, whereas the trend in accretion of C3-C and total C had P of 0.004 and 0.001 (Kaye et al. 2001).

plots (including the same plots used by Kaye et al.), and found no difference in aboveground litterfall between species, but 30% greater belowground production by the N-fixing *Falcataria.* The total input of detrital C was about 280 g m^{-2} yr^{-1} greater under the N fixer (Process 1 in Figure 3), or about three times greater than the gain of tree-derived C. The effect of *Falcataria* on soil C appeared to result from an increase in detrital inputs (Process 1), a decrease in output of stabilized soil C (Process 2b), perhaps an increase in formation of stabilized soil C (Process 3), and not to any decrease in the rate of decomposition of recent detritus (Process 2b).

Increased inputs of detrital C may have played a role in the accretion of C under *Acacia mearnsii* in Australia. The rate of divergence between the soils under *Acacia* and *Eucalyptus globulus* averaged 130 g C m^{-2} yr^{-1} for 10 years (Pares 2002), and litterfall was about 40 g C m^{-2} yr^{-1} greater under *Acacia* (Forrester 2004). The net production of fine roots may have been about 30 g C m^{-2} yr^{-1} greater under *Acacia*, but the difference was not significant. These components of the production budget suggest that increased inputs of detrital C (Process 1 in Figure 3) may have been important, but not sufficient to account for the full rate of C accretion. Forrester (2004) found that decomposition in the O horizon was more than twice as rapid under *Acacia*, so Process 2a was not lower under *Acacia*. Therefore, the higher soil C under the *Acacia* likely included a substantial increase in formation of stabilized organic matter (Process 3 in Figure 3), or by reduced decomposition of the stabilized pool (Process 2b in Figure 3).

The contribution of processes in Figure 3 to C accretion were examined by Resh et al. (2002) using the C4-C and C3-C approach used by Kaye et al. (2000). They examined the retention of old soil C and gain of new C under N fixers at two locations in Hawaii (one was the same as used by Kaye et al. 2000), and two locations in Puerto Rico. About 55% of the greater C found under the N-fixers across all 4 locations (110 g m^{-2} yr^{-1}) resulted from greater retention of older soil C, and 45% from C fixed by the current generation of trees. About 60% of the variance among plots in retention of older soil C (derived from C4 plants) was associated with the net accretion of N ($P <$ 0.001). This study could not identify cause and effect: did high rates of N accretion result from higher N retention when C was also retained, or was C retention higher when the N supply was higher?

In one of the Puerto Rico N-fixation sites used by Resh et al. (2002), Kaye et al. (2002) examined the effects of tree species on the long-term retention of added ^{15}N. These stands of *Eucalyptus robusta*, N-fixing *Casuarina equisetifolia*, and N-fixing *Leucaena leucocephala* were labeled with highly enriched ^{15}N at 2 years of age. Seven years later, Kaye et al. (2002) quantified the amount of ^{15}N retained in the soil, and the proportion that remained in stabilized organic matter. The stabilized proportion was determined by incubating soil samples under optimal conditions for microbial activity; any N that could not be accessed by microbes within 13 months was defined as non-labile N (stabilized in humus). They found that about 45% of the added N remained in the soil, and that 20% to 25% more of the retained N was incorporated in stabilized pools under the two N-fixing species than under the *Eucalyptus*, indicating greater formation of stabilized organic matter (Process 3 in Figure 3).

What mechanisms might account for the increase in stabilized soil organic matter under N-fixing species? The mechanisms could operate by increasing the rate of formation of stabilized soil organic matter, or reducing the rate of decomposition of stabilized organic matter (as indicated in the natural abundance of ^{13}C experiments; Process 2b in Figure 3).

Increased stabilization of soil C under N-fixing trees could be driven by strictly chemical processes. The greater availability of N could lead to increased humification by covalently bonding with C atoms in low-molecular-weight compounds to form larger, more recalcitrant molecules. Alternatively, the N could inhibit the formation of lignin-degrading enzymes by fungi (Berg 2000, Carrierro et al. 2000). This chemical mechanism was tested in Hawaii on the same soil type as in the *Eucalyptus* and *Falcataria* experiment of Kaye et al. (2000) by repeated additions of inorganic fertilizer N to a pure *Eucalyptus* plantation. Over a period of 6 years, a total of 160 g N/m^2 was added (in 4 applications/yr), with no effect on either the total amount of C in the soil, or on the retention of older, cane-derived soil C (Binkley et al. 2004).

Laboratory incubations of soils from the fertilized and not-fertilized plots of *Eucalyptus* showed no differences in rates of CO_2 release, and additions of N during the course of the incubations also failed to influence CO_2 release. This single case study showed that increased stabilization of C under the influence of N-fixing *Falcataria* was not a simple chemical reaction following increased availability of inorganic N.

Increased stabilization of organic matter under N-fixing trees is probably a function of the biological changes that cascade from the input of N-rich organic matter. Intensive research will be needed to characterize the importance of microbial changes (such as the dominance of bacteria and fungi) and changes in the soil invertebrate communities. Creative experimental approaches will be needed, perhaps including the application of special treatments that reduce the influence of particular components, such as fungicides, electricity to repel earthworms, and reciprocal transplanting of soil blocks from one forest type to another.

The patterns found here would clearly benefit from an increase in the number of case studies. The value of future case studies would be substantially increased by a focus on hypothesis testing about processes that drive any observed changes. For example, one might hypothesize that the increases in soil C depend on the activity of earthworms. This hypothesis could be tested by treatments that reduce worm densities under N-fixing trees (such as use of electric probes). Another test of this hypothesis could examine patterns in soil C and worm densities across the boundaries between stands of N-fixing and non-N-fixing trees; tight covariation would support the hypothesis, whereas a lack of correlation would refute the hypothesis.

REFERENCES

Berg B 2000 Litter decomposition and organic matter turnover in northern forest soils. For. Ecol. Manage. 133, 13-22.

Binkley D and Sollins P 1990 Acidification of soils in mixtures of conifers and red alder. Soil Sci. Soc. Am. J. 54, 1427-1433.

Binkley D and Giardina C 1998 Why trees affect soils in temperate and tropical forests: the warp and woof of tree/soil interactions. Biogeochemistry 42, 89-106.

Binkley D and Ryan M 1998 Net primary production and nutrient cycling in replicated stands of *Eucalyptus saligna* and *Albizia falcataria*. For. Ecol. Manage. 112, 79-85.

Binkley D, Giardina C and Bashkin M 2000 Soil phosphorus pools and supply under the influence of *Eucalyptus saligna* and nitrogen-fixing *Albizia falcataria*. For. Ecol. Manage. 128, 241-247.

Binkley D 1983 Interaction of site fertility and red alder on ecosystem production in Douglas-fir plantations. For. Ecol. Manage. 5, 215- 227.

Binkley D, Sollins P, Bell R, Sachs D and Myrold D 1992 Biogeochemistry of adjacent conifer and alder/conifer ecosystems. Ecology 73, 2022-2034.

Binkley D, Lousier J D and Cromack K Jr 1984 Ecosystem effects of Sitka alder in a Douglas-fir plantation. For. Sci. 30, 26-35.

Binkley D 1982 Nitrogen fixation and net primary production in a young Sitka alder ecosystem. Can. J. Bot. 60, 281-286.

Binkley D, Kaye J, Barry M and Ryan M G 2004 First rotation changes in soil carbon and nitrogen in a *Eucalyptus* plantation in Hawaii. Soil Sci. Soc. Am. J. 68, 1713-1719.

Bormann B T and DeBell D S 1981 Nitrogen content and other soil properties related to age of red alder stands. Soil Sci. Soc. Am. J. 45, 428-432.

Carrierro M, Sinsabaugh R, Repert D and Parkhurst D 2000 Microbial enzyme shifts explain litter decay responses to simulated nitrogen deposition. Ecology 81, 2359–2365.

Compton J E and Cole D W 1998 Phosphorus cycling and soil P fractions in Douglas-fir and red alder stands. For. Ecol. Manage. 110, 101-112.

Forrester D I 2004 Mixed-species plantations of nitrogen-fixing and non-nitrogen-fixing trees. PhD dissertation. Australian National University, Canberra. 196 pp.

Homann P S, Van Miegroet H and Cole D W 1992 Cation distribution, cycling and removal from mineral soil in Douglas-fir and red alder forests. Biogeochemistry 16, 121-150.

Johnson D W 1995 Soil properties beneath ceanothus and pine stands in the Eastern Sierra Nevada. Soil Sci. Soc. Am. J. 59, 918-924.

Johnson D W and Lindberg SE 1992 Atmospheric deposition and forest nutrient cycling. Springer-Verlag, New York. 707 pp.

Kaye J P, Resh S C, Kaye M W and Chimner R A 2000 Nutrient and carbon dynamics in a replacement series of *Eucalyptus* and *Albizia* trees. Ecology 81, 3267-3273.

Kaye J, Binkley D, Zou X and Parrotta J 2002 Non-labile ^{15}Nitrogen retention beneath three tree species in a subtropical plantation. Soil Sci. Soc. Am. J. 66, 612-619.

Mailly D and Margolis H A 1992 Forest floor and mineral soil development in *Casuarina equisetifolia* plantations on the coastal sand dunes of Senegal. For. Ecol. Manage. 55, 259-278.

Pares A 2002 Soil organic carbon sequestration in mixed and monospecific plantations of *Eucalyptus globulus* ssp. *Pseudoglobulus* and *Acacia mearnsii.* Honours thesis. Australian National University, Canberra. 79 pp.

Resh S, Binkley D and Parrotta J 2002 Greater soil carbon sequestration under nitrogen-fixing trees compared with *Eucalyptus* species. Ecosystems 5, 217-231.

Rhoades C and Binkley D 1996 Factors influencing decline in soil pH in Hawaiian *Eucalyptus* and *Albizia* plantations. For. Ecol. Manage. 80, 47-56.

Rothe A, Cromack K Jr, Resh S C, Makineci E and Son Y 2002 Soil carbon and nitrogen changes under Douglas-fir with and without red alder. Soil Sci. Soc. Am. J. 66, 1988-1995.

Zou X 1993 Species effects on earthworm density in tropical tree plantations in Hawaii. Biol. Fert. Soils. 15, 35-38.

Chapter 9

EFFECTS OF TREE SPECIES ON C- AND N-CYCLING AND BIOSPHERE-ATMOSPHERE EXCHANGE OF TRACE GASES IN FORESTS

H. Papen[1], P. Rosenkranz, K. Butterbach-Bahl, R. Gasche, G. Willibald & N. Brüggemann

*Forschungszentrum Karlsruhe GmbH, Institute for Meteorology and Climate Research, Atmospheric Environmental Research (IMK-IFU), Kreuzeckbahnstr. 19, D-82467 Garmisch-Partenkirchen, Germany. [1]Corresponding author*E-mail: hans.papen@imk.fzk.de*

INTRODUCTION

Forest soils exchange a variety of gases with the atmosphere, including greenhouse gases that influence climate (CO_2, CH_4, NO, N_2O). The net fluxes of these gases depend on microbiological processes of production and consumption of these gases within the soil (e.g. Firestone and Davidson, 1989; Conrad, 1996). These processes include nitrification, denitrification, methanogenesis, methane oxidation, respiration and CO_2 fixation by microbial autotrophs. The magnitude of trace gas exchange between soils and the atmosphere depends on a multitude of interacting environmental factors including pH, substrate availability, soil temperature, soil moisture, and soil texture (e.g. Firestone and Davidson, 1989). In recent years, the importance of biotic factors has been demonstrated, especially forest type and tree species. Indeed, the net exchange of N-trace gases may depend more heavily on these biotic factors than on the abiotic factors (e.g. Butterbach-Bahl et al., 1997; Papen and Butterbach-Bahl, 1999; Gasche and Papen, 1999; Menyailo and Huwe, 1999; Butterbach-Bahl and Papen, 2002; Erickson et al., 2002; Menyailo and Hungate, this volume).

In this paper we report results obtained from long term measurements of N- and C-trace gas fluxes from different temperate forest ecosystems dominated by beech, spruce, pine or oak. We also examine the key factors that may account for the biotic effects on trace gas flux, especially litter

D. Binkley and O. Menyailo (eds.),
Tree Species Effects on Soils: Implications for Global Change, 165–172.

quality (e.g. Stump and Binkley, 1993; Scott and Binkley, 1997; Binkley and Giardina, 1998; Papen and Butterbach-Bahl, 1999; Gasche and Papen, 1999; Priha and Smolander, 1999; Coté et al., 2000; Augusto et al., 2002; Menyailo et al., 2002; Bengtsson et al., 2003).

STUDY SITES AND METHODS

Our studies on the effect of different tree species/forest types on N- and C-trace gas emissions were performed at three temperate forest sites:

1. Höglwald experimental site, Bavaria, Germany, with pure spruce and pure beech stands;
2. In the Northeastern German Lowlands, near the city of Eberswalde, with pure pine, pure oak, and pure beech stands; and
3. In Western Jutland, Denmark, with pure stands of larch, oak, spruce, pine, beech.

Details about location, stands and soil characteristics are given in detail by Papen and Butterbach-Bahl (1999) and Kreutzer and Weis (1998) for the Höglwald site, by Butterbach-Bahl et al. (2001) for the Northeastern Lowland sites and by Brüggemann et al. (2004) for the forests in West Jutland, Denmark.

Details about methods used for continuous and fully automated measurements of N- and trace gas fluxes (using both static and dynamic chambers) are given by Butterbach-Bahl et al. (1997), Papen and Butterbach-Bahl (1999) and Butterbach-Bahl et al. (2001). Gross mineralization (ammonification) rates in soil samples from the organic layer (O_h horizon) were determined by the ^{15}N isotope pool dilution method (Davidson et al., 1991). Gross nitrification rates and respiration rates of soil samples were determined using the barometric process separation method (Ingwersen et al., 1999). N_2 emissions from soil stemming from actual denitrification were quantified using the newly developed soil core incubation method as described by Butterbach-Bahl et al. (2002).

RESULTS AND DISCUSSION

Figure 1 and 2 show the results obtained from long-term measurements at the Höglwald experimental site (Bavaria, Germany), where -amongst other trace gases- the N_2O- and NO/NO_2-emission rates from the soil into the atmosphere have been monitored with high temporal resolution (1-2 hourly) over several consecutive years (1994-1997; for details see: Papen and Butterbach-Bahl, 1999; Butterbach-Bahl and Papen , 2002). Given are the results for a c. 95 year old pure spruce stand and a 120 year old pure beech stand at the Höglwald experimental site. Most important is to note that at this

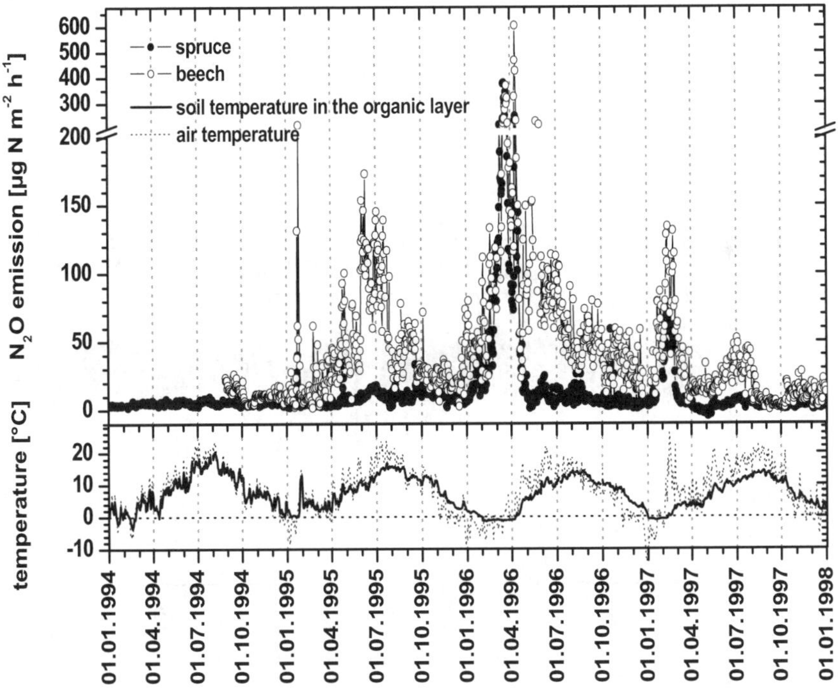

Figure 1. Effect of forest type (spruce vs. beech) on magnitude of N_2O emission rates at the Höglwald experimental site (Bavaria, Germany). In addition the time course of soil (organic layer) and air temperature is given over the entire observation period (1994-1997). Adapted from: Papen and Butterbach-Bahl (1999).

site both spruce and beech stock on identical soil type and are exposed in general to identical edaphic and climatic conditions (Papen and Butterbach-Bahl, 1999). The interannual variation in fluxes of N-trace gases showed very large interannual variation for both stands (Figures 1 and 2), and the loss of nitrous oxide was much greater from the beech soil than from the spruce soil (Figure 1). The effect of tree species was reversed for nitric oxide emissions, where the soil under spruce was a much larger source for nitric oxide than the soil under beech (Figure 2). For NO_2 deposition rates to the forest floors, no significant differences between spruce and beech could be determined indicating that the tree species had no detectable effect on NO_2 deposition.

The annual mean (1994-1996) rate of N_2O emission from the spruce stand was relatively low (16.4 µg N_2O-N m^{-2} h^{-1}), whereas the emissions from the beech stand were more than 3-fold greater (54 µg N_2O-N m^{-2} h^{-1}).

Extrapolated to a full year, these average flux rates would indicate an N loss of about 0.14 g m^{-2} yr^{-1} from spruce, and 0.47 g m^{-2} yr^{-1} from beech. On the other hand, the long-term annual mean for NO emission from soil under spruce (91.8 µg NO-N m_{-2} h_{-1}; about 0.8 g N m_{-2} yr_{-1}) was about 3.2 times

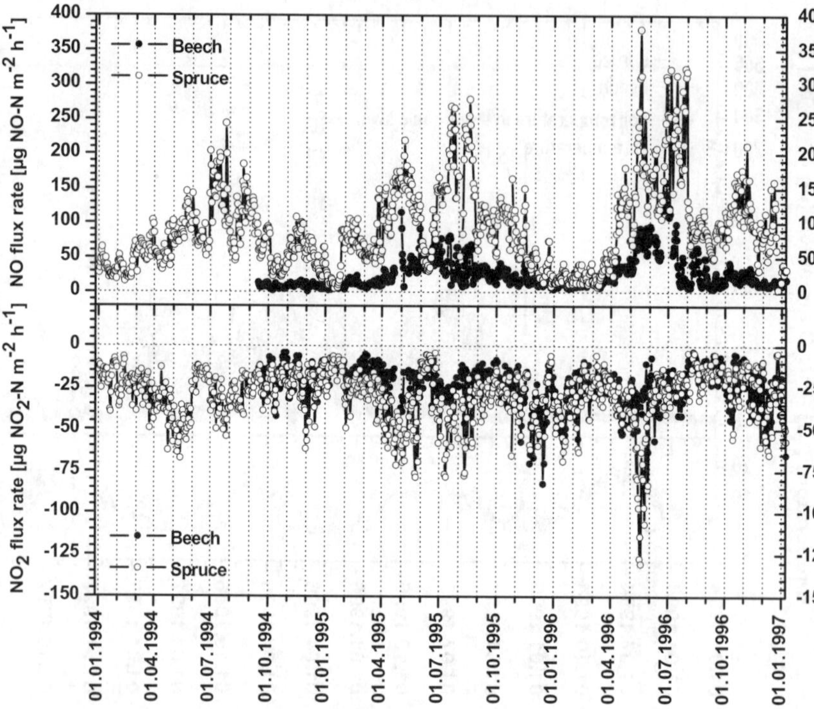

Figure 2: Effect of forest type (spruce vs. beech) on magnitude of NO and NO_2 flux rates at the Höglwald experimental site (Bavaria, Germany) during the observation period 1994-1997. Adapted from Gasche and Papen (1999).

higher than NO release from soil under beech (28.4 µg NO-N m^{-2} h^{-1}; about 0.25 g N m^{-2} yr^{-1}; Table 1). These stands appear to be so similar in soil type and climate that the differences in N gas fluxes are likely the result of the influence of the dominant tree species. Moreover, since the N_2O to NO emission ratio for beech was 2.0, but was 0.17 for spruce, the tree species apparently determines the dominant N-trace gas species primarily emitted from the soil into the atmosphere. Interestingly, the stands differed only by about 30% in the sum of both gases, despite the order of magnitude difference in the ratios of the two gases.

Could the differences between these two stands be extrapolated in general to conifers vs. broadleaved species? We studied N-trace gas fluxes from forest soils in the Northeastern German Lowlands stocked with pine, beech and oak on comparable soil. Our preliminary results showed that the ratio of N_2O:NO emission for pine was about 0.08, compared with about 5 for oak (Table 1). However, this ratio was 0.4 for beech, far lower than the 2.0 ratio from the beech stand at Höglwald. Our tentative results indicate that the ratio of N_2O:NO emissions may be much lower for conifers (primarily NO producers) than for broadleaved species (primarily N_2O producers), but that the actual ratios may vary substantially across sites.

In order to elucidate any tree species effect on N_2 emissions via denitrification, we measured N_2 emissions from all sites with a recently developed soil-core incubation system developed in our laboratory. This system allows quantification of the N_2 losses from forest soils into the atmosphere stemming from actual denitrification (for details see: Butterbach-Bahl et al., 2002). At the Höglwald and Northeastern German Lowland sites, the deciduous species showed higher losses of N_2 than the conifers, by factors of 1.7 to 2.5 (Table 1).

The beech stand at Högwald also showed 2.7-fold greater rates of net methane oxidation (or consumption) than the spruce stand. At the Northeastern Lowlands the soil beneath with oak (deciduous tree species) showed significantly higher methane oxidation rates than the pine soil, but the beech soil may have had even lower rates than the pine. We need more data to support any final conclusion from this site.

With regard to CO_2 release from forest soils, our field studies at the pine, beech and oak sites in the Northeastern German Lowlands indicate increasing CO_2 emission in the following sequence: pine, beech, oak (Table 1). This result is in line with results obtained from determinations of gross mineralization rates, which were found to be more than 2 times higher in the soil under oak as compared to beech and pine (Table 1). However, the CO_2 release from intact forest soils includes material respired by both plants and saprophytic microbes, so the interpretation of the CO_2 release is not as simple as the emissions of the N trace gases.

Litter quality may be a key factor by which tree species might modulate soil-microbial processes involved in N- and C-trace gas production, consumption and emission from soil. In order to get more insight in the role litter quality plays in this context we took soil samples from the organic layer of different forest stands (larch, oak, beech, spruce, pine) in Western Jutland (Denmark) which were stocking on the same soil type and with the same age of trees. The tree species exerted a clear effect on the magnitude of the soil microbial respiration, mineralization, and nitrification (Table 2). The sequence of decreasing rates for all processes was spruce, beech, larch, oak and pine. However, there was no obvious relation between the magnitude of the microbial processes and N_2O production; the soil from the beech site showed the highest rates followed by spruce, pine, oak and larch, which is in line with results obtained at the other field sites studied (Höglwald, NE German Lowlands, respectively: see above). Menyailo and Huwe (1999) also detected lowest rates of N_2O production under larch (*Larix sibirica* Ledeb.), followed by Scots pine (*Pinus sylvestris* L.) and Norway spruce in Siberia. We do not know if soil under larch and oak would show the same shifted from N_2O to NO emission as was found beneath spruce and pine, or if nitrification and denitrification would tend to proceed all the way to nitrate and dinitrogen. Experiments are in progress to resolve these open questions.

Table 1: Long term annual mean ± SE (1994-1996) emission rates of N_2O (Papen and Butterbach-Bahl 1999), NO (Gasche and Papen 1999), and N_2 (Butterbach-Bahl et al. 2002), and CH_4 oxidation rates (Butterbach-Bahl and Papen 2002) of soils stocked with spruce and beech at the Höglwald experimental site, mean N_2O, NO-, CH_4- (Butterbach-Bahl et al. 2001) CO_2 and N_2-flux rates from soils stocked with beech, pine and oak at the NE German Lowland forest sites as well as mean gross nitrification and ammonification rates in the respective organic soil horizon (O_h-layer).

	N_2O	NO	N_2	Gross nitrification	Gross ammonification	CO_2	CH_4
	-------(μg N m^{-2} h^{-1})-----			(mg N kg^{-1} soil d^{-1})		(mg C m^{-2} h^{-1})	(μg CH_4 m^{-2} h^{-1})
Höglwald Forest							
Picea abies	16.4 ± 1.2	91.8 ± 1.9	82	n.d.	n.d.	n.d.	27.7 ± 0.4
Fagus sylvatica	58.4 ± 2.3	28.4 ± 0.9	142	n.d.	n.d.	n.d.	73.6 ± 2.2
NE German Lowlands							
Pinus sylvestris	2.7 ± 1.2	34.2 ±11.7	28.8	3.3	6.0	88.7	202.6 ± 27.4
Fagus sylvatica	6.3 ± 1.8	14.8 ± 5.8	72.9	4.5	5.0	122.1	188.9 ± 32.3
Quercus robur	1.6 ± 1.4	0.3 ± 0.2	52.2	0.1	11.3	160	259.9 ± 34.1

Table 2: Effect of different tree species (via litter quality) on N_2O production, respiration, gross nitrification and mineralization (compiled from: Brüggemann et al., 2004 (submitted).

Western Jutland, Denmark	N_2O release (μg N kg^{-1} soil d^{-1})	Respiration (mg C kg^{-1} soil d^{-1})	Gross mineralization	Gross nitrification
			-----(mg N kg^{-1} soil d^{-1})-----	
Picea abies	72	345	25	40
Fagus sylvatica	176	240	19	32
Larix leptolepsis	1.0	160	18	18
Quercus robur	8.0	120	12	14
Pinus mugo	101	99	9	10

CONCLUSIONS

Tree species differ in effects on the magnitude of N_2O-, NO-, CO_2-emissions from soils, as well as methane oxidation within soils. Tree species determine not only the magnitude of the fluxes of N-trace gases, but also the individual N-trace gas compounds. Soil under beech primarily emits N_2O and less NO, whereas soil under spruce and pine primarily emit NO and less N_2O. Whether this difference applies in general for all deciduous and coniferous tree species urgently needs more rigorous research. Litter quality probably plays a dominant role in determining rates of trace gas flux, along with species effects on soil temperature and moisture regimes. As illustrated in other chapters in this volume, the biotic communities in soils differ strongly under the influence of different species.

ACKNOWLEDGEMENTS

The work was supported by the Bundesministerium für Bildung, Wissenschaft, Forschung und Technologie (BMBF), Bonn, Germany, under crontract numbers: KFA-BEO 0339615, KFA-BEO 0339628, BMBF-BEO 0339728, BMBF BEO 0339729A as well as BEO 0330534C.

REFERENCES

Augusto L, Ranger J, Binkley D and Rothe A 2002 Impact of several common tree species of European temperate forests on soil fertility. Ann. For. Sci. 59, 233-253.

Bengtsson G, Bengtson P. and Mansson K F 2003 Gross nitrogen mineralization-, immobilization-, and nitrification rates as a function of soil C/N ratio and microbial activity. Soil Biol. Biochem. 35, 143-154.

Binkley D and Giardina C 1998 Why do tree species affect soils? The Warp and Woof of tree-soil interactions. Biogeochem. 42, 89-106.

Brüggemann N, Rosenkranz P, Papen H, Pilegaard K and Butterbach-Bahl K (2004) Pure stands of temperate forest tree species modify soil respiration and N turnover. Soil Biol. Biochem (submitted).

Butterbach-Bahl K, Gasche R, Breuer L and Papen H 1997 Fluxes of NO and N_2O from temperate forest soils: impact of forest type, N deposition and of liming on the NO and N_2O emissions. Nutr. Cycl. Agroecos. 48, 79-90.

Butterbach-Bahl K, Breuer L, Gasche R, Willibald G and Papen H 2001 Exchange of trace gases between soils and the atmosphere in Scots pine forest ecosystems of the northeastern German Lowlands. 1. Fluxes of N2O, NO/NO2 and CH4 at forest sites with different N-deposition. For. Ecol. Manage. 167, 123-134.

Butterbach-Bahl K and Papen H 2002 Four years continuous record of CH_4-exchange between the atmosphere and untreated and limed soil of a N-saturated spruce and beech forest ecosystem in Germany. Plant Soil 240, 77-90.

Butterbach-Bahl K, Willibald G and Papen H 2002 Soil core method for direct simultaneous determination of N_2 and N_2O emissions from forest soils. Plant Soil 240, 105-116.

Conrad R 1996 Soil microorganisms as controllers of atmospheric trace gases (H_2, CO, CH_4, OCS, N_2O, and NO). Microbiol. Rev. 60, 609-640.

Coté L, Brown S, Pare D, Fyles J and Bauhus J 2000 Dynamics of carbon acid nitrogen mineralization in relation to stand type, stand age and soil texture in the boreal mixedwood. Soil Biol. Biochem. 32, 1079-1090.

Davidson, E A, Hart S C, Shanks C A and Firestone M K 1991 Measuring gross nitrogen mineralization, immobilization, and nitrification by N-15 isotopic pool dilution in intact soil cores. J. Soil Sci. 42, 335-349.

Erickson H, Davidson E A and Keller M 2002 Former land-use and tree species affect nitrogen oxide emissions from a tropical dry forest. Oecologia 130, 297-308.

Firestone M K and Davidson E A 1989 Microbiological basis of NO and N_2O production and consumption in soil. In: Andreae, M.O., Schimel, D.S. (Eds.), Exchange of Trace Gases Between Terrestrial Ecosystems and the Atmosphere. John Wiley, Chichester, pp. 7-21.

Gasche R and Papen H 1999 A 3-year continuous record of nitrogen trace gas fluxes from untreated and limed soil of a N-saturated spruce and beech forest ecosystem in Germany - 2. NO and NO_2 fluxes. J. Geophys. Res. 104, 18505-18520.

Ingwersen J, Butterbach-Bahl K, Gasche R, Richter O and Papen H 1999 Barometric process separation: New method for quantifying nitrification, denitrification, and nitrous oxide sources in soils. Soil Sci. Soc. Am. J. 63, 117-128.

Kreutzer K and Weis W 1998 The Höglwald field experiments – aims, concept and basic data. Plant Soil 199, 1-10.

Menyailo O V and Huwe B 1999 Activity of denitrification and dynamics of N_2O release in soils under six tree species and grassland in central Siberia. J. Plant Nutr. Soil Sci. 162, 533-538.

Menyailo O V, Hungate B A and Zech W 2002 The effect of single tree species on soil microbial activities related to C and N cycling in the Siberian artificial afforestation experiment - Tree species and soil microbial activities. Plant Soil 242, 183-196.

Papen H and Butterbach-Bahl K 1999 A 3-year continuous record of nitrogen trace gas fluxes from untreated and limed soil of a N-saturated spruce and beech forest ecosystem in Germany - 1. N_2O emissions. J. Geophys. Res. 104, 18487-18503.

Priha O and Smolander A 1999 Nitrogen transformations in soil under *Pinus sylvestris*, *Picea abies* and *Betula pendula* at two forest sites. Soil Biol. Biochem. 31, 965-977.

Scott N A and Binkley D 1997 Foliage litter quality and annual net N mineralization - comparison across North American forest sites. Oecologia 111, 151-159.

Stump L M and Binkley D 1993 Relationships between litter quality and nitrogen availability in Rocky-Mountain forests. Can. J. For. Res. 23, 492-502.

Chapter 10

SIGNIFICANCE OF FORESTS AS SOURCES FOR N_2O AND NO

Klaus Butterbach-Bahl and Ralf Kiese
Institute for Meteorology and Climate Research, Atmospheric Environmental Research, Forschungszentrum Karlsruhe, Kreuzeckbahnstr. 19, 82467 Garmisch-Partenkirchen, Germany

ABSTRACT

Forest ecosystems cover approximately 20% of the Earth's land area, and forests are a major source of gases such as N_2O and NO that influence Earth's climate. Tropical forests are one of the most important sources of N_2O in the global atmospheric budget. Recent results from field experiments indicate that temperate forests may be more important sources than previously expected, owing to underestimations of N_2O emissions. The higher than expected emissions of N_2O may result from increased availability of N in temperate forests from atmospheric N deposition and its stimulating effect on N_2O but also NO emissions. High rates may also be driven by repeated freezing and thawing of forest soils. Some non-tropical forests may also have a weak capacity to act as a net sink for N_2O, though the mechanism is not currently known. Tropical forest soils are major sources of NO, and the NO emission from temperate forests may increase following increased atmospheric N deposition. The large uncertainties in the budgets for these gases in forest ecosystems may be reduced if process-oriented models are developed and applied to simulate all major C and N cycling and biosphere-atmosphere exchange processes, in a spatially explicit approach that will allow scaling from points to regions to the globe.

D. Binkley and O. Menyailo (eds.),
Tree Species Effects on Soils: Implications for Global Change, 173–191.

INTRODUCTION

Forests cover approximately 29 million km^2 of Earth's surface (GLCCD, 1998), or about 20% of Earth's land area. The greatest cover of forests is found in high-latitude (10.3 million km^2) and low-latitude regions (12.8 million km^2); forests cover only 5.9 million ha in the mid-latitude regions (GLCCD, 1998). Forests contain about 613 to 938 Gt of carbon (C), with more than half stored in soils, along with substantial quantities of organically bound nitrogen (Batjes, 1996). The huge pools of C and N in the forests are by no means static, with turnover rates varying with climate, site properties and management activities. Turnover times may range from a few decades for some sites, to hundreds or even thousands of years for others. Microbially mediated processes determine the rates of C and N turnover; complex organic molecules are broken down (Figure 1) in processes of mineralization (releasing CO_2 and NH_4^+), nitrification (producing NO_3^-, and N trace gases), and denitrification (producing N oxides and N_2). Two N gases, nitrous oxide (N_2O) and nitric oxide (NO; both facultative or obligate intermediates of the microbial processes of nitrification and denitrification) influence atmospheric chemistry and global climate due to a) their direct involvement in tropospheric O_3 chemistry (NO) (e.g. Crutzen, 1970, 1995) and b) the significant capability to absorb infrared radiation (N_2O) (IPCC, 1997).

Rates of production of N_2O and NO are still highly uncertain for forests. The IPCC (1997) estimated that tropical forests generated from 2.2 to 3.7 Tg N_2O-N yr^{-1}. The mean estimate of 3.0 Tg N_2O-N yr^{-1} accounted for about 18% of all atmospheric N_2O sources, second only to agricultural soils as a source of N_2O. Temperate forests were estimated to emit 0.1 to 2.0 Tg N_2O-N yr^{-1} (mean value: 1.0 Tg N_2O-N yr^{-1}; IPCC 1997). The huge ranges of uncertainties for N_2O production are similar to the uncertainty for the secondarily active trace gas NO. Lee et al. (1997) and Davidson and Kingerlee (1997) estimated that soils contribute between 7 and 21 Tg NO-N yr^{-1} (15 to 47 % of the global total). Forest soils may release less NO than agricultural systems owing to redeposition on forest canopies, which may range up to 50% of emitted NO compared with 20% for agricultural systems (Lee et al., 1997).

In this chapter we provide an overview of our current knowledge on N trace gas emissions from forest soils worldwide, summarize the underlying mechanisms, and provide perspectives on ways to improve our knowledge of the role of forests as sources and sinks for atmospheric NO and N_2O.

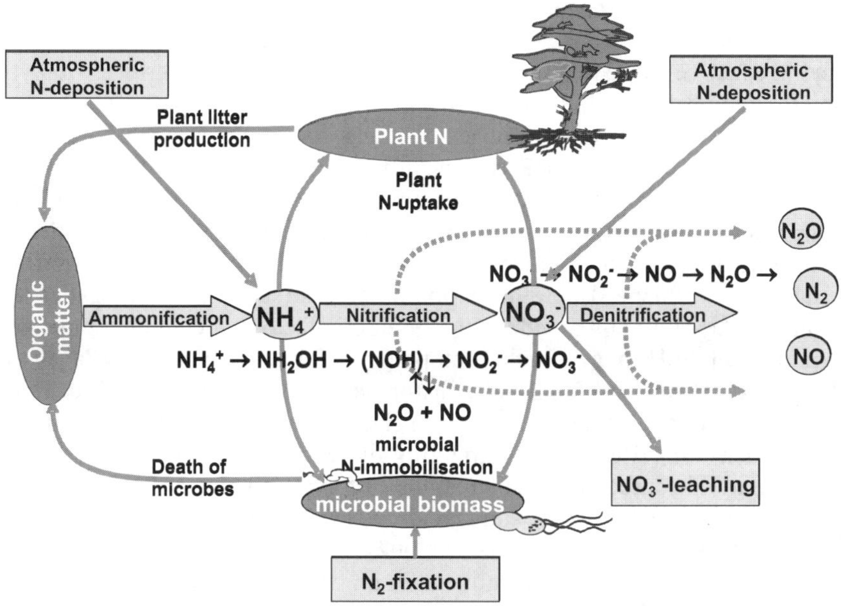

Figure 1. Processes involved in N-cycling and N-trace gas production and consumption..

PROCESSES INVOLVED IN N-TRACE GAS PRODUCTION, CONSUMPTION AND EMISSION IN FOREST SOILS.

The exchange of N_2O and NO between soils and the atmosphere depends on simultaneous, opposing processes of production and consumption (Conrad, 1996, 2002). Production of N_2O or NO in soils is predominantly due to the microbial processes of nitrification and denitrification. Nitrification is an oxidative process during which NH_3/NH_4^+ is oxidised to NO_2^-/NO_3^-. This process requires the availability of molecular oxygen. In contrast, denitrification is a reductive process, which mainly occurs in oxygen depleted soil zones (Zumft, 1997). Under anaerobic conditions, some microbes can use NO_3^-/NO_2^- as an alternative electron acceptor, thereby reducing NO_3^-/NO_2^- sequentially to NO, N_2O and finally to N_2 (Conrad, 1996). Both processes can occur simultaneously in soils, although the rates of the two processes depend on the soil aeration and the microsite availability of substrates. At low soil pH-values (<4.0) nitrite, which is an intermediate of nitrification and denitrification, can also be chemically decomposed to NO via chemo-denitrification (e.g. Van Cleemput and Baert, 1984). The importance of other soil processes remains poorly known, including any role of dissimilatory nitrate reduction to ammonium in the production of NO and N_2O (Silver et al.,

2001). A recently discovered consumption process is the further reduction of N_2O to N_2 by denitrifying bacteria (Conrad, 2002). For NO, both oxidative and reductive consumption pathways have been described for the microbiological processes of nitrification and denitrification (e.g. Dunfield and Knowles, 1998). The rates of N-trace gas consumption in soils are related to the diffusivity in the soil, which influences oxygen availability and the residence time of N gases. Soil diffusivity depends on soil moisture and other properties such as texture and organic matter content. This complex pattern of processes involved in N-trace gas production (Figure 1), consumption and emission is influenced by other biotic processes (such as mineralization and plant uptake of N) and a wide range of abiotic factors (including temperature and moisture, pH, N availability, cropping or forest management (Papen and Butterbach-Bahl, 1999). All these factors can vary by several fold (or more) across space and through time (see Davidson et al., 1998; Papen and Butterbach-Bahl, 1999). This variability of driving factors explains the huge uncertainty in source and sink strengths of forest soils for N-trace gases. For further details, see Firestone and Davidson, 1989; Davidson, 1991; Granli and Boeckman, 1994; and Conrad, 1996, 2002).

The key environmental influences on fluxes of N_2O and NO differ between major climate zones. Moisture variations may be most important under constant temperature conditions in the tropics, whereas freeze/thaw cycles may be especially important at higher latitudes (Papen and Butterbach-Bahl, 1999; Gasche and Papen, 1999; Butterbach-Bahl et al., 2004).

BOREAL FORESTS

Boreal forests cover a nearly continuous belt of coniferous forests across the upper parts of North America and Eurasia. Long, severe winters (up to six months with mean temperatures below freezing) and short summers (50 to 100 frost-free days) are characteristic, with wide ranges of temperatures between winter lows and summer highs. Decomposition rates may be slow, resulting in large accumulations of soil organic matter, acidic conditions, and podzolized profiles. Only a few data on N trace gas exchange are available. Kim and Tanaka (2003) measured mean N_2O summer fluxes for different forest sites in interior Alaska in a range of 0 to 45 μg N_2O-N m^{-2} h^{-1} (mean value: 23.3 μg N_2O-N m^{-2} h^{-1}), whereas Huttunen et al. (2003) reported for forested peatlands in northern Finland mean N_2O summer emissions in the range of 3.6 to 11.9 μg N_2O-N m^{-2} h^{-1}. A somewhat lower range of N_2O fluxes -6 - + 13 μg N_2O-N m^{-2} h^{-1} was observed in summer in boreal forests of Ontario, Canada (Schiller and Hastie, 1996). The latter observation was similar to N_2O flux rates from a boreal forest site at Hyytiälä, Finland (NOFRETETE project, Vesala et al., personal communications), which seem to be the only flux measurements at present which have covered an entire year. We expect that freeze/thaw cycles are very important in determining gas fluxes in boreal forests, but this remains largely speculative (see section on temperate forests). Fires and draining of wetlands have large effects on gas fluxes; Martikainen et al. (1995) and Regina et al. (1996) showed that N_2O fluxes from drained forested peatland sites can be as high as 5.2 kg N_2O-N ha^{-1} yr^{-1}, whereas Kim and Tanaka (2003) found that forest fires reduce summer N_2O emissions in a range of 10 to 50 %.

Reports about NO emissions from boreal forest ecosystems are even more scarce. Regina et al. (1996) measured NO fluxes in a forested boreal peatland in Finland and found that fluxes are < 15 μg NO-N m^{-2} h^{-1}. The same observation was also made by Vesala et al. for the Hyytiälä site in Finland (personal communications). Obviously, more measurements are urgently needed to clarify the role of boreal forest soils as sinks or sources for NO.

TEMPERATE FORESTS

In the last two decades many reports about N_2O emissions from temperate forest soils have been published (e.g. Goodroad and Keeney, 1984; Bowden et al., 1990; Ambus and Christensen, 1995; Borken and Brumme, 1997; Papen and Butterbach-Bahl, 1999; Butterbach-Bahl et al., 2002a; Jungkunst et al., 2004). Most of the work has been carried out in Europe and the USA, and data are sparse for other regions including Asia, Australia/New Zealand and South America (e.g. Oura et al., 2001). These studies have demonstrated that temperate forests are a major source for N_2O, but rates are low on a hectare basis (<1 kg N ha^{-1} yr^{-1}).

In a few cases, temperate forests can be net sinks for N_2O consumption from the atmosphere, at rates of 1-0.2 kg N ha^{-1} yr^{-1} (e.g. Butterbach-Bahl et al., 1998; Papen et al., 2001). Most forest soils are net producers of N_2O into the atmosphere, and these fluxes may or may not follow clear seasonal patterns. Nitrogen-rich sites may show a pronounced increase in emissions in late spring or early summer (e.g. Brumme and Beese, 1992; Papen and Butterbach-Bahl, 1999; Zechmeister-Boltenstern et al., 2002).

Most studies have measured emissions of N_2O at intervals of 2 to 4 weeks, and these long intervals may miss episodic pulses related to rewetting and freeze/thaw events. Soil freezing and thawing is important in N_2O effluxes from agricultural soils (Duxbury et al., 1982; but see also e.g. Christensen and Tiedje, 1990) and in forest soils (Figures 2 and 3, Papen and Butterbach-Bahl 1999, Teepe et al., 2000). For the Höglwald Forest site in Southwest Germany, the only site worldwide where N_2O emissions are followed continuously in sub-daily resolutions since end of 1993, we found that N_2O emissions during freezing and thawing contributed to the annual sum of N_2O emissions between 0 to 82 % at the spruce site and 0 - 73.5 % at the beech site (average for 1994-1997: spruce: 58.3%; beech: 47.6%). The variability of the contribution of freeze-thaw events to the annual N_2O emission depended on soil moisture status and on the severity and length of the soil freezing period. Papen and Butterbach-Bahl (1999) hypothesized that increased N_2O emissions during freeze-thaw events result from the interaction of several processes that occur near freezing:

a) dieback of the microbial biomass, which increases the availability of easily decomposable organic material for surviving microbes;
b) expansion of the anaerobic zones in the frozen soil due to a reduced pore volume by freezing of water and, thus, reduced gas diffusion;
c) sensitivity to low temperature of the N_2O reductase enzyme, the last enzyme of the denitrification chain, which leaves N_2O as the major end product of denitrification (Holtan-Hartwig et al., 2002); and
d) a tight coupling of nitrification and denitrification processes in the frozen/ thawing soils.

The contributions of these (and other) factors to increased N_2O emissions during freezing and thawing are the focus of ongoing research (e.g. Koponen and Martikainen, 2004).

Increased soil N supply from atmospheric deposition of N may increase N_2O production in temperate forests. Most of Europe receives at least 5 kg N ha^{-1} yr^{-1} in deposition; some parts of central Europe receive more than 20 kg N ha^{-1} yr^{-1}; and the Netherlands, parts of Northern Germany, and the UK receive up to 50 kg N ha^{-1} yr^{-1} (e.g. Van der Eerden et al., 1998). These large inputs may alleviate N limitation to plant growth, and the fluxes of radiatively active trace gases (e.g. NO, N_2O, CO_2, CH_4; Aber et al., 1989, 1992; Butterbach-Bahl et al. 1997, 2002a; Fenn et al., 1998; Maghill et al., 1997; Pilegaard et al., 1999; Priha et al., 1995; Sitaula et al., 1995; Van Dijk and Duyzer, 1999). The high-deposition environment of the Höglwald Forest beech site showed emission rates of 3.8 kg N_2O-N ha^{-1} yr^{-1} (Butterbach-Bahl et al., 2002b), similar to the 4.0 kg N_2O-N ha^{-1} yr^{-1} emission from a beech forest in Schottenwald near Vienna, Austria (Zechmeister et al., 2002). The most extreme rate reported comes from a high-N-deposition forest of oak, beech and hornbeam in the Netherlands, where Tietema et al. (1991) reported loss rates of 20 kg N_2O-N ha^{-1} yr^{-1}. Research in the EU-funded project NOFRETETE (Nitrogen Oxide Emissions from European Forest Ecosystems; http://195.127.136.75/nofretete/) supports the role of high N deposition as a driver of N trace gas emission.

Measurements of NO flux are much more challenging than those of N_2O, so data on NO emissions remain scarce. Reported NO emission rates for a wide variety of forest types range on short time scales between 0 and 380 μg NO-N m^{-2} h^{-1} (= 0 and 8 kg N ha^{-1} yr^{-1} (e.g. Skiba et al., 1994; Fenn et al., 1996; Gasche and Papen, 1999, 2002; Van Dijk and Duyzer, 1999; Pilegaard et al., 1999; Butterbach-Bahl et al., 2002a). NO emissions from forest soils show a clear seasonal pattern with highest emission rates observed during the summer months (e.g. Gasche and Papen, 1999; van Dijk and Duyzer, 1999).

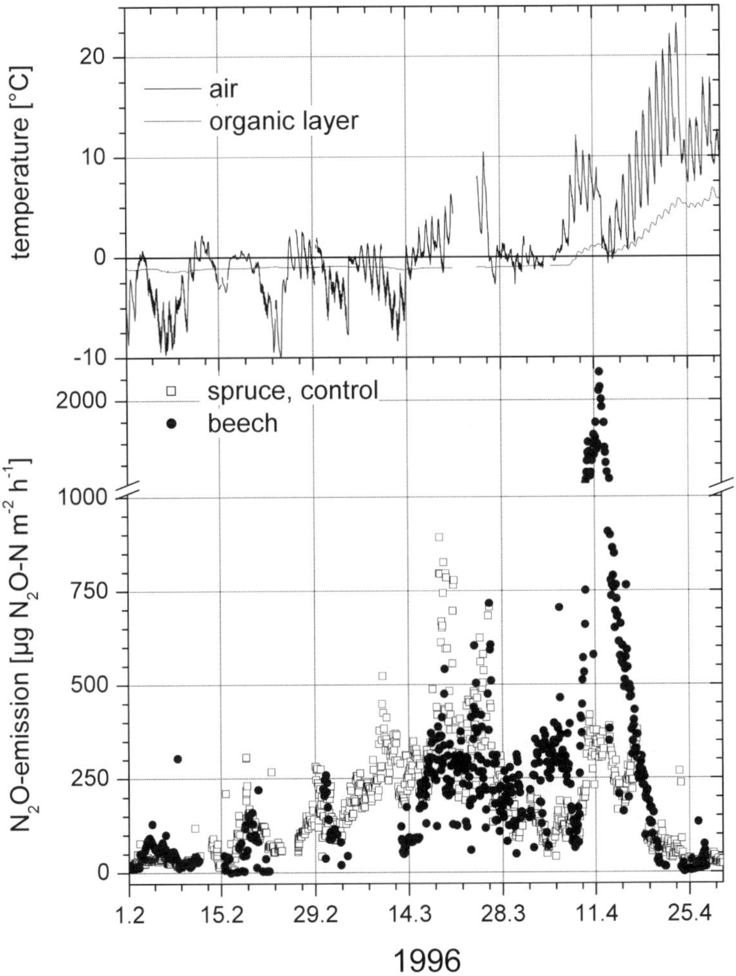

Figure 2. Time course of N_2O emission from single measuring chambers at the beech and spruce site of the Höglwald Forest, in relation to forest floor and air temperature in the period February 1 to April 30. The majority of N_2O emissions were associated with relatively brief periods of freezing and thawing (see Papen and Butterbach-Bahl 1999 for more detail).

However, as was already pointed out for N_2O, long-term measurements are still scarce and only at the Höglwald Forest site in Southeastern Germany, NO fluxes have been followed in sub-daily resolution for several years (Gasche and Papen, 1999). Freezing and thawing do not appear to be important for NO emissions at Höglwald, as NO emissions are more triggered by nitrification activity rather than by denitrification activity (Conrad, 1996, 2002). The NOFRETETE project will increase the available information on NO emissions.

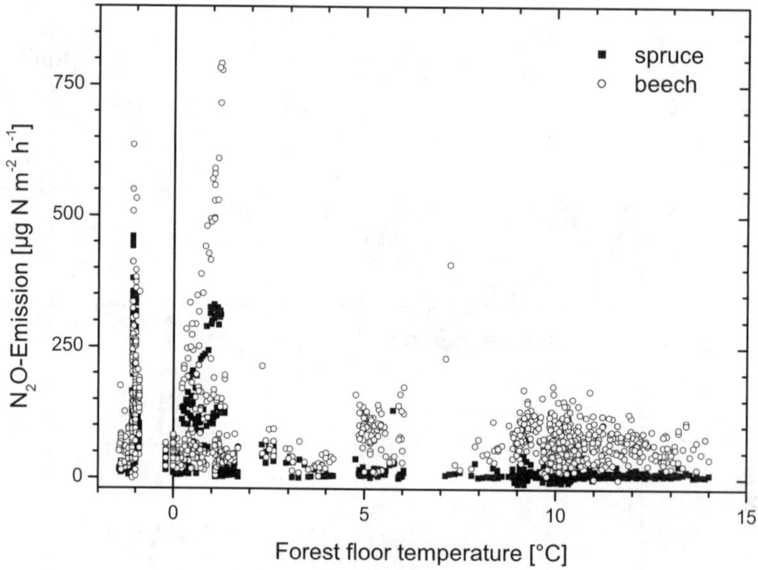

Figure 3. Nitrous oxide emissions from the Höglwald Forest spruce and beech sites in relation to temperature (during 1996) The huge N_2O emissions occurring around 0°C resulted from freezing and thawing (after Papen and Butterbach-Bahl 1999).

Emission of NO also appears to increase with increasing deposition of N from the atmosphere. Rates from high deposition sites include: 10 kg N ha^{-1} yr^{-1} for the Speulderbos sites in the Netherlands (Van Dijk and Duyzer, 1999), 8.0 kg N ha^{-1} yr^{-1} at Höglwald in southern Germany (Gasche and Papen, 1999, 2002). Medium and low deposition sites show lower rates: 3.1 kg N ha^{-1} yr^{-1} from a beech forest at Soro, Denmark (Pilegaard et al., 1999), and <0.8 kg N ha^{-1} yr^{-1} for some pine forests in northeastern German Lowlands (Butterbach-Bahl et al., 2002a). Emissions of NO from temperate forests may increase in the future in response to sustained high rates of N deposition.

MEDITERRANEAN FORESTS

The Mediterranean climate has mild moist winters and hot dry summers. This combination occurs in 5 major regions of the world: the Mediterranean, south-central and southwestern Australia, the fynbos of southern Africa, the Chilean matorral, and the Mediterranean ecoregions of California. Emissions of N-trace gases are low from Mediterranean forests relative to temperate and tropical forests, largely because of low annual precipitation. Reported rates of

N_2O emission from Mediterranean forests are mostly lower than 5µg N_2O-N m^{-2} h^{-1} (Figure 4), extrapolating to < 0.5 kg N_2O-N ha-^{1} yr^{-1} (Fenn et al., 1996; Bernal et al., 2003).

Microbial activity and associated N-trace gas emissions are substantial only after precipitation events in summer, spring and through fall (Hart and Firestone, 1991; Davidson et al., 1991, 1992; Jacinthe and Lal, 2004; Musacchio et al., 1996). Rewetting of the soil after long dry periods can lead to high pulses of NO and N_2O emissions as the increased availability of readily decomposed organic matter and inorganic nitrogen stimulate microbial activity. These pulsed emissions depend on the adaptation of nitrifying and denitrifying bacteria to survive severe desiccation and starvation. The speed of the stepwise activation of the enzyme changes within the denitrification process may significantly influence the partitioning between NO and N_2O- emission during these events (Otte et al., 1996). Mediterranean forests show maximum N_2O emissions when soil moisture is higher than field capacity. In contrast, NO emissions dominate below field capacity (Anderson et al., 1988; Fenn et al., 1996; Fenn et al., 1998). This pattern indicates that nitrification may be the major driver of NO emissions, whereas denitrification is the major process leading to N_2O emission.

Mediterranean forests can occasionally consume N_2O under very dry conditions (Figure 4; Butterbach-Bahl et al., 1998; Papen et al., 2001; Fenn et al., 1996). Since a N_2O-uptake can only be explained by denitrification, this phenomenon may be explained as adaptation of denitrifier bacteria to the lack of nitrate during dry conditions and the alternative use of N_2O as metabolism substrate. The lack of nitrate is further supported by the general low nitrogen availability in Mediterranean forest soils.

Fire plays a major role in the development of Mediterranean forests, with N losses of 100 kg/ha or more from each fire (Johnson et al., 1998). Unfortunately, no studies have documented the effects of fires on N-trace gas emissions.

TROPICAL RAIN FORESTS

Tropical rain forests are found in a belt around the equator and in the humid subtropics. These forests are characterized by warm, humid climates with high annual rainfall. Tropical rain forests currently cover less than 5% of the Earth's land surface, in tropical Africa, India and Southeast Asia, the New World tropics, and Australia and many Pacific islands.

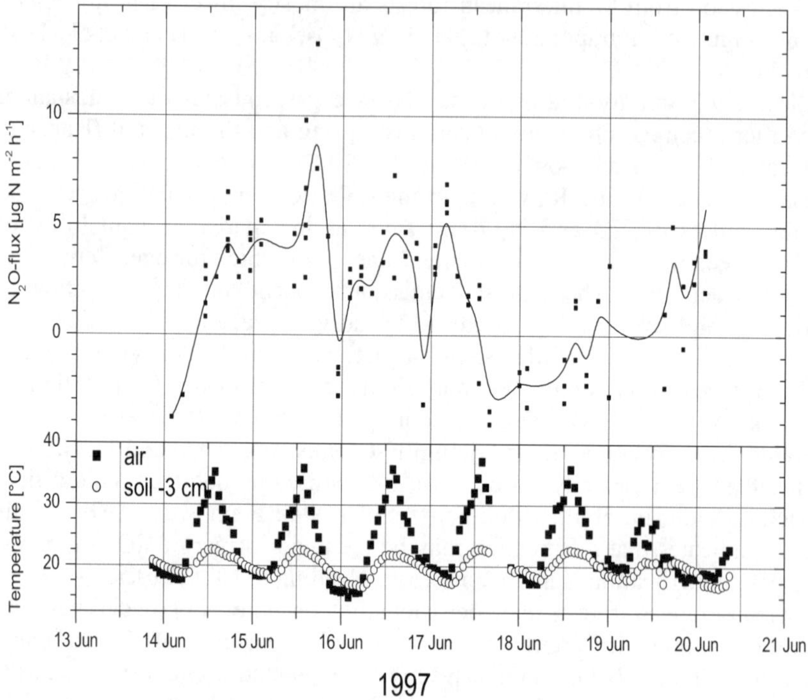

Figure 4. Summer N_2O fluxes in a Mediterranean mixed hardwood stand in the Toscana region of Italy (after Bernal et al., 2003).

Moist and warm climates favour decomposition, and the high rates of microbial N-transformations lead to very high emissions of N-trace gas emissions, ranging from 2.2 to 3.7 Tg N yr^{-1} (IPCC, 2001). Despite the magnitude of these fluxes, N_2O emission rates have been estimated for only a few tropical rainforests, and most of these estimates are based on sporadic sampling that may not capture the spatial and temporal dynamics shown in Figure 5 for an Australian rain forests (Kiese and Butterbach-Bahl, 2002; Kiese et al., 2003). The paucity of good spatial and temporal information accounts for the current high uncertainty in the source strength for N_2O for tropical rain forest soils (Breuer et al., 2000; Kroeze and Mosier, 2002).

Reported N_2O fluxes are in a range of 4.2 to 80.0 µg N m^{-2} h^{-1} for New World tropical rain forests (i.e. Goreau and DeMello, 1988; Verchot et al., 1999; Erickson et al., 2002; Garcia-Montiel et al., 2003), 1.7 to 207.0 µg N m^{-2} h^{-1} for one African rain forest site in the Congo (Serca et al., 1994), <1.0 to 15 µg N m^{-2} h^{-1} for a Southeast Asian rain forest (Ishizuka et al., 2002), and < 1.0 to 570.8 µg N m^{-2} h^{-1} for Australian rain forests (Breuer et al., 2000; Kiese and Butterbach-Bahl et al., 2002; Kiese et al., 2003). These hourly rates scale to annual emissions on the order of 0.3 to 7.5 kg N ha^{-1} yr^{-1}.

The emission of N_2O from tropical forest soils is influenced by the same suite of factors as those from temperate forest soils: vegetation, soil texture, soil C/N ratio, soil pH, substrate availability, and especially soil moisture (e.g. Keller and Reiners, 1994; Davidson et al., 1993; Steel et al., 2004; Kiese and Butterbach, 2002; Breuer et al., 2000; Erickson et al., 2002). Sites with seasonal patterns in rainfall typically show higher N_2O-emission during the wetter parts of the year (30-500 μg N_2O-N m^{-2} h^{-1}) than during the drier seasons (< 15 μg N_2O-N m^{-2} h^{-1}) due to limitations of soil microbial C and N turnover processes by soil moisture (Figure 5; Garcia Méndez et al., 1991; Steudler et al., 1991, Verchot et al., 1999; Breuer et al., 2000; Kiese and Butterbach-Bahl, 2002; Kiese et al., 2003;). Temporal variations in rates tend to be low when rates are low (Figure 5; Keller and Reiners, 1994, Kiese et al., 2003). Rates may also differ among years in response to major differences in annual climate. For example, the lowland rain forest ecosystem at Australia emitted 0.95 N_2O-N ha^{-1} yr^{-1} in a dry year, and 7.45 kg N_2O-N ha^{-1} yr^{-1} in a wet year (Figure 5; Kiese et al., 2003).

The existing data on NO emissions indicate rates that are similar to those for N_2O, totalling about 1.1 Tg NO-N yr^{-1} (Davidson and Kingerlee, 1997). Again, the available information for NO is much more sparse than for N_2O, and only a few studies have adequately characterized spatial and temporal variation (Butterbach-Bahl et al., 2004; Verchot et al., 1999; and Gut et al., 2002; Serça et al.,1994; Keller and Reiners, 1994). The temporal variability of NO emissions is pronounced and mainly driven by changes in soil moisture due to rainfall and subsequent drying out of the soil (Butterbach-Bahl et al., 2004). Highest NO emissions were observed with the re-wetting of the soil with the first rainfalls at the end of the dry season. Depending on the duration and severity of antecedent dry periods, the pulses in NO emission can be as high as 500 μg N m^{-2} h^{-1} (Davidson et al., 1991, 1993; Butterbach-Bahl et al., 2004). However, wetting events during the wet season result only in small increases of NO emissions or even in a decline of NO fluxes with higher values of soil moisture (Butterbach-Bahl et al., 2004). This lack of detailed measurements of pulse emissions of NO could mean that NO emissions are underestimated by a factor of 3 for tropical rain forests, and the actual total emission could be as high as 3 Tg NO-N yr^{-1}. Clearly we need more long-term measurements at high temporal and spatial resolution for many tropical forest sites.

CONCLUSIONS

Forest soils are characterized by high spatial and temporal variation in rates of emission of N trace gases, and this variation limits the precision of global rates estimates. Laboratory studies have shown that soil water content (followed by soil temperature) was the most important factor determining NO emissions (Gasche and Papen, 1999; Yang

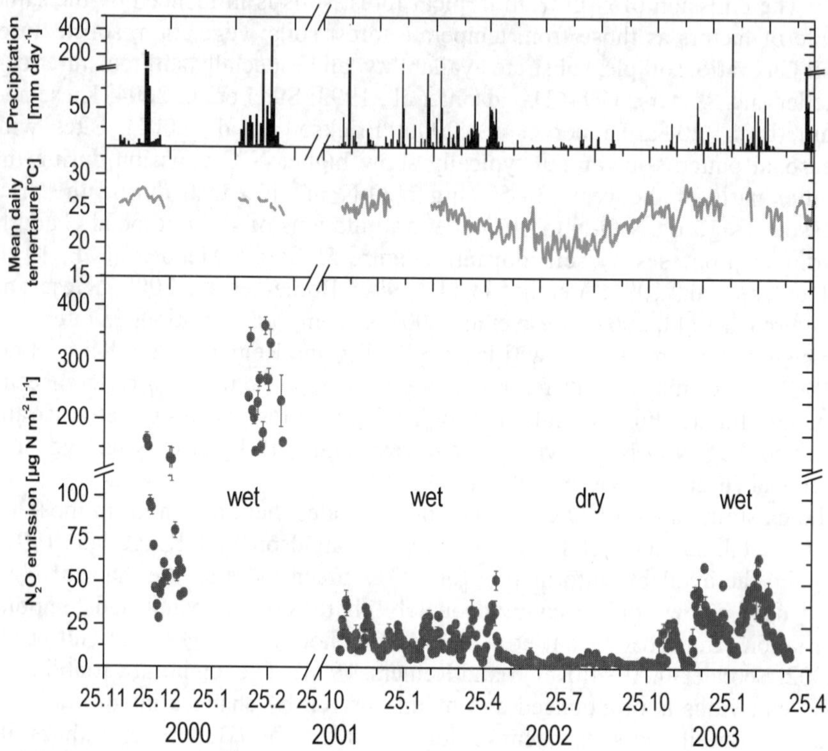

Figure 5. Seasonal and interannual variability of N_2O fluxes at the lowland tropical rain forest site Bellenden Ker, Australia (after Kiese and Butterbach-Bahl, 2002).

and Meixner, 1997; Butterbach-Bahl et al., 2004). For this reason, we believe that a promising strategy to overcome these problems is to combine field and laboratory experiments with the further development and application of process oriented models, which are able to describe the major processes in forest soils involved in N-trace gas production and emission (e.g. mineralisation, nitrification, denitrification, plant-microbe competition for inorganic nitrogen) and their dependency on changing environmental conditions. Such models may then be linked to geographic information systems to extrapolate to regional and global estimates. Several models of this sort have been developed recently (e.g. Potter et al., 1996; Butterbach-Bahl et al., 2001, 2004; Rastetter et al., 2002; Del Grosso et al., 2002). For example, C dynamics have been examined in grassland soils with the processed-based CENTURY model (Parton et al. 1987, 1998) and intensively validated for a number of grassland and other ecosystem types world-wide (e.g. Parton et al. 1993, Del Grosso et al. 2002). Another detailed process oriented model is the DNDC model, which has its original focus on the simulation of N dynamics in soils and on the prediction of the biosphere-atmosphere exchange of N trace gases in agricultural and forest ecosystems, accounting for detailed mechanisms driving microbial N turnover processes such as nitrification and denitrification (Li et al. 1992, 2000; Butterbach-Bahl et al. 2001). The DNDC model also allows consideration of management options and their potential effects on N cycling. Biogeochemical models have already been used in the recent past for the regionalization of site results, including providing a framework for inventories of N trace gas emissions from agricultural, forest and grassland soils (e.g. Potter et al., 1996; Li et al., 1996, Butterbach-Bahl et al. 2001, 2004; Brown et al. 2002). An example for the use of one biogeochemical model to estimate regional fluxes of N_2O is provided in Figure 6 for tropical forests in Australia (for further details see Kiese et al., 2004). We stress that point-based models have obvious weaknesses in characterizing the spatial variability of C and N cycling, including effects of landscape position on water availability and soil properties (Pennock and Frick, 2001). These weaknesses can be overcome only if biogeochemical models are embedded in or networked with landscape models such as regional hydrological models and dynamic vegetation models. This approach would allow consideration of the effects of spatial and temporal influences of factors such as distance to the groundwater table, or matter input and losses via wind and water erosion.

ACKNOWLEDGEMENTS

The work was supported by the European Commission in the NOFRETETE project of the fifth framework program.

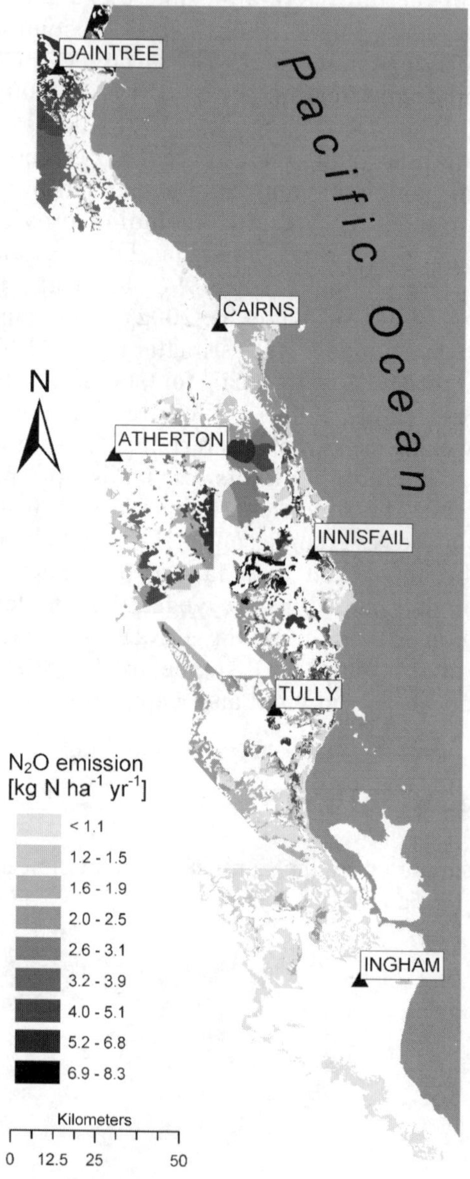

Figure 6. Regional distribution of annual sums of N_2O emissions from tropical rainforests of Australia (July 1996 to June 1997). The emission inventory was derived by coupling of the biogeochemical model PnET-N-DNDC with a detailed GIS databases holding all relevant input driving variables in spatial and temporal resolution (after see Kiese et al., 2004).

REFERENCES

Aber J D, Nadelhoffer K J, Steudler P and Melillo J M 1989 Nitrogen saturation in northern forest ecosystems. BioScience 39, 378-386.

Aber J D 1992 Nitrogen cycling and nitrogen saturation in temperate forest ecosystems. Tree 7, 220-224.

Ambus P and Christensen S 1995 Spatial and seasonal nitrous oxide and methane fluxes in Danish forest-, grassland-, and agroecosystems. J. Environ. Qual. 24, 993-1001.

Anderson I C, Levine J S, Poth M A and Riggan P J 1998 Enhanced biogenic emissions of nitric oxide and nitrous oxide following surface biomass burning. Journal of Geophysical Research 93, 3893-3898.

Batjes N H 1996 Total carbon and nitrogen in the soils of the world. Europ. J. Soil Sci. 47, 151-163.

Bernal S, Butturini A, Nin E, Sabater F and Sabater S 2003 Leaf Litter dynamics and nitrous oxide Emissions in a Mediterranean Riparian Forest: Implications for soil nitrogen dynamics. Journal of Environmental Quality 32, 191-197.

Borken, W and Brumme R 1997 Liming practice in temperate forest ecosystems and the effects on CO_2, N_2O and CH_4 fluxes. Soil Use Management 13, 251-257.

Breuer L, Papen H and Butterbach-Bahl 2000 N_2O emission from tropical forest soils of Australia. Journal of Geophysical Research 105, 26353-26367.

Brown L, Syed B, Jarvis S C, Sneath R W, Phillips V R, Goulding K W T and Li C 2002 Development and application of a mechanistic model to estimate emission of nitrous oxide from UK agriculture. Atmos. Environm. 36, 917-928.

Brumme R and Beese F 1992 Effects of liming and nitrogen fertilization on emissions of CO_2 and N_2O from a temperate forest. J. Geophys. Res. 97, 12,851-12,858.

Butterbach-Bahl K, Gasche R, Breuer L and Papen H 1997 Fluxes of NO and N_2O from temperate forest soils: impact of forest type, N deposition and of liming on the NO and N_2O emissions. Nutr. Cycl. Agroecos. 48, 79-90.

Butterbach-Bahl K, Gasche R, Huber C, Kreutzer K and Papen H 1998 Impact of nitrogen input by wet deposition on N-trace gas fluxes and CH_4-oxidation in spruce forest ecosystems of the temperate zone in Europe. Atmospheric Environment 32, 559-564.

Butterbach-Bahl K, Stange F, Papen H and Li C 2001Regional inventory of nitric oxide and nitrous oxide emissions for forest soils of Southeast Germany using the biogeochemical model PnET-N-DNDC. J. Geophys. Res. 106, 34155-34166.

Butterbach-Bahl K, Breuer L, Gasche R, Willibald G and Papen H 2002a Exchange of trace gases between soils and the atmosphere in Scots pine forest ecosystems of the North Eastern German Lowlands, 1. Fluxes of N_2O, NO/NO_2 and CH_4 at forest sites with different N-deposition. Forest Ecology and Management 167, 123-134.

Butterbach-Bahl K, Willibald G, Papen H and Gasche R 2002b Exchange of N-gases at the spruce and beech sites at the Höglwald Forest – A summary. Plant and Soil 240, 117-123.

Butterbach-Bahl, K, Kock M, Willibald G, Hewett B, Buhagiar S, Papen H and Kiese R. Temporal variations of fluxes of NO, NO_2, N_2O, CO_2 and CH_4 in a tropical rain forest ecosystem. Global Biogeochem. Cycl. 18, doi:10.1029/ 2004GB002243.

Butterbach-Bahl K, Kesik M, Miehle P, Papen H and Li C 2004 Quantifying the regional source strength of N-trace gases across agricultural and forest ecosystems with process based models. Plant and Soil 260, 311-329.

Christensen S and Tiedje J M 1990 Brief and vigorous N_2O production by soil at spring thaw. J. Soil Sci. 41, 1-4.

Conrad R 1996 Soil microorganisms as controllers of atmospheric trace gases (H_2, CO, CH_4, OCS, N_2O and NO). Microbiological Reviews 60, 609-640.

Conrad R 2002 Microbiological and biochemical background of production and consumption of NO and N_2O in soil, in Trace Gas Exchange in Forest Ecosystems, edited by R. Gasche et al., pp. 3-33. Kluwer Academic Publishers, Dordrecht, Boston, London.

Crutzen P J 1970 The influence of nitrogen oxides on the atmospheric ozone content, Q. J. R. Meteorol. Soc. 96, 320-325.

Crutzen P J 1995 Ozone in the troposphere, in Composition, Chemistry, and Climate of the Atmosphere, H. B. Sing, editor, pp. 349-393. Van Nostrand-Reinold, New York.

Davidson E A 1991 Fluxes of nitrous and nitric oxide from terrestrial ecosystems, in Microbial Production and Consumption of Greenhouse Gases: Methane, Nitrogen Oxides, and Halomethanes, edited by J. E. Rogers and W. B. Whitman, pp. 219-235. American Society for Microbiology, Washigton, D.C.

Davidson E A, Vitousek P M, Matson P A, Riley R, Garcia-Mendez G and Maas J M 1991 Soil emissions of nitric oxide in a seasonally dry tropical forest of Mexico. Journal of Geophysical Research 96, 15439-15445.

Davidson E A, Hart S C and Firestone M K 1992 Internal cycling of nitrate in soils of a mature coniferous forest. Ecology 73, 1148-1156.

Davidson E A, Matson P A, Vitousek P M, Riley R, Dinkin K, Garcia-Méndez, G and Maass J M 1993 Processes regulating soil emission of NO and N_2O in a seasonally dry tropical forest. Ecology 74, 130-139.

Davidson E A and Kingerlee W 1997 A global inventory of nitric oxide emissions from soils. Nutrient Cycling in Agroecosystems 48, 37-50.

Davidson E A, Potter C S, Schlesinger P and Klooster S A 1998 Model estimates of regional nitric oxide emissions from soils of the Southeastern United States. Ecol. Applic. 8, 748-759.

Del Grosso S, Ojima D, Parton W, Mosier A, Peterson G and Schimel D 2002 Simulated effects of dryland cropping intensification on soil organic matter and greenhouse gas exchanges using the DAYCENT ecosystem model. Environm. Pollut. 116, S75-S83.

Dunfield P F and Knowles R 1998 Organic matter, heterotrophic activity, and NO consumption in soils. Global Change Biol., 4, 199-207.

Duxbury J M, Bouldin D R, Terry R E and Tate R L 1982 Emissions of nitrous oxide from soils. Nature 298, 462-464.

Erickson H, Davidson E A and Keller M 2002 Former land-use and tree species affect nitrogen oxide emissions from a tropical dry forest. Oecologia 130, 297-308.

FAO 1995 Digital Soil Map of the World and Derived Soil Properties, Version 3.5, CD-ROM. Food and Agriculture Organization of the United Nations. Rome, Italy.

Fenn M E, Poth M A, and Johnson D W 1996 Evidence for nitrogen saturation in the San Bernadino Mountains in southern California. Forest Ecol. Manage. 82, 211-230.

Fenn M E, Poth M A, Aber J D, Baron J S, Bormann B T, Johnson D W, Lemly A D, McNulty S G, Ryan D F and Stottlemeyer R 1998 Nitrogen excess in North American ecosystems: predisposing factors, ecosystem responses, and management strategies, Ecol. Appl. 8, 706-733.

Firestone M K, Davidson E A 1989 Microbiological basis of NO and N_2O production and consumption in soil, in Exchange of Trace Gases between Terrestrial Ecosystems and the Atmosphere, edited by M. O. Andreae and D. S. Schimel, pp. 7-21. John Wiley & Sons Ltd., Chichester, United Kingdom.

Garcia-Montiel D C, Steudler P A, Piccolo M, Neill C, Melillo J and Cerri CC 2003 Nitrogen oxide emissions following wetting of dry soils in forest and pastures in Rondônia, Brazil. Biogeochemistry 64, 319-336.

Gasche R and Papen H 1999 A 3-year continuous record of nitrogen trace gas fluxes from untreated and limed soil of a N-saturated spruce and beech forest ecosystem in Germany 2. NO and NO_2 fluxes. J. Geophys. Res. 104, 18505-18520.

GLCCD 1998. Global Land Cover Characteristics Database, Version 1.2. Data available at http://edcdaac.usgs.gov/glcc/glcc.html.

Goodroad L L and Keeney D R 1984 Nitrous oxide emission from forest, marsh, and prairie ecosystems. J. Environ. Qual. 13, 448-452.

Goreau T J and De Mello W Z 1988 Tropical deforestation: Some effects on atmospheric chemistry. Ambio 17, 275-281.

Granli T and Bockmann O C 1994 Nitrous oxide from agriculture. Norwegian J. Agri. Sci. 12, 7-128.

Gut A, van Dijk S M, Scheibe M, Rummel U, Welling M, Ammann C, Meixner F X, Kirkman G A, Andreae M O and Lehmann B E 2002 NO emission from an Amazonian rain forest soil: Continuous measurements of NO flux and soil concentration. J. Geophys. Res. 107, 8057, doi: 10.1029/2001JD000521.

Hart S C and Firestone M K 1991 Forest floor - mineral soil interactions in the internal nitrogen cycle of an old-growth forest. Biogeochemistry 12, 103-127.

Holtan-Hartwig L, Dörsch P and Bakken L R 2002 Low temperature control of soil denitrifying communities: kinetics of N_2O production and reduction. Soil Biol. Biochem. 34, 1797-1806.

Huttunen J T, Nykänen H, Martikainen P J and Nieminen M 2003 Fluxes of nitrous oxide and methane from drained peatlands following forest clear-felling in southern Finland. Plant and Soil 255, 457-462.

IPCC 1997 Intergovernmental Panel on Climate Change guidelines for national greenhouse gas inventories. Chapter 4: Agriculture: Nitrous oxide from agricultural soils and manure management. OCDE, Paris.

IPCC 2001 Intergovernmental Panel on Climate Change, Climate Change 2001: The Scientific Basis. Contribution of Working Group I to the Third Assessment Report of the Intergovernmental Panel on Climate Change [Houghton JT, Ding Y, Griggs DJ, Noguer M, van der Linden PJ, Dai X, Maskell K, Johnson CA (eds.)]. Cambridge University Press, Cambridge, United Kingdom and New York, NY, USA, 881pp.

Ishizuka S, Tsuruta H and Murdiyarso D 2002 An intensive field study on CO_2, CH_4, and N_2O emissions from soils at four land-use types in Sumatra, Indonesia. Global Biogeochemical Cycles 16, 1049, doi:10.1029/2001GB001614.

Jacinthe P A and Lal R 2004 Effects of soil cover and land-use on the relations flux-concentration of trace gases. Soil Science 169, 243-259.

Johnson D W, Susfalk R B, Dahlgren R A and Klopatek J M 1998 Fire is more important than water for nitrogen fluxes in semi-arid forests. Environmental Science Policy 1, 79-86.

Jungkunst H F, Fiedler S and Stahr K 2004 N_2O emissions of a mature Norway spruce (Picea abies) stand in the Nlack Forest (southwest Germany) as differentiated by the soil pattern. J. Geophys. Res. 109, 7302.

Kiese R and Butterbach-Bahl K 2002 N_2O and CO_2 emissions from three different tropical forest sites in the wet tropics of Queensland, Australia. Soil Biology and Biochemistry 34, 975-987.

Kiese R, Hewett B, Graham A and Butterbach-Bahl K 2003 Seasonal variability of N_2O-emissions and CH_4-uptake from/ by a tropical rainforest soil of Queensland, Australia. Global Biogeochemical Cycles 17, 1043, doi: 10.1029/2002GB002014.

Kiese R, Li C, Hilbert W, Papen H and Butterbach-Bahl K 2004 Regional application of PnET-N-DNDC for estimating the N_2O source strength of tropical rainforests in the Wet Tropics of Australia. Global Change Biology in press.

Kim Y, and Tanaka N 2003 Effect of forest fire on the fluxes of CO_2, CH_4, and N_2O in boreal forest soils, interior Alaska. J. Geophys. Res. 108, D1, doi:10.1029/2001JD00063.

Kroeze C and Mosier A 2002 Progress in source strength estimates of nitrous oxide (N_2O). In: Non-CO_2 greenhouse gases: Scientific understanding, control options and policy aspects, (eds van Ham J, Baede AP, Guicherit R, Williams-Jacobse JG), pp. XXIX–XXXIII, Millpress, Rotterdam.

Koponen H T and Martikainen P T 2004 Soil water content and freezing temperature affect freeze-thaw related N_2O production in organic soil. Nutr. Cycl. Agroecosys. 69, 213-219.

Laine J, Silvola J, Tolonen K, Alm J, Nykänen H, Vasander H, Sallantaus T, Savolainen I, Sinisalo J and Martikainen P J 1996 Effects of water-level draw down on global climatic warming : northern peatlands. Ambio 25, 179-184.

Lee D S, Köhler I, Grobler E, Rohrer F, Sausen R, Gallardo-Klenner L, Olivier J G J, Dentener F J and Bouwman A F 1997 Estimations of global NO_x-emissions and their uncertainties. Atm. Environ. 31, 1735-1749.

Li C, Frolking S and Frolking TA 1992 A model of nitrous oxide evolution from soil driven by rainfall events: 1. Model structure and sensitivity. J. Geophys. Res. 97, 9759-9776.

Li C, Aber J, Stange F, Butterbach-Bahl K and Papen H 2000 A process-oriented model of N_2O and NO emissions from forest soils: 1. Model development. J. Geophys. Res. 105, 4369-4384.

Magill A H, Aber J D, Hendricks J J, Bowden R D, Melillo J M and Steudler P A 1997 Biogeochemical response of forest ecosystems to simulated chronic nitrogen deposition. Ecol. Appl. 7, 402-415.

Martikainen P J, Nykänen H, Alm J and Silvola J 1995 Change in fluxes of carbon dioxide, methane and nitrous oxide due to forest drainage of mire sites of different trophy. Plant and Soil 168-169, 571-577.

Musacchio A, Fioretto A, Palermo A M, Guzzo S, Pellegrino G and DeSanto A V 1996 Contribution of the litter and superficial soil layer to N-2 fixation, denitrification and N2O and CH4 emissions in a Pinus laricio Poiret forest (Sila mountains, Italy). Pedobiologia 40, 302-310.

Otte S, Grobben N G, Robertson L A, Jetten M S, Kuenen J G 1996 Nitrous oxide production by Alcaligenes faecalis under transient and dynamic aerobic and anaerobic conditions. Applied and Environmental Microbiology 62, 2421-2426.

Oura N, Shindo J, Fumoto T, Toda H and Kawashima H 2001 Effects of nitrogen deposition on nitrous oxide emissions from the forest floor. Water, Air and Soil Pollution 130, 673-678.

Papen H and Butterbach-Bahl K 1999 A 3-year continuous record of nitrogen trace gas fluxes from untreated and limed soil of a N-saturated spruce and beech forest ecosystem in Germany, 1. N_2O emissions. J. Geophys. Res. 104, 18487-18503.

Papen H, Daum M, Steinkamp R and Butterbach-Bahl K 2001 N_2O- and CH_4-Fluxes from Soils of a N-limited and N-fertilized Spruce Forest Ecosystem of the Temperate Zone. Journal Applied Botany 75, 159-163.

Parton W J, Schimel D S, Cole C V and Ojima D S 1987 Analysis of factors controlling soil organic matter levels in great plain grasslands. Soil Sci. Soc. Am. J. 51, 1173–1179.

Parton W J, Scurlock J M O, Ojima D S, Gilmanov T G, Scholes R J, Schimel D S, Kirchner T, Menaut J C, Seastedt T, Garcia Moya E, Kamnalrut A and Kinyamario J I 1993 Observations and modeling of biomass and soil organic matter dynamics for the grassland biome worldwide. Glob. Biogeochem. Cycle. 7, 785-809.

Parton W J, Hartman M, Ojima D S and Schimel D S 1998 DAYCENT and its land surface submodel: description and testing. Global and Planetary Change 19, 35-48.

Pennock D J and Frick A H 2001 The role of field studies in landscape-scale applications of process models: an example of soil redistribution and soil organic carbon content using CENTURY. Soil and Tillage Res. 58, 183-191.

Pilegaard K, Hummelshøj P and Jensen N O 1999 Nitric oxide emission from a Norway spruce forest floor. J. Geophys. Res. 104, 3433-3445.

Potter C S, Matson P A, Vitousek P M and Davidson E A 1996 Process modeling of controls on nitrogen trace gas emissions from soils worldwide. J. Geophys. Res. 101, 1361-1378.

Priha O and Smolander A 1995 Nitrification, denitrification and microbial biomass in soil from two N-fertilized and limed Norway spruce forests. Soil Biol. Biochem. 27, 305-310.

Rastetter E B, Aber J D, Peters D P C, Ojima D S and Burke I C 2003 Using mechanistic models to scale ecological processes across space and time. BioScience 53, 68-76.

Regina K, Nykänen H, Maljanen M, Silvola J and Martikainen P J 1998 Emissions of N2O and NO and net nitrogen mineralization in a boreal forested peatland treated with different nitrogen compounds. Can. J. For. Res. 28, 132-140.

Riley H R and Vitousek P M 1995 Nutrient dynamics and nitrogen trace gas flux during ecosystem development in montane rain forest. Ecology 76, 292-304.

Schiller C L and Hastie D R 1996 Nitrous oxide and methane fluxes from perturbed and unperturbed boreal forest sites in northern Ontario. J. Geophys. Res., 101, 22767-22774.

Serca D, Delmas R, Jambert C and Labroue L 1994 Emissions of nitrogen oxides from equatorial rain forest in central Africa: Origin and regulation of NO emission from soils. Tellus 46B, 243-254.

Silver W L, Herman D J and Firestone M K 2001 Dissimilatory nitrate reduction to ammonium in upland tropical forest soils. Ecology 82, 2410-2416.

Skiba U, Fowler D and Smith K A 1994 Emissions of NO and N_2O from soils. Environ. Mon. Assess. 31, 153-158.

Sitaula B K, Bakken L R and Abrahamsen G 1995 N-fertilization and soil acidification effects on N_2O and CO_2 emission from temperate pine forest soil. Soil Biol. Biochem. 27, 1401-1408.

Steel S, Zarin D J, Capanu M, Littell R, Davidson E A, et al. 2004 Moisture and substrate availability constrain soil trace gas fluxes in an Eastern Amazonian regrowth forest. Global Biogeochemical Cycles 18, GB2009, doi: 10.1029/ 2003GB002210.

Steudler P A, Melillo J M, Bowden R D, Castro M S and Lugo A E 1991 The effect of natural and human disturbance on soil nitrogen dynamics and trace gas fluxes in a Puerto Rican wet forest. Biotropica 23, 356-363.

Teepe R, Brumme R and Beese F 2000 Nitrous oxide emissions from soil during freezing and thawing. Soil Biol. Biochem. 32, 1807-1810.

Tietema A, Bouten W and Wartenbergh P E 1991 Nitrous oxide dynamics in an oak-beech forest ecosystem in the Netherlands. For. Ecol. Manag. 44, 53-61.

Van Cleemput O and Baert L 1984 Nitrite: A key compound in N loss processes under acid conditions? Plant and Soil 76, 233-241.

Van Dijk S M, Duyzer J H 1999 Nitric oxide emissions from forest soils. J. Geophys. Res. 104, 15955-15961.

Van der Eerden L J M, deVisser P H B and vanDijk C J 1998 Risk of damage to crops in the direct neighbourhood of ammonia sources. Environm. Pollut. 102, 49-53.

Verchot L V, Davidso E A, Cattânnio J H, Ackermann I L, Erickson H E and Keller M 1999 Land use andbiogeochemical controls of nitrogen oxide emissions from soils in eastern Amazonia. Global Biogeochemical Cycles 13, 31-46.

Yang W X and Meixner F 1997 Laboratory studies on the release of nitric oxide from subtropical grassland soils: The effect of soil temperature and moisture, in Gaseous nitrogen emissions from grasslands, Jarvis S C and Pain B F (eds.), pp. 67-71, CAB International, Wallingford, Oxon, U.K.

Zechmeister-Boltenstern S, Hahn M, Meger S and Jandl R 2002 Nitrous oxide emissions and nitrate leaching in relation to microbial biomass dynamics in a beech forest soil. Soil Biol. Biochem. 34, 823-832.

Zumft W G 1997 Cell biology and molecular basis of denitrification. Microb. Molec. Biol. Rev. 61, 533-616.

Chapter 11

MICROBIAL PROCESSING OF HUMIC SUBSTANCES FROM MEADOW AND FOREST SOILS

Z. Filip[1] and M. Tesařová[2]
[1]*Marie Curie Chair, Dept. of Biochemistry and Microbiology, Institute of Chemical Technology, Technická 5, CZ-16628 Prague, Czech Republic, Phone +420 220 445 174, Zdenek.Filip@vscht.cz*
[2]*Department of Soil Science and Microbiology, Mendel University of Agriculture and Forestry, Zemědělská 1, 613 00 Brno, Czech Republic. Phone/Fax: +420-545133054; E-mail: tesarova@mendelu.cz*

ABSTRACT

Soil organic matter represents the main carbon reservoir of the biosphere. Recent concern for increasing atmospheric CO_2 level and global warming results in growing interest in the sequestration of organic C in soil organic matter. This is true especially for humic substances which represent differently extractable fractions of soil organic matter. Humic acids account for recalcitrant components of humic substances, but even these compounds may become subject to degradation and transformation by soil microorganisms. The contributions of various species of fungi, actinomycetes, and also non-mycelial aerobic or anaerobic bacteria to these processes have been studied mainly under laboratory conditions. The degradation of humic acids can be followed by optical, gravimetric and chemical methods, and also by using microbiological and biochemical procedures. We report on the value of a multi-factorial approach that includes both quantitative and qualitative parameters. This approach is illustrated by analysis of humic acids extracted from two soils under permanent meadow, and from a forest soil. When added as either supplemental source of nutrients or as the sole source of carbon or nitrogen, between 9 % and 63 % of humic acids were microbially utilized and structurally altered. The formation of microbial biomass was enhanced up to

D. Binkley and O. Menyailo (eds.),
Tree Species Effects on Soils: Implications for Global Change, 193–212.

4-fold in comparison to cultures without humic acids. Humic acids from soils under permanent meadow were more resistant to microbial degradation than those from a forest soil. A complex microbial assemblage from forest soil could mineralize up to 74% of C present in liquid cultures enriched with humic acids. Our studies combine with other experiments in the literature to demonstrate that humic acids play a major role in both the sequestration of soil C, and as substrates for the soil community.

INTRODUCTION

Soils form the largest reservoir of carbon (C) in terrestrial ecosystems; the pool of 1200 to 1500 x 10^9 Mg C is five times the mass of C in vegetation (Batjes, 1996; Breuer, 1999). The pools of C in forests are very dynamic; in New Zealand, about half the annual CO_2 emissions from energy and industrial uses are offset by new forests that have been converted from large areas of pasture (Parfit et al. 2003). A small fraction of the total organic matter found in soil comes from recently added plant detritus, but most of the soil organic matter is comprised of highly altered compounds known as humic substances. The bulk humic substances are commonly divided into humic acids (alkali-soluble but acid-insoluble), fulvic acids (soluble in both alkali and acids), and insoluble humins. In total, humic substances usually comprise between 60 to 80 % of the soil organic matter. Their main characteristic is a high molecular weight polymeric structure based mainly on different phenolic precursors originating from lignin degradation and cell pigments released from decaying plant and/or microbial cells (Haider et al., 1975). Humic substances are generally regarded as highly resistant to biological and chemical degradation (Brady and Weil, 2002). Schulten and Schnitzer (1997) presented a complicated structural model of humic substances, which could give a reason for a high degree of their stability. Nevertheless, even relatively stable sinks of matter and energy have finite rates of turnover, depending in part on environmental and biological conditions. For example, evidence and opinions differ regarding the decline in soil organic matter following disturbances such as clearcutting (e.g., Beyer et al., 2001; Johnson and Curtis, 2001; Rusco et al., 2001, Yanai et al., 2003).

In general, C storage in terrestrial ecosystems changes with shifts in the balance between primary production and the decomposing activity of heterotrophic microorganisms (Post et al. 1997). The stability and degradability of humic substances in soils are a fundamental part of understanding the role of soils in the global CO_2 budget. In this chapter, we review the state of knowledge on the dynamics of humic substances, and then summarize some experiments with additions of humic substances to microbial cultures.

METHODS FOR INVESTIGATING MICROBIAL PROCESSING OF HUMIC SUBSTANCES

Growth of microorganisms

Estimation of microbial growth on media containing humic substances represents a simple way of how to demonstrate microbial degradation of humic substances. Growth rates can be very slow unless other sources of C (such as glucose) are also provided (Murzakov et al., 1989; Kontchou and Blondeau, 1992).

In addition to serving as a substrate for microbial metabolism, humic substances may also affect the physiology of microbial cells. Martin et al. (1976) reported a growth stimulation in cultures of some actinomycetes in which nutritive effects were probably not involved as the cultures were supplied with readily available glucose. The authors discussed the possibility of (i) absorption of some inhibitory molecules by humic substances, (ii) release of stimulatory organic molecules, and (iii) the provision of trace elements from the humic substances which enhance the microbial growth. Humic substances appear to influence the metabolism of some microbes by acting as surfactants (Visser, 1985). According to Vallini et al. (1997) humic acids stimulated growth and activity of axenic cultures of soil nitrifying bacteria *Nitrosomonas europaea* and *Nitrobacter agilis* by increasing the permeability of cell membranes, enhancing nutrient utilization.

The potential importance of humic acids can be illustrated with the results of a 1-year incubation experiment that involved additions of humic acids extracted from fertilized and manured soils (Table 1, Filip and Kubat, 2001). In a full strength nutrient broth, microbial biomass was 5-fold greater when enriched with humic acids. The biomass was only 20% greater if humic acids served as the only source of C. The degradation of humic acids in these incubations was 23% when humic acids represented the only C source for microbes, but just 2.7% if other C sources were present.

Decolorization of media

Many authors use a loss in color of nutrient media containing humic substances, and caused by growth and metabolic activities of microorganisms as an evidence of humic substances degradation (e.g., Bhardway and Gaur, 1971; Khandewal and Gaur, 1980). A light absorption at 370 nm has been recommended for the respective measurements (Blondeau, 1989; Dari et al., 1995). Color intensity depends on pH, so adjustments in pH are crucial. Also, the adsorption of different humic substances fractions on the microbial biomass can result in a loss in color of microbial cultures, and thus need not be consistent with humic substances degradation (Behmel, 1988; Dari et al., 1995; Mishustin and Nikitin, 1961). Therefore, repeated washing or gentle extraction of microbial biomass from cultures containing humic substances is necessary before measuring optical density. In long-term incubated cultures the microbial production of dark colored pigments could become a source of additional problems (Gordienko and Kunz, 1984). These factors illustrate the limitations of using optical measurements for evaluation the degradation of humic substances.

Table 1. Carbon contents (mg C/L), yield of biomass (C), and percentage of utilization of humic acids extracted from Orthic Luvisol (OL), Orthic Luvisol with fertilizer (OL+F), and Distric Cambisol with fertilizer (DC+F) (after Filip and Kubat 2001).

Cultures[a]	Carbon distribution in		Biomass C (% of control)	Humic acid utilization (% of added)
	Fresh cultures	Biomass yield		
Controls				
FB, Inoculum OL	12000	847	100.0	--
FB, Inoculum OL+F	12000	1153	100.0	--
FB, Inoculum DC+F	12000	473	100.0	--
With humic acids from OL				
FB+humic acids	12506	2183	257.7	16.3
FB-C+humic acids	506	240	28.3	47.3
FB-N+humic acids	12506	697	82.3	35.3
With humic acids from OL+F				
FB+humic acids	12533	1827	158.5	18.0
FB-C+humic acids	533	250	21.7	41.3
FB-N+humic acids	12533	733	63.6	21.3
With humic acids from DC+F				
FB+humic acids	12500	2547	519.4	2.7
FB-C+humic acids	500	97	20.5	22.7
FB-N+humic acids	12500	603	127.5	20.0

[a]FB=full broth, FB-C=FB deficient in carbon (glucose), FB-N=FB deficient in nitrogen ($NaNO_3$)

Weight loss

Weight loss is a clear indication of the processing of a substrate, but the resolution of this approach may be limited by the low amounts of humic substances recovered from the microbial cultures. Humic acid degradation could be overestimated if some humic substances adsorb onto microbial biomass and are removed prior to weighing (Dehorter et al., 1992). Nevertheless, we commonly use this approach in combination with respective pretreatments (as suggested here). It basically allows one to obtain microbially processed humic substances for consecutive analyses (Filip and Alberts, 1994; Filip and Bielek, 2002; Filip and Kubat, 2001, 2003; Filip et al., 1999). In brief, after incubation the microbial cultures should be centrifuged at a high speed. The biomass pellets should be treated (such as with 0.05 M NaOH) to remove any adsorbed humic substances. After that the biomass should be washed several times with deionized water and centrifuged again. All supernatants are collected and filtered (with a glassfiber filter with a 1 µm pore size). The filter is also rinsed (with 0.02 M NaOH) and all supernatants composited and acidified to pH 1.5. Humic acids precipitate over night, and the precipitate can be separated by centrifugation, dialyzed against several changes of deionized water, freeze-dried and finally weighed before individual analyses. To recover sufficient amounts of humic acids for analysis, several parallel culture solutions may need to be composited.

CO_2 collection

Microbial utilization of humic substances and especially the mineralization rate obtained in the respective cultures can be estimated also by measuring the release of CO_2 (Monib et al., 1981; Steinbrenner and Mundstock, 1975). If humic substances are not the sole source of C in the culture, isotopic labeling of C is necessary for determining the contribution of humic substances (Haider and Martin, 1988). The interpretation of these experiments may be limited by the uneven isotopic labeling of the humic substances, as typically only the aliphatic side chains of the complex molecules are labeled. Their relatively easy microbial degradation could be misleadingly interpreted as a complete degradation of humic substances. A possibility exists to correlate CO_2 evolution with the concentration of different carbon isotopes (^{12}C, ^{13}C, ^{14}C), which could be attributed to specific fractions of soil organic matter. Using this method, Bol et al. (2003) found the rate of C mineralization in soil increased with temperature (10, 20, and 35 °C); more of the C from the relatively stable soil organic matter pool was mineralized at the highest temperature.

In general, precise quantitative estimations of the degradation of humic substances are difficult to obtain. Many authors (including ourselves) combine both quantitative and qualitative approaches to obtain a more complete picture of microbial processing of humic substances. Common microbiological procedures include elemental analysis, and spectroscopy using ultraviolet light (UV), visible light, Fourier Transform Infrared (FTIR), and ^{13}C-NMR (Dehorter et al., 1992; Filip and Kubat, 2001, 2003; Filip et al. 1999; Gordienko and Kunz, 1984; Rammuni et al., 1987; Solntseva, 1994).

DEGRADATION OF HUMIC SUBSTANCES BY COMPLEX POPULATIONS OF SOIL MICROORGANISMS AND SOME INDIVIDUAL HETEROTROPHIC BACTERIA

Mishustin and Nikitin (1961) published an important, basic study on microbial degradation of humic substances related to many soils in the former Soviet Union. They examined the loss of color in cultures containing humic acids, mineral salts and glucose, and inoculated cultures with aqueous soil suspensions. Optical density changed markedly within 1 to 2 months. When humic acids were the only source of C for microbes, the loss of color was low even after 12 to 14 months. From the microbial isolates obtained, bacteria of the genus *Pseudomonas* were most active in a humate decomposition. An enhanced activity of peroxidases was detected in the respective microbial cultures. Marthur and Paul (1967a) found *Bacillus* sp., *Pseudomonas* sp., *Agrobacterium* sp., and *Arthrobacter* sp. could utilize humic substances as the only C source. Later work by Bhardway and Gaur (1971) found a decolorization capacity towards humic substances in soil isolates of *Bacillus subtilis* and *Bacillus brevis* only when an easily utilized C source was present in the culture medium.

Tuyev and Emtsev (1984) investigated microbial processing of a highly purified humic acids preparation (0.7% ash) originating from a podzol soil. Under anaerobic conditions, *Clostridium sporogenum* reduced the N content of humic acids by 40% and that of carboxyl (COOH) groups by 63%. Apparently, the hydrolytical cleavage of humic acids aliphatic structures was accompanied by an increase in aromatic structures. Kontchou and Blondeau (1990) found the activity of heterotrophic bacteria to be similar under both anaerobic and aerobic conditions, though humic acids degradation was low in 100 days. *Bacillus* and *Pseudomonas* spp. were the dominant bacteria in both aerobic and anaerobic cultures. Solntseva (1994) reported a mixed heterotrophic microflora was unable to degrade humic substances in a liquid medium, but that *Bacillus subtilis, Bacillus megaterium, Flavobacterium rigense, Nocardia opaca* and *Pseudomonas alcaligenes* (all isolated from a podzol soil) were capable of utilizing humic acids as a sole C and N sources in shaken cultures incubated at 24 °C. The humic acids utilization included splitting of aliphatic structures, and a subsequent cleavage of aromatic structures. *B. subtilis* removed only aliphatic parts of humic acids molecules, while *N. opaca* utilized aromatic compounds. Coates et al. (2002) demonstrated that reduced humic substances could be reoxidized by anaerobic bacteria such as *Geobacter*, *Geothrix*, and *Wolinella*. The anaerobic bacteria could be isolated from aquatic and terrestrial environments (including forest soils), and were capable of utilizing humic substances as an energy source.

Hüttmann and Beyer (1996) found 97 heterotrophic types of soil microorganisms (bacteria and fungi) capable of utilizing humic acids as a sole C source. They assumed this microbial capacity would be important in the natural development of podzolized soils. In our short-term laboratory experiments (21 days, shaken cultures) and long-term experiments (1 year, static cultures), humic acids from various agricultural soils were partially degradable by mixed assemblages of microorganisms originating from the same soils. About 25% of the humic acids were utilized if humic acids served as a supplemental nutrient source, and up to 47 %, and 41 % of the humic acids were degraded if humic acids served as the sole source of C or N, respectively (Filip and Kubát, 2001, 2003). In other laboratory experiments, *Variovorax (Alcaligenes) paradoxus, Pseudomonas fluorescens,* and a yeast *Cryptococcus* sp. have been identified in a BIOLOG® system as the dominant microbial species capable of utilizing humic acids from soils with various contents of metals (Filip and Bielek, 2002).

DEGRADATION OF HUMIC SUBSTANCES BY FUNGI

The role of fungi in degrading humic substances was also reported in the 1960s. Hurst (1963) reported a decolorization of humic substances in cultures of *Polystrictus versicolor* accompanied by decreases in COOH groups in humic substances. *Penicillium frequentans* showed a capacity to degrade N contained in side chains of humic acids, and also within aromatic groups (Marthur and Paul, 1967a,b). Salicyl alcohol and salicyl aldehyde were detected in culture filtrates as the main degradation products. *Aspergillus fumigatus*, *Aspergillus niger* and *Aspergillus oryzae* were shown to decompose humic substances in experiments by Bhardwaj and Gaur (1971), based on the criterion of decolorization of sodium humate added in a liquid medium. Similarly, a decolorization was induced in a semi-solid agar medium by *A. fumigatus*, *Fusarium solani* and *Penicilium roseopurpureum* (Khandelwal and Gaur, 1980). Mishra and Srivastava (1986) reported a 20 to

30 % decomposition of humic acids from a forest soil by *Aspergillus awamori*, *Humicola insolense* and *Penicillium* sp. Experiments that rely on changes in color may be confounded by the ability of many soil fungi (including *Aspergillus* spp.) to form dark-colored polymers in the mycelium, especially in cultures sufficiently supplied with glucose (Filip et al., 1972, 1974, 1976; Haider et al. 1975). These substances resemble soil humic acids in many respects, and may confound experimental results.

Several authors investigated the decomposition of humic substances by lignin-degrading white-rot fungi. Haider and Martin (1988) reported a mineralization rate of 35% for ^{14}C labeled humic acids after only 18 days in cultures of *Phanerochaete chrysosporium*, and the process was accompanied by a loss in color in the medium. Only 5 to 8 days were needed for this fungus (and also *Trametes versicolor)* to mineralize 30% of soil humic acids (Dehorter et al., 1992); the total mineralization rate reached 60% based on NMR spectra of the splitting of aromatic structures in the humic acids preparations. The authors also observed the presence of some novel microbial products involved in humic acids recovered from fungal cultures. In our experiments, different lignolytic fungi demonstrated the capacity to transform the structure and nutritional utilization of aquatic humic acids, especially if some additional sources of C and N were present. A laccase-producing fungus, *Cladosporium cladosporoides*, appeared highly effective. The transformations of humic acids comprised some losses in C, gains in N and O, losses in aromacity, and increases in aliphatic structures such as amino acids and carbohydrates (Claus and Filip, 1998; Claus et al., 1997).

Grams et al. (1999) reported a range of reductions in absorbance at 340 nm (loss in color intensity) after 21 days for humic substances from a black calcareous forest mull in cultures of different wood-degrading basidiomycetes (57% reduction), terricolous basidiomycetes (28% reduction), soil-born microfungi (26% reduction) and ectomycorrhizal fungi (19% reduction). The more active degraders reduced the overall molecular weight of humic acids and fulvic acids. The degradation of hydroxylated aromatic molecules was suggested, and peroxidase and a β-glucosidase appeared to be important.

DEGRADATION OF HUMIC SUBSTANCES BY ACTINOMYCETES

Actinomycetes (filamentous bacteria) are both producers and degraders of complex organic substances, including humic substances (Küster, 1955, 1975; Kutzner 1968). Hayakawa and Nonomura (1987) isolated actinomycete species *Dactylosporangium*, *Micromonospora*, *Microtetraspora*, *Nocardia*, *Streptomyces*, *Streptosporangium*, and *Thermomonospora* from agar plates containing humic substances as the only C and N source. Khandelwal and Gaur (1980) reported a decrease in humic substances extracted from a manure by 44 to 48 %, and humic substances from soil to 26 % by after 40 days if the humic substances were added as supplemental nutrient sources to cultures of nine individual strains of *Streptomyces* spp. In this respect, *Streptomyces candidus* appeared the most effective species. Monib et al. (1981) estimated degradation of humic acids by 28 isolates of *Streptomyces*; during 120 days of incubation, humic acids decomposition followed a cyclic pattern of increases and decreases before 19 to 34% were finally decomposed. A high proportion of the evolved CO_2 (up to 66 %) was released within three days, and 16 isolates showed the highest activity even during the first 24 hours. The actinomycetes apparently grew at the expense of carbohydrates, amino acids and other easily decomposable structural units of humic acids. Steinbrenner and Mundstock (1975) followed the decomposition of humic substances over 200 days in a pure culture of *Nocardia erythropolis*, and found 10 % of humic acids and 14 % of fulvic acids, respectively, to be decomposed according to amounts of CO_2 released from the cultures.

Kontchou and Blondeau (1992) demonstrated biodegradation of humic acids from a forest soil by *Streptomyces viridosporus* growing in a medium containing glucose and mineral salts. The solution was 60% decolored in 48 hours, but about one quarter of the decolorization resulted from adsorption of the humic acids onto cells, rather than actual breakdown of the humic acids. Peroxidase activity was measured in culture filtrates of *S. viridosporus*, but this actinomycete could not use humic acids as the sole source of C. Dari et al. (1995) found fifteen strains of *Streptomyces* sp. isolated from soil could decolorize humic acids in the presence of glucose. Degradation properties studied in batch cultures showed that the catabolic activity towards humic acids (maximum 47 %) was stimulated by incubation with oxygen added and was cell surface-associated. Dehorter et al. (1992) reported a loss of about 30% humic acids in cultures of *S. viridosporus* and *Streptomyces* sp.; no substantial structural changes were found using ^{13}C NMR.

The literature clearly documents that many soil bacteria (including actinomycetes) and fungi can decompose humic substances. The diverse assemblages of microbes in soils may show high activity because of the complex enzymatic system they develop. Therefore, in the experiments we describe below we used diluted suspensions of complex microbial assemblages originating in the same soil samples from which the humic substances were extracted.

EXPERIMENTS ON MICROBIAL PROCESSING OF HUMIC SUBSTANCES FROM MEADOW AND FOREST SOILS

In these experiments, humic acids were obtained from: (i) an umbrice-planosol under permanent natural meadow (Meadow I) located in Bohemian-Moravian Uplands of the Czech Republic; (ii) the same soil under permanent meadow but over 6 years fertilized with 180 kg N, 30 kg P, 60 kg K ha^{-1} yr^{-1} (Meadow II); (iii) a haplic podzol under spruce forest in the Beskydy Mountains of the Czech Republic (Forest). For more detail, see Filip and Tesarova, 2004.

A Czapek-Dox nutrient broth was prepared, containing glucose and $NaNO_3$ as C and N sources, mineral nutrients, and buffered at pH 7. Media were prepared to be C-deficient by omitting glucose and N-deficient by omitting $NaNO_3$. Portions of broth were sterilized by autoclaving before addition of freeze-dried humic acids (1 g/L). The nutrient broth with or without humic acids added (controls) was inoculated with 1 ml soil suspension diluted 10^{-3} and originating from the respective soils. The average counts of colony forming units, CFU g^{-1} dry soil, were:

Meadow I: $290 x 10^6$ for total bacteria, $28 x 10^4$ for actinomycetes and $3 x 10^4$ for fungi;

Meadow II: $292 x 10^6$ for total bacteria, $15 x 10^4$ for actinomycetes and $4 x 10^4$ for fungi;

Forest: $93 x 10^6$ for total bacteria, $4 x 10^4$ for actinomycetes and $100 x 10^4$ for fungi.

Triplicate samples of microbial cultures were incubated in the dark for 12 months at room temperature. Individual replicates that did not differ optically from each other were combined to provide sufficient amounts of residual humic acids for analyses. The re-isolation of humic acids was done in the same way as described earlier in this chapter (see in *Loss in weight).* Various analyses were performed in triplicate, and standard deviations were always less than 5% of the means of the triplicates.

Microbial growth became visible a few weeks after inoculation in the culture flasks without humic acids as increased turbidity in the nutrient broth. Microbial biomass appeared diffusely distributed in cultures developing from Meadow I and II suspensions, whereas coarse particles could be observed in the cultures developing from the Forest inoculum. The highest yield of biomass (671 mg in the 100 ml culture vessel) and the maximum increase (412 %) relative to the corresponding control was obtained in a culture enriched with humic acids from Meadow I soil (Table 2). The biomass yield was very low when humic acids from Meadow I or II served as sole C-source, generating only 3.7% (Meadow I) and 4.2% (Meadow II) of the corresponding control values. The Forest humic acids provided more carbon and nitrogen for microbial biomass, but still only 26% or 71.8% of the mass in the comparable controls. Because also the percentage of carbon in the microbial biomass was remarkably low when humic acids served the only C-source, the majority of carbon from the Forest humic acids was undoubtedly mineralized.

The individual preparations differed in utilization by soil microorganisms (Table 3). The mean utilization was only 10.6 % for humic acids preparations from Meadow I and II soils when serving as a supplementary nutrient source, compared with 45 % for humic acids from Forest soil. All humic acids preparations were utilized in significantly larger quantities when serving as sole C- or N-sources. In this respect, the microbial utilization of humic acids from Forest soil was 2 to 3 times higher in comparison to humic acids from Meadow I and II soils.

Microbial processing of humic acids resulted in distinct changes in the elemental composition of the individual preparations (Table 4). The carbon contents were enhanced by 4% on average for Meadow I, 5.5% for Meadow II and 7.6% for Forest preparations. The humic acids preparations serving as sole N-sources were depleted of nitrogen, whereas those added as a supplement to the full-strength nutrient broth showed an increased nitrogen content. The C:N ratio reflected these alterations. The H:C and O:C ratios mostly declined the re-isolated humic acids. An H:C ratio of 2:1 indicated that the carbon atoms in humic acids are aliphatic in character, whereas a ratio of 1:1 indicates aromatic structures (Thurman, 1985). Similarly, Yonebayashi and Hattori (1988, 1989) attributed the lower H:C ratio to an enhanced condensation process in humic acids, and a progressive humification of soil organic matter. The increased carbon content and the decrease in H:C ratios in our humic acids samples indicated that aliphatic structural units were preferentially utilized by microbes, while more condensed aromatic structures remained resistant to biodegradation. The extinction curves in the visible range of light also indicated a slight increase in optical density of the Forest humic acids that served as either C- or N- sources in microbial cultures (Figure 1).

A "Novel humic acid" was formed in the full-strength broth inoculated with a diluted suspension of Forest soil, and it demonstrated an elemental composition similar in part to soil humic acids but an extremely low optical density (Table 4; Figure 1).

Table 2. Effect of humic acids on production of microbial biomass, composition of biomass, and conversion of C and N supplied in cultures into biomass after 12 month incubation.

Cultures	Biomass yield		Biomass C		C in full broth		Biomass N		N in full broth	
	mg/L	% of control	%	mg/L	mg/L	Conversion %	%	mg/L	mg/L	Conversion %
Controls										
FB, Inoculum OL	1630	100	46.1	751	12000	6.3	3.7	60	500	12.0
FB, Inoculum OL+F	1660	100	45.2	767	12000	6.4	3.4	56	500	11.2
FB, Inoculum DC+F	1420	100	47.5	674	12000	5.6	4.0	57	500	11.4
With humic acids from OL										
FB+humic acids	6710	412	41.2	2764	12480	22.1	2.2	148	541	27.4
FB-C+humic acids	60	4	21.6	13	480	2.7	5.9	73	541	0.6
FB-N+humic acids	670	41	47.3	317	12480	2.5	1.0	7	41	17.0
With humic acids from OL+F										
FB+humic acids	2460	148	41.4	1018	12480	8.2	2.4	59	542	10.9
FB-C+humic acids	70	4	22.6	16	480	3.3	6.2	4	542	0.7
FB-N+humic acids	480	29	47.2	227	12480	1.8	1.2	8	42	19.0
With humic acids from DC+F										
FB+humic acids	2580	182	37.2	960	12470	7.7	3.9	101	540	18.7
FB-C+humic acids	370	26	10.7	40	470	8.5	3.5	13	540	2.4
FB-N+humic acids	1020	72	35.9	366	12470	2.9	1.2	12	40	30.0

[a]FB=full broth, FB-C=FB deficient in carbon (glucose), FB-N=FB deficient in nitrogen ($NaNO_3$)

Table 3. Recovery of humic acids from microbial cultures and percentage of humic acids utilized after 12 month incubation (mg humic acids/L added).

Cultures[a]	Humic acids recovered (mg/L)	Humic acids utilized (%)
With humic acids from Meadow I		
FB, Inoculum OL	837	12.3
FB, Inoculum OL+F	737	26.3
FB, Inoculum DC+F	827	17.3
With humic acids from Meadow II		
FB+humic acids	910	9.0
FB-C+humic acids	703	29.7
FB-N+humic acids	780	22.0
With humic acids from Forest		
FB+humic acids	550	45.0
FB-C+humic acids	373	62.7
FB-N+humic acids	440	56.0

[a]FB=full broth, FB-C=FB deficient in carbon (glucose), FB-N=FB deficient in nitrogen ($NaNO_3$)

Table 4. Elemental composition (% ash free) and atomic ratios of original humic acids (humic acids) and humic acids reisolated from microbial cultures after 12 month incubation.

Cultures[a]	C	N	H	O	C:N	H:C	O:C
With humic acids from Meadow I							
Original	48.5	4.1	5.1	42.3	13.9	1.3	0.6
From FB	51.9	4.8	5.0	38.3	12.7	1.2	0.5
From FB-C	52.0	3.7	4.5	29.8	16.6	1.0	0.6
From FB-N	53.6	3.3	5.1	38.0	18.6	1.1	0.5
With humic acids from Meadow II							
Original	48.4	4.2	5.1	42.3	13.4	1.3	0.7
From FB	54.0	5.4	5.4	35.2	11.5	1.2	0.5
From FB-C	53.6	3.7	4.3	38.4	17.2	1.0	0.5
From FB-N	54.2	3.3	4.8	37.7	18.8	1.1	0.5
With humic acids from forest							
Original	46.8	4.0	5.6	43.6	13.4	1.4	0.7
From FB	52.6	5.1	5.6	36.7	12.2	1.3	0.5
From FB-C	55.9	4.3	5.2	34.8	15.0	1.1	0.5
From FB-N	54.6	3.2	5.2	37.0	19.8	1.1	0.5
Novel humic acids[b]	51.9	5.2	7.4	35.6	11.7	1.7	0.5

[a]FB=full broth, FB-C=FB deficient in carbon (glucose), FB-N=FB deficient in nitrogen ($NaNO_3$).
[b]Amorphous precipitate at pH 1.5.

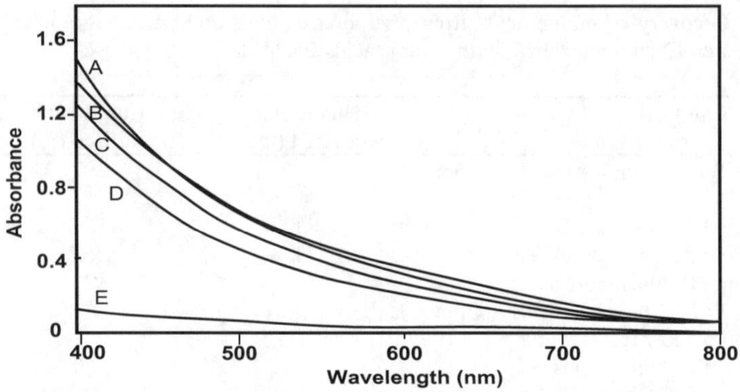

Figure 1. Visible spectra of humic acids from a Forest soil, and incubated (A) as a sole Source of carbon; (B) humic acids incubated as a sole source of nitrogen; (C) original humic acids; (D) humic acids incubated in a full-strength nutrient broth; (E) microbially produced "Novel humic acids" from a full-strength nutrient broth without humic acids added. (after Filip and Tesarova 2004).

The mineralization activity of the complex microbial population from Forest soil was much higher (up to 100 %) that that from Meadow soils I and II as shown in cultures without humic acids added (Table 5). The same capacity was found in cultures enriched with humic acids from Forest soil: The total carbon mineralization amounted 57% in cultures additionally supplied with humic acids, and 45% and 74% respectively, in cultures with humic acids from Forest as sole sources of either C or N.

In a similar comparison of pasture and forest soils, the total C mineralization (after 250 days) in mineral soil samples supplemented by litter additions was higher in samples originating from a pasture than forest (Parfit et al. 2003). The authors see this feature as consistent with a low lignin concentration in pasture litter in comparison to that from forest. Humification of plant residues, however, means biochemical transformations including cleavage of links in lignin structures by fungal enzymes, and resulting in a relative decrease of phenolic units resistant to degradation (Haider, 1999; Haider et al., 1975). Similar processes may occur preferentially in a forest soil while the humic substances structures under permanent meadow could become more stabilized by either chemical alterations or physical binding to inorganic soil structures. Well-designed and replicated experiments are needed to clarify generalizable features of grassland and forest soils.

FTIR spectroscopy offered insight into structural characteristics of humic acids preparations. We based our attribution of IR spectra on data published by Silverstein and Bassler (1964), Bellamy (1975) and Stevenson (1982). Typical FTIR spectra of a humic acids preparation are shown in Figure 2. In the original humic acids from Forest soil (Figure 2A), the following molecular groupings could be detected: H-bonded O–H groups (also partially N–H stretch) at 3423 cm^{-1}, C–H stretching vibrations in CH_3 and CH_2 groups of aliphatics at 2925-2850 cm^{-1}, C=O stretching in carboxyls at 1713 cm^{-1}, amide I and II stretch at 1640 and 1548 cm^{-1}, respectively, C–O stretching (in polysaccharides) at 1033 cm^{-1}, and C–O–C bonding in aromatics at 528 and 470 cm^{-1} (eventually O and/or OH bond in metallic-ion containing groups). The carboxyl- and amide-related bands disappeared in humic acids preparations exposed to microbial activity, and H-bonded C=O groups (usually attributed to quinines) became dominant at 1599 cm^{-1}. Similarly, OH-deformations and C=O stretch in phenolics were expressed strongly. The polysaccharide-related IR-band at 1033 cm^{-1} disappeared almost completely from the spectrum when humic acids had been sole C-source (Figure 2B). The same IR-band was strongly expressed when humic acids were the sole N-source in cultures supplied with an easily utilizable C-source (glucose). In

Table 5. Elemental composition (% ash free) and atomic ratios of original humic acids (humic acids) and humic acids reisolated from microbial cultures after 12 month incubation.

Cultures[a]	Carbon (mg/L culture) distributed in				Mineral-ization[b] (%C)
	Biomass (1)	Recovered humic acids (2)	Remaining broth (3)	Total (1+2+3)	
Controls					
FB, Inoculum Meadow I	751	--	6793	7544	37.1
FB, Inoculum Meadow II	767	--	7877	8644	28.0
FB, Inoculum Forest	674	67[c]	4233	4974	58.5
With humic acids from Meadow I					
From FB	2764	455	3410	6629	44.8
From FB-C	13	383	14	410	14.6
From FB-N	317	443	2563	3323	73.4
With humic acids from Meadow II					
From FB	1018	491	7877	9386	21.8
From FB-C	16	377	14	407	15.2
From FB-N	227	423	2847	3497	72.0
With humic acids from forest					
From FB	960	289	3903	5152	56.7
From FB-C	40	208	12	260	44.7
From FB-N	366	240	2637	3243	74.0

[a]FB=full broth, FB-C=FB deficient in carbon (glucose), FB-N=FB deficient in nitrogen ($NaNO_3$).[b]Calculated from the total C in culture (see Table 1). [c]Amorphous precipitate at pH 1.5.

the same humic acids, the 528 and 470 cm^{-1} bands remained also visible (Figure2C). The novel humic acids formed during the 12-month incubation in control cultures inoculated with a suspension of Forest soil strongly expressed IR bands of C–H and NH_2/NH groups in aliphatics (Figure 2D). The formation of some novel microbially produced humic acids-like substances was observed also by Ramunni et al. (1987) as a side effect of humic acids degradation after a 12-month aerobic incubation. Nevertheless, in our experiments we do not assume any strong quantitative coincidence from this feature, since we obtained significant losses in humic acids amounts in the individual humic acids enriched cultures.

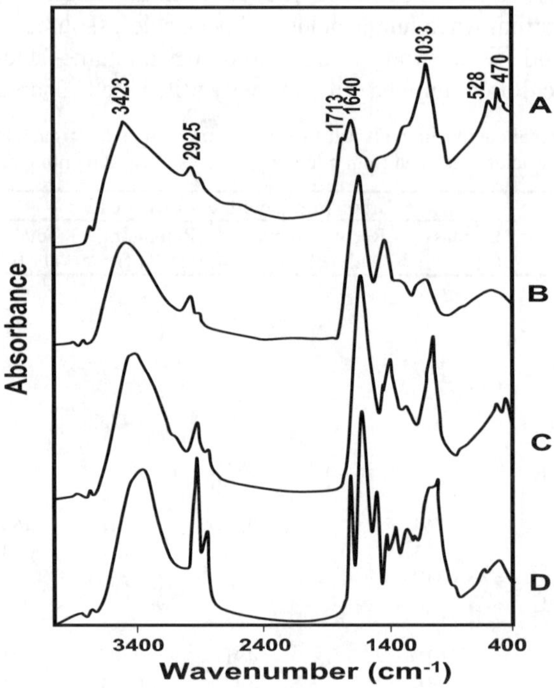

Figure 2. Fourier-transform infra-red (FTIR) spectra of humic acids from a Forest soil. (A) Original humic acids; (B) humic acids incubated as a sole source of carbon; (C) humic acids incubated as a sole source of nitrogen; (D) microbially produced "Novel humic acids" from a full strength nutrient broth without humic acids added (after Filip and Tesarova 2004).

CONCLUSIONS

Humic substances are the major pool of C sequestered in terrestrial ecosystems. Field observations have shown changes in both quantity and quality of soil organic matter as result of long-term climatic influences and/or human impact in many regions. Here we demonstrated that the losses in humic substances result from the activity of complex populations of soil microorganisms. Bacteria, actinomycetes and fungi can utilize humic substances as either supplemental or sole nutrient sources. Humic acids from a single forest soil were utilized more easily than those from a meadow soil (with or without a history of fertilization). These conclusions are based on results of 12-month incubation experiments in a laboratory, and their possible implication to environmental issues, such as CO_2 release from soil organic matter to the atmosphere, should be further investigated with a much broader representation of soils.

ACKNOWLEDGEMENTS

The senior author (Z F) gratefully acknowledges a Visiting Professorship at the Mendel University of Agriculture and Forestry in Brno (Czech Rep.) granted by courtesy of the Johann-Gottfried-Herder Foundation, Bonn (Germany). The co-author (M T) wishes to acknowledge a Grant No. 432100001/MSM from the Czech Ministry of Education and Youth, Prague.

REFERENCES

Batjes N H 1996 Total carbon and nitrogen in the soils of the world. Eur. J. Soil Sci. 47, 151-163.

Behmel P 1988 Sorption von Huminsäuren durch Bakterienzellwände. *In* Organische Inhaltsstoffe des Bodens. Wolfgang Ziechmann zum Geburtstag. Ed. U. Müller-Wegener. pp. 1-13. Göttingen.

Bellamy L J 1975 Infrared spectra of complex molecules. Chapman and Hall, London.

Beyer L, Kahle P, Kretschmer H and Wu Q 2001 Soil organic matter composition of man-impacted sites in North Germany. J. Soil Sci. Plant Nutr. 164, 359-364.

Brady N C and Weil R R 2002 The nature and properties of soils. Prentice Hall, Upper Saddle River, NJ.

Breuer G 1999 Die Zukunft der CO_2-Senken. Naturw. Rdsch. 52, 434436.

Bhardway K K R and Gaur A C 1971 Isolation and characterization of some humic acid decomposing bacteri and fungi from soil. Zbl. Bakt II 126, 307-312.

Blondeau R 1989 Biodegradation of natural and synthetic humic acids by the white rot fungus *Phanerochaete chrysosporium*. Appl. Environ. Microbiol. 55, 1282-1285.

Boll R Bolger T Cully R and Little D 2003 Recalcitrant soil organic materials mineralize more efficiently at higher temperatures. J. Plant Nutr. Soil Sci. 166, 300-307.

Claus H and Filip Z 1998 Degradation and transformation of aquatic humic substances by laccase-producing fungi *Cladosporium cladosporioides* and *Polyporus versicolor*. Acta Hydrochim. Hydrobiol. 26, 180-185.

Claus H Filip Z and Alberts J J 1997 Microbial utilization and transformation of riverine humic substances. *In* Proc. 8th Meeting. Int. Humic Substances Soc. pp. 561-566, Wroclaw.

Coates J D, Cole K A, Chakraborty R, O'Conor S M and Achenbach L A 2002 Diversity and ubiquity of bacteria capable of utilizing humic substances as electron donors for anaerobic respiration. Appl. Environ. Microbiol. 68, 2445-2452.

Dari K Bechet M and Blondeau R 1995 Isolation of soil *Streptomyces* strains capable of degrading humic acids and analysis of their peroxidase activity. FEMS Microbiol. Ecol. 16, 115-122.

Dehorter B Kontchou C Y and Blondeau R 1992 ^{13}C NMR spectroscopic analysis of soil humic acids recovered after incubation with some white rot fungi and actinomycetes. Soil Biol. Biochem 24, 667-673.

Filip Z and Alberts J J 1994 Microbial utilization resulting in early diagenesis of salt-marsh humic acids. Sci. Total Environ. 144, 121-135.

Filip Z and Bielek P 2002 Susceptibility of humic acids from soils with various contents of metals to microbial utilization and transformation. Biol. Fertil. Soils 36, 426-433.

Filip Z Haider K Beutelspacher H and Martin J P 1974 Comparison of IR-spectra from melanins of microscopic soil fungi, humic acids and model phenol polymers. Geoderma 11, 37-52.

Filip Z Haider K and Martin J P 1972 Influence of clay minerals on the formation of humic substances by *Epicoccum nigrum* and *Stachybotrys chartarum*. Soil Biol. Biochem 4, 147-154.

Filip Z and Kubat J 2001 Microbial utilization and transformation of humic substances extracted from soils of long-term field experiments. Eur. J. Soil Biol. 37, 167-174.

Filip Z and Kubat J 2003 Aerobic short-term microbial utilization and degradation of humic acids extracted from soils of long-term field experiments. Eur. J. Soil Biol. 39, 175-182.

Filip Z Pecher W and Berthelin J 1999 Microbial utilization and transformation of humic acids extracted from different soils. J. Plant Nutr. Soil Sci. 162, 215-222.

Filip Z Semotan and Kutilek M 1976 Thermal and spectrophotomertic analysis of some fungal melanins and soil humic compounds. Geoderma 15, 131-142.

Filip Z and Tesarova M 2004 Microbial degradation and transformation of humic acids from permanent meadow and forest soils. Int. Biodeter. Biodegr. 54, 225-231.

Gordienko S A and Kunz F 1984 Role of microorganisms in humic acid transformation. Ecology 2, 90-95.

Grams G, Diegenhagen D and Sorge S 1999 Degradation of soil humic extracts by wood- and soil-associated fungi, bacteria and commercial enzymes. Microb. Ecol. 37, 140-151.

Haider K 1999 Von der toten organischen Substanz zum Humus. Z. Pflanzenernähr. Bodenkd. 162, 363-371.

Haider K M and Martin J P 1988 Mineralization of ^{14}C labeled humic acids and of humic acid bound ^{14}C-xenobiotics by *Phanerochaete chrysosporium*. Soil Biol. Biochem. 20, 425-429.

Haider K Martin J P and Filip Z 1975 Humus biochemistry. *In* Soil biochemistry. Eds. E A Paul and A D McLaren. pp. 195-244. Dekker, New York.

Hayakawa M and Nonomura H 1987 Humic acid-vitamine agar, a new medium for the selective isolation of soil actinomycetes. J. Ferment. Technol. 65, 501-509.

Hurst H M 1963 Aromatic acid-reducing system in fungi. *In* Enzyme chemistry of phenolic compounds. pp. 121-128. Pergamon, New York.

Hüttmann S and Beyer L 1996 Mikrobielle Verwertung verschiedener Huminstoff-Fraktionen eines Podsols in Schleswick-Holstein. Mitteilgn. Dtsch. Bodenkundl. Gesellsch. 81, 197-200.

Johnson DW, Curtis PS. 2001 Effects of forest management on soil C and N storage: meta analysis. For. Ecol. Manag. 140:227–38.

Khandewal K C and Gaur A C 1980 Degradation of humic acids extracted from manure and soil by some streptomycetes and fungi. Zbl. Bakt. II 135, 119-122.

Konchou C Y and Blondeau R 1990 Effect of heterotrophic bacteria on different humic substances in mixed batch cultures. Can. J. Soil Sci. 70, 51-59.

Kontchou C Y and Blondeau R 1992 Biodegradation of soil humic acids by *Streptomyces viridosporus*. Can. J. Microbiol. 38, 203-208.

Küster E 1955 Humusbildung und Phenoloxidasen bei Streptomyceten. Z. Pflanzenernähr. Bodenkd. 69, 137-142.

Küster E 1979 Bedeutung der Aktinomyceten für den Abbau von Cellulose, Lignin und Huminstoffen im Boden. Z. Pflanzenernähr. Bodenkd. 142, 365-374.

Kutzner H J 1968 Über die Bildung von Huminstoffen durch Streptomyceten. Landwirtsch. Forsch. 21, 48-61.

Martin J P, Filip Z and Haider K 1976 Effects of montmorillonite and humate on growth and metabolic activity of some actinomycetes. Soil Biol. Biochem 8, 409-413.

Marthur S P and Paul E A 1967a Microbial utilization of soil humic acids. Can. J. Microbiol. 13, 573-580.

Marthur S P and Paul E A 1967b Partial characterization of soil humic acids through biodegradation. Can. J. Microbiol. 13, 581-586.

Mishra B and Srivastava L L 1986 Degradation of humic acids of a forest soil by some fungal isolates. Plant Soil 96, 413-416.

Mishustin E N and Nikitin D I 1961 Susceptibility of humic acids to the soil microflora. Microbiology 30, 687-694.

Monib M, Hosny I, Zohdy L and Khalafallah M 1981 Studies on humic acid decomposing Streptomycetes. Zbl. Bakt II 36, 15-25.

Murzakov B G, Semenov A M, Evdokimova M D, Ykashina V M and Dorofeeva I K. 1989 Decomposition of soil humus compounds by microorganisms. Microbiology 58, 88-93.

Murzakov, B G, Semenov A M, Gorchakova A V and Ermekbaeva G A 1990 Effect of fractions of humic compounds on soil microflora. Microbiology 59, 63-67.

Parfit R L, Scott N A, Ross D J, Salt G J and Tate K R 2003 Land use change effects on soil C and N transformations in soil of high N status: comparison under indigenous forest, pasture and pine plantation. Biogeochemistry 66, 203-221.

Post W M, King A W and Wullschleger S D 1997 Historical variations in terrestrial biospheric carbon. Global Biogeochem. Cy. 11, 99-109.

Ramunni A, Scialdone R and Felleca D 1987 Decomposition of humic acid by incubation in a soil water extract under various conditions oxygen availability. Plant Soil 97, 15-23.

Rusco E, Jones R and Bidoglio G 2001 Organic matter in the soils of Europe: Present status and future trends. European Soil Bureau, 14 pp. JRC/IES Ispra.

Schulten H R and Schnitzer M 1997 Chemical model structures for soil organic matter and soils. Soil Sci. 162, 115-130.

Silverstein R and Bassler G C 1964 Spectrometric identificatin of organic compounds. Wiley, New York.

Solntseva I E 1994 Certain energy aspects of microorganism-induced transformation of humic substances. *In* Humic substances in the global environment and implications on human health. Eds. N. Senesi and T.M. Miano. pp. 373-379. Elsevier, Amsterdam.

Steinbrenner K and Mundstock I 1975 Untersuchungen zum Huminstoffabbau durch Nokardien. Arch. Acker- u. Pflanzenbau u. Bodenk. 19, 243-255.

Stevenson F J 1982 Humus chemistry. Wiley, New York.

Thurman E M 1985 Humic substances in groundwater. *In* Humic substances in soil, sediment and water. Eds. G R Aiken, D M Mc Knight, R L Wershaw and P MacCarthy. pp. 87-103. Wiley, New York.

Tuyev N A and Emtsev V T 1984 Changes in the chemical composition of humus compounds during their decomposition by *Clostridium* microorganisms. *In* Soil biology and conservation of the biosphere. Ed. J Szegi. pp. 333-338. Budapest.

Vallini G, Pera A, Agnolucci M and Valdrighi M M 1997 Humic acids stimulate growth and activity of in vitro tested axenic cultures of soil autotrophic nitrifying bacteria. Biol. Fertil. Soils 24, 243-248.

Visser S A 1985 Physiological action of humic substances on microbial cells. Soil Biol. Biochem. 17, 457-462.

Yanai RD, Currie WS, and Goodale CL. 2003. Soil carbon dynamics after forest harvest: an ecosystem paradigm reconsidered. Ecosystems 6:197-2003.

Yonebayashi K and Hattori T 1988 Chemical and biological sturies on environmental humic acids. I. Composition of elemental and functional groups of humic acids. Soil Sci. Plant Nutr. 34, 524-571.

Yonebayashi K and Hattori T. 1989 Chemical and biological studies on environmental humic acids. II. ^{1}H-NMR and IR spectra of humic acids. Soil Sci. Plant Nutr. 35 383-392.

Chapter 12

PHOSPHORUS COMPOUNDS UNDER DIFFERENT PLANTS IN AN ARTIFICIAL SOIL FORMATION EXPERIMENT

M.I. Makarov[1], T.I. Malysheva
Department of Soil Science, Moscow State University, 119992 Moscow, Russia
[1]Corresponding author. Tel: +7 095 9393774; Fax: +7 095 9391716.E-mail: mmakarov@soil.msu.ru

ABSTRACT

The phosphorus (P) status of noncalcareous loam soils developed in large lysimeters was evaluated after 30 years of (i) crop rotation, (ii) permanent grasses, and (iii) several types of forest vegetation. The soils of forest treatments (spruce, oak/maple and mixed) were characterized by the higher concentration of organic C and N, while the accumulation of organic P (P_o) was not pronounced because of low P concentration in the soil organic matter. The C to P ratio in the primitive soils under trees was high and varied insignificantly for different forest types. The fractionation of P in the primitive soils showed that the content of inorganic and organic P compounds did not point to an increase of P_o mineralization in soils under forests (including conifers). Inorganic fractions extractable with $NaHCO_3$ and NaOH were lower in the Ah horizon of spruce and mixed forest soils compared with grassland soil, while extractable organic fractions were higher in spruce and oak/maple forest soils. ^{31}P NMR spectroscopy indicated that the large concentration of inorganic pyrophosphate in the direct alkaline extract from the spruce and mixed forest soils may be due to the increase of P compounds produced by fungi in these soils. Another specific characteristic of P species extracted from the spruce and mixed forest soils was lower proportion and concentration of relatively unstable phosphate diesters (phospholipids and DNA) that can indicate higher P_o mineralization, although it was not revealed

D. Binkley and O. Menyailo (eds.),
Tree Species Effects on Soils: Implications for Global Change, 213–228.

by P fractionation procedures. At the same time, ^{31}P NMR spectroscopy indicated that at the initial stage of soil formation the effect of vegetation type on the composition of soil organic phosphates was not significant. However, the organic P compounds in these primitive soils differed from those found in the majority of developed, mature soils. The organic P compounds in the lysimeter soils were not resistant to alkaline hydrolysis, and contained high proportion of relatively labile diester phosphates that indicate the possible important role of organic phosphates in plant P nutrition at the initial stage of soil formation.

INTRODUCTION

Changes in global climate and carbon (C) cycles could change forest ecosystems and their role as C sinks. Soil nutrient cycles may play a critical role in these ecosystem responses, and phosphorus (P) is one of the most important limiting nutrients in ecosystems. Many ecosystems are first nitrogen limited, however the tendency towards P limitation can be increased by nitrogen deposition from the atmosphere, while only soil resources limit P availability.

Soil P occurs in a variety of chemical forms, including inorganic orthophosphate and condensed phosphates, as well as organic monoesters, diesters, phosphonates and condensed phosphates. Different forms of soil phosphates are associated with biological and geochemical processes, and the relative proportions of P compounds are controlled by the combined effect of these processes; it is difficult to separate the contribution of each of these interacting factors (Magid et al., 1996). However, the problem is simplified by analyzing soils that have formed under similar climatic conditions, but with marked difference in one of the soil-forming factors (parent material, vegetation, relief or time). The study of vegetation's influence on soil nutrients can hardly be performed for natural ecosystems, because changes of natural plant communities depend on a variety of environmental parameters. To solve this problem, some authors have compared soils under artificially altered vegetation, in so-called common garden experiments. Many studies have examined the effects of clear-cutting, deforestation, and afforestation on soil P status (Sharpley and Smith, 1985; Condron et al., 1996; Williams, 1996; Saikh et al., 1998; Ross et al., 1999; Chen et al., 2000; Turrion et al., 2000).

The combined results from these studies provide insight on the transformation of soil P resulting from changed vegetation, in situations where long-term pedogenesis has already developed a soil system under relatively stable environmental conditions. Information is lacking on the initial dynamics of P in soil formation, and on the influence of vegetation on this early development.

The evolution of the P status in the course of pedogenesis is illustrated by a scheme, compiled on the basis of numerous studies of soil chronosequences developed from noncalcareous parent materials in New Zealand (Williams and Walker, 1969; Adams and Walker, 1975; Walker and Syers, 1976). According to this scheme, the availability of inorganic phosphates is low in the initial stage of soil formation, since P remains unweathered from primary minerals, and the role of organic phosphates in P nutrition is more important.

Soil fractions of inorganic and organic P compounds have variable biological availability depending their behaviour in the soil and therefore at the different degrees involved in nutrient cycling. Two main alternative approaches exist to characterize the availability of soil phosphates. One is based on phosphorus fractionation using extractants of varying strength and composition (most popular was proposed by Hedley et al. 1982). Another involves characterization of the chemical nature of soil phosphates, including the ^{31}P NMR spectroscopy studies proposed by Newman and Tate (1980).

The large lysimeter experiment through the Soil Science Department of Moscow State University provides a unique opportunity for observation of the soil formation on a noncalcareous loam under different vegetation. Our objective was to estimate inorganic and organic P compound transformation at the initial stage of soil formation (about 30 years) under model vegetation cover simulating different types of south-taiga forest compared with permanent grasses and crop rotation using P fractionation and ^{31}P NMR spectroscopy.

MATERIALS AND METHODS

Phosphorus compounds were studied in the surface horizons of primitive soils formed in the upper part of the artificial loam column under several types of vegetation in the long-term lysimeter experiment. The experiment was started in 1967, when 18 m^3 of noncalcareous loam was placed into the 9 m^2 pots to a depth of 2 m. Major forest types found in the South-taiga were simulated in this experiment, with two replicate lysimeters (pots) for: 1) Norway spruce (*Picea abies*), 2) deciduous forest with oak (*Quercus robur*) and maple (*Acer platanoides*), and 3) mixed forest with spruce, oak and maple. Three non-tree treatments were: 1) control fallow, 2) permanent grasses (mostly timothy (*Phleum pratense*), foxtail (*Alopecurus pratensis*), brome (*Bromus inermis*) and clovers (*Trifolium pratense* and *T. repens*)), and 3) crop rotation with cereals and tilled crops and perennial grasses.

The loam material in all lysimeters was fertilized once in 1967 before planting trees and grasses, applying 6 g N m^{-2} as NH_4NO_3 and 3.5 g P m^{-2} as superphosphate. No more fertilizers were added to the soils of control, meadow and forest treatments for the whole time. The crop-rotation treatment received a total of 131 g N m^{-2} and 89 g P m^{-2} over 31 years (1967–1997). The soil samples were collected in 1998. Soil horizons were not recognizable in the loam of control and crop rotation treatments, but grasses and trees favored a more intense transformation of the loam (primitive soil profiles have been formed). Gray-colored Ah horizons (5 cm thick) were apparent under permanent grasses, and O and Ah horizons had developed under all trees. The thickness of these horizons was 0–1 cm (O) and 1–5 cm (Ah) under oak/maple, and 0–3 cm (O) and 3–6 cm (Ah) under conifers and the mixed-species treatments. Gray tongues of organic substance along root channels were noted up to 10 cm depth. Therefore, the surface loam layer of 0–5 cm was investigated in the control and crop rotation treatments, while Ah and AC horizons were characterized for the grass and forest treatments. Soil samples were collected in two replicates from each individual lysimeter (= 4 replicates for each treatment). The mixed sample of the initial loam was analyzed in two replicates.

Total organic C and N were determined by dry combustion on an Elementar Vario EL elemental analyzer. Total P (P_t) was determined after digestion in a mixture of concentrated H_2SO_4 and $HClO_4$ at 250 °C (Guggenberger et al., 1996a), and P_o by the method of Saunders and Williams (1955). Inorganic P (P_i) was calculated by differences between P_t and P_o.

Soil P was fractionated using a slight modification of the Tiessen and Moir (1993) procedure. Soil samples (0.5 g) were sequentially extracted by 0.5 *M* $NaHCO_3$ (pH 8.7), 0.1 *M* NaOH and 1.0 *M* HCl at a soil-to-solution ratio of 1:60 and shaken for 16 h. In the first two extracts, P_t was determined after digestion of the solution in a mixture of concentrated H_2SO_4 and $HClO_4$, and P_i was measured after the solution lightening by acidification to pH 1.5 and centrifugation. P_o was calculated by the difference. In the HCl extract only P_i was determined directly. Unextractable (insoluble) P_o and P_i were calculated by differences between total amounts of these forms in the soil and the sum of the extractable fractions of P_o and P_i, respectively. The P contents were determined by the ammonium molybdate-ascorbic acid method (John, 1970).

For ^{31}P NMR spectroscopy, P compounds were sequentially extracted with 0.1 *M* NaOH, 0.1 *M* HCl and 0.1 *M* NaOH at a soil-to-solution ratio of 1:10. The samples were shaken with NaOH for 1 h and left to stand overnight, while samples were extracted with HCl for 30 min. Then all extracts were centrifuged for 30 min at 2,560 *g*. The acidic supernatant was thrown out, while the alkaline were dialyzed against deionized water using SERVAPOR® dialysis tubing with a molecular weight cut off of 12,000–14,000 and freeze dried. Dry material was dissolved in 0.5 *M* NaOH to create P concentrations of 0.5–1.0 mg ml^{-1}. One ml of D_2O was added to 2 ml each NaOH solution, and the samples were transferred to 10-mm NMR tubes. ^{31}P NMR spectra were obtained on a Bruker DRX 500 NMR spectrometer under conditions described elsewhere (Makarov et al., 2002a). Interpretation of spectra was based on literature assignments (Newman and Tate, 1980; Condron et al., 1990; Makarov et al., 2002b; Turner et al., 2003).

Mean values were compared by one-way ANOVA. All differences with

$P < 0.05$ are reported as significant. Analyses were performed using STATISTICA 5.0 for Windows (Statsoft, 1995).

RESULTS

Total organic matter and associated elements

The initial loam contained 0.16% C, 0.02% N and 416 mg kg^{-1} P. Organic compounds represented only 7% of P_t. The C:N and C:P were 8.0 and 59, respectively (Table 1). In the control (unvegetated) treatment, soil C and N were three times the initial concentrations (0.52% C, 0.06% N), whereas P had changed little (436 mg kg^{-1} P, 15% of P_t occurring in the organic form). The unvegetated controls had a C:N of 8.6, and C:P of 79.

In all experimental treatments with plants, the accumulation of organic matter and related elements (C, N and P) was pronounced. The lowest increase of the C and N concentrations was typical of the crop rotation, reaching 0.86% C and 0.09% N. However, P_t content in these fertilized plots rose to 673 mg kg^{-1}, which was higher than in the treatments of permanent grasses, spruce and mixed forests. Insignificant contribution of biological P accumulation is evident by low contents of the P_o, both concentration (69 mg kg^{-1} – at the control level) and proportion (10% – below the control level), while the P_i have risen by the factor of 1.6. The C:N and C:P constituted 9.6 and 124, respectively.

The accumulation of C, N, and P_o was higher in the soils of permanent grasses and all forest treatments than in the crop rotation treatment. The concentrations of C in the Ah horizon were 2.28, 4.07, 4.54 and 7.70% in the soil of permanent grasses, mixed, spruce and oak/maple forest, respectively. The corresponding N concentrations were 0.19, 0.19, 0.21 and 0.36%. The content of P_t in the Ah horizon under spruce and mixed forests remained at the control level (432–458 mg kg^{-1}), while it rose to 518 mg kg^{-1} under permanent grasses and reached the maximum (830 mg kg^{-1}) under oak/maple trees. The amount of P_o increased in the soil of permanent grasses, spruce and mixed forests to 164–178 mg kg^{-1} and to 337 mg kg^{-1} in the soil of oak/maple forest. The proportion of P_o constituted 33–41% of the P_t. The concentration of P_i decreased in the Ah horizon of all grass and forest soils (with exception of oak/maple forest) due to its biological transformation into P_o. Oak/maple trees have favored the highest transfer of P into the Ah horizon from the underlying layers. As a result, the concentration of P_o increased to maximal value, while the concentration of P_i also increased considerably.

The C:N and C:P ratios were higher in the Ah horizon of grass and forest soils compared with the control and crop rotation treatments. The organic matter accumulated under permanent grasses appeared to be more saturated with N and P than that in the soils under all forest treatments. In the Ah

Table 1. Some properties of the primitive soils (standard deviation in parentheses; different letters within column indicate significant differences $P < 0.05$).

Treat-ment	Depth (cm)	pH (H_2O)	C	N	P_t	P_o	P_o	C:N	C:P_o
			----(g kg^{-1})----		---(mg kg^{-1})----		% P_t		
Control	0–5	6.78^{a} (0.09)	5.2^{a} (1.2)	0.6^{a} (0.1)	436^{a} (20)	65^{a} (10)	15^{a} (2)	8.6^{a} (0.7)	79^{af} (7)
Crop rotation	0–5	6.29^{b} (0.15)	8.6^{b} (1.2)	0.9^{b} (0.1)	673^{b} (43)	69^{ad} (11)	10^{b} (1)	9.6^{a} (0.6)	124^{b} (5)
Perma-nent grasses	0–5	6.15^{b} (0.07)	22.8^{c} (3.3)	1.9^{c} (0.3)	518^{c} (30)	169^{b} (16)	33^{c} (1)	11.9^{b} (1.1)	135^{b} (7)
	5–10	5.88^{c} (0.11)	7.1^{ab} (1.0)	0.8^{be} (0.1)	425^{a} (29)	92^{c} (10)	22^{d} (1)	8.6^{a} (0.4)	77^{a} (4)
Spruce forest	3–6	5.98^{bc} (0.14)	45.4^{d} (4.6)	2.1^{c} (0.3)	432^{a} (40)	178^{b} (23)	41^{e} (2)	22.0^{c} (1.7)	256^{c} (11)
	6–10	5.64^{d} (0.11)	14.2^{e} (3.0)	0.9^{be} (0.2)	405^{a} (20)	99^{cd} (21)	24^{d} (4)	15.3^{d} (1.5)	144^{d} (2)
Oak/maple forest	1–5	6.30^{b} (0.17)	77.0^{f} (6.4)	3.6^{d} (0.5)	830^{d} (50)	337^{e} (39)	40^{e} (3)	21.8^{c} (1.3)	229^{e} (13)
	5–10	5.63^{d} (0.05)	8.3^{b} (1.9)	0.6^{a} (0.1)	431^{a} (39)	87^{ac} (13)	20^{d} (2)	14.4^{de} (1.3)	95^{f} (11)
Mixed forest	3–6	6.24^{b} (0.11)	40.7^{d} (3.3)	1.9^{c} (0.3)	458^{a} (43)	164^{b} (25)	36^{ce} (5)	21.6^{c} (1.8)	251^{ce} (28)
	6–10	5.53^{d} (0.09)	9.0^{b} (1.6)	0.7^{ae} (0.1)	392^{a} (39)	93^{cd} (17)	24^{d} (2)	12.5^{be} (1.2)	97^{f} (10)
Initial loam		5.45	1.6	0.2	416	27	7	8.0	59

horizon under permanent grasses, the C:N was 11.9, and C:P was 135. These ratios in the soils of forest treatments ranged from 21.6 to 22.0 (C:N), and from 229 to 256 (C:P). The different tree species exerted no pronounced effect on the amount of N and P in soil organic matter.

The concentrations of C and N in the underlying AC horizon (from 5 or 6 cm to 10 cm) of the grass and forested soils were close to the Ah horizon of the control and crop rotation treatments. The higher C concentration reached 1.42% in the AC horizon under spruce, which may attest to a higher mobility of organic matter formed under conifers. The content of P_t in the AC horizon did not exceed the control, whereas the concentration (87–99 mg kg^{-1}) and proportion (20–24%) of P_o were higher in all meadow and forest soils. The concentration of P_i decreased not only in the Ah horizon but also in the underlying AC horizon as compared to the control. The composition of organic matter being formed in the AC horizon differed among the tree species treatments; the depletion of soil organic matter N (C:N 12.5 to 15.3) and P (C:P 95 to 144) being greatest under spruce.

Inorganic and organic phosphorus fractions

The sequential extraction of the initial and control (unvegetated) loams with $NaHCO_3$, NaOH and HCl dissolved 57–58% of P_t. The extractability of P_i and P_o was 59–61% and 33–38%, respectively (Tables 2 and 3). The maximum P_i was extracted with HCl, and the concentration of P_i in the $NaHCO_3$ and NaOH extracts was 5 and 9–16 times higher than of P_o in the control and initial loam, respectively.

The extractability of the inorganic (71%) and organic (51%) P compounds increased in crop rotation treatment which had received nearly 90 g fertilizer P/m^2. The concentrations of all extractable P fractions increased to various extents. Among inorganic forms, the concentration of the most available P (extracted with $NaHCO_3$) increased by a factor of 2.5, while the content of the less soluble Al- and Fe-phosphates (extracted by NaOH), as well as Ca-phosphates (extracted by HCl), increased by factors of 1.6 to 1.9. In contrast, the content of the less labile fraction of organic phosphates extracted with NaOH increased to a greater extent.

Table 2. Inorganic phosphorus fractions of the primitive soils (standard deviation in parentheses, different letters within columns indicate differences $P < 0.05$).

Treat-ment	Depth	---------------P extractable with--------------						Nonextractable P	
		--$NaHCO_3$--		---NaOH---		----HCl----			
	(cm)	mg kg^{-1}	% P_i	mg kg^{-1}	% P_i	mg kg^{-1}	% P_i	mg kg^{-1}	% P_i
Control	0–5	58^{a} (12)	16^{a} (3)	67^{ad} (9)	18^{ab} (2)	102^{a} (12)	27^{a} (1)	144^{ab} (18)	39^{a} (1)
Crop rotation	0–5	144^{b} (18)	24^{b} (2)	125^{b} (14)	21^{a} (2)	161^{b} (19)	27^{a} (2)	174^{a} (16)	29^{b} (4)
Perma-nent grasses	0–5	52^{a} (8)	15^{ad} (2)	65^{ad} (13)	19^{ab} (4)	105^{a} (13)	30^{ab} (3)	127^{bc} (19)	36^{abc} (4)
	5–10	48^{ad} (8)	14^{ac} (3)	65^{ad} (11)	20^{ab} (4)	97^{a} (16)	29^{ab} (3)	123^{bc} (19)	37^{acd} (4)
Spruce forest	3–6	26^{c} (5)	10^{c} (2)	40^{c} (8)	16^{b} (3)	91^{a} (17)	36^{bc} (7)	97^{c} (13)	38^{ad} (3)
	6–10	37^{cd} (8)	12^{cd} (2)	51^{ac} (9)	17^{b} (2)	92^{a} (13)	30^{ab} (4)	126^{b} (10)	41^{ad} (2)
Oak/ma-ple forest	1–5	117^{b} (14)	24^{b} (1)	81^{d} (9)	16^{b} (2)	187b (26)	38^{c} (3)	108^{bc} (17)	22^{b} (4)
	5–10	59^{a} (11)	17^{a} (3)	72^{d} (12)	21^{ab} (3)	96^{a} (16)	28^{ab} (4)	117^{bc} (14)	34^{c} (1)
Mixed forest	3–6	29^{c} (10)	10^{c} (3)	38^{c} (8)	13^{b} (2)	97^{a} (14)	33^{bc} (4)	130^{b} (14)	44^{d} (4)
	6–10	31^{c} (6)	10^{c} (2)	51^{ac} (9)	17^{ab} (3)	93^{a} (10)	31^{b} (2)	124^{b} (12)	41^{ad} (3)
Initial loam		45	12	64	16	121	31	159	41

Table 3. Organic phosphorus fractions of primitive soils (standard deviation in parentheses; different letters within column indicate differences at $P<0.05$).

Treatment	Depth (cm)	P extractable with $NaHCO_3$		P extractable with NaOH		Nonextractable P	
		mg kg^{-1}	% P_o	mg kg^{-1}	% P_o	mg kg^{-1}	% P_o
Control	0–5	12^{a}	19^{a}	13^{af}	20^{a}	40^{a}	62^{ae}
		(4)	(4)	(4)	(5)	(7)	(9)
Crop	0–5	14^{a}	20^{ab}	21^{b}	31^{bd}	34^{a}	49^{ab}
rotation		(2)	(4)	(4)	(4)	(7)	(6)
Permanent	0–5	36^{bd}	21^{ab}	58^{c}	34^{bd}	75^{b}	44^{b}
grasses		(8)	(5)	(11)	(7)	(16)	(8)
	5–10	15^{a}	16^{ac}	16^{ab}	17^{ae}	61^{bc}	67^{ae}
		(3)	(3)	(5)	(6)	(8)	(6)
Spruce	3–6	49^{bc}	27^{b}	83^{d}	47^{cd}	46^{ac}	26^{c}
forest		(6)	(4)	(7)	(4)	(8)	(3)
	6–10	14^{a}	14^{ac}	27^{b}	27^{abf}	58^{bc}	59^{ae}
		(3)	(4)	(7)	(5)	(8)	(5)
Oak/maple	1–5	59^{c}	17^{a}	133^{e}	39^{d}	145^{d}	43^{b}
forest		(10)	(3)	(25)	(6)	(15)	(6)
	5–10	10^{a}	12^{c}	10^{a}	12^{e}	67^{b}	77^{d}
		(3)	(2)	(2)	(2)	(9)	(4)
Mixed	3–6	31^{d}	19^{a}	57^{c}	35^{b}	76^{b}	47^{b}
forest		(7)	(3)	(10)	(4)	(11)	(4)
	6–10	14^{a}	15^{ac}	20^{bf}	22^{af}	59^{bc}	63^{e}
		(3)	(4)	(5)	(5)	(8)	(6)
Initial loam		5	18	4	15	18	67

The phosphate status depended significantly on the type of vegetation in the Ah horizon of the grass and forest treatments (Tables 2 and 3). For the permanent grass treatment, the total extractability and the composition of the extractable P_i were nearly the same as for the control soil. At the same time, the solubility of P_o increased to 56%, and the concentrations of P_o extractable with $NaHCO_3$ and NaOH and nonextractable P_o increased by a factor of 3.0, 4.5 and 1.9, respectively.

The inorganic P status of the Ah horizon under trees showed different changes, and the effects differed among the forest types. Spruce and mixed-forest changed the total extractability of P_i insignificantly, whereas the concentrations and proportions of different extracted fractions were significantly modified. The amount of P_i extracted with $NaHCO_3$ and NaOH decreased by a factor of 1.7–2.2, and the proportions of these fractions in the total P_i decreased to 10% and 13–16%, respectively. The concentration of the acid-soluble P_i changed negligibly, but proportion became somewhat higher. The extractability of P_i was the highest (to reach 78%) under oak/maple forest. The concentrations of all extractable fractions increased (from insignificant increase of P_i extractable with NaOH to twofold increase of P_i extractable with $NaHCO_3$).

The extractability of P_o, as well as the concentration and proportion of its different fractions in the Ah horizon under mixed forest were close to those under permanent grasses (Table 3). The soils of two other forest treatments showed higher concentrations of extractable P_o fractions. The total extractability and the proportion of extractable fractions depended on the type of vegetation. The most labile P_o was found in the spruce soil, where 74% of P_o was extracted, and the proportions of P_o extractable with $NaHCO_3$ and NaOH constituted 27 and 47%, respectively.

The trends in the change of P_i status indicated in the Ah horizon under spruce and mixed forests (a decrease of the concentrations of P_i extractable with $NaHCO_3$ and NaOH) extended to a depth of at least 10 cm. The P_i status of the AC horizon under permanent grasses and oak/maple forest did not differ from the surface layer of the control loam. Additional accumulation of P_o in the AC horizon of all experimental treatments was predominantly in the nonextractable fraction. The most labile P_o extractable with $NaHCO_3$ remained at the control level in the AC horizon of all soils, while the concentration of the NaOH-extractable P_o increased only in the spruce forest treatment.

^{31}P NMR spectroscopy of alkaline extracts

Phosphorus-31 NMR spectra of NaOH extracts from primitive soils are shown in Figs. 1 and 2. A range of organic and inorganic P compounds were found in the dialyzed alkaline extracts:

1) monoesters identified by resonance with the chemical shift from 3.0 to 5.5 ppm;
2) diesters represented by DNA (0 ppm) and
3) sum of phospholipids and teichoic acids (0.8–1.9 ppm);
4) inorganic orthophosphate (5.8 ppm) and
5) pyrophosphates (–5.2 ppm).

No peaks of phosphonates (about 20 ppm) or polyphosphates (about –20 ppm) were registered even though these forms are commonly found in alkaline extracts of many soils (i.e. Newman and Tate, 1980; Tate and Newman, 1982; Makarov et al., 1997; Makarov et al., 2004). The P concentrations in the dialyzed NaOH extracts from the soil of the control treatment were too low for adequate NMR spectroscopy.

Most of the dialyzed NaOH extracts contained an unusually high concentration of P_i. The direct alkaline extracts from the soils of permanent grasses and oak/maple forest only contained 12–13% of P as inorganic orthophosphate and pyrophosphates, which appears to be close to typical values of dialyzed extracts from different soils (Table 4). Inorganic P proportion reached 26–41% in the direct NaOH extracts from the soils of other experimental treatments and ranged from 56 to 79% in the alkaline extracts obtained from all soils after acidic pretreatment.

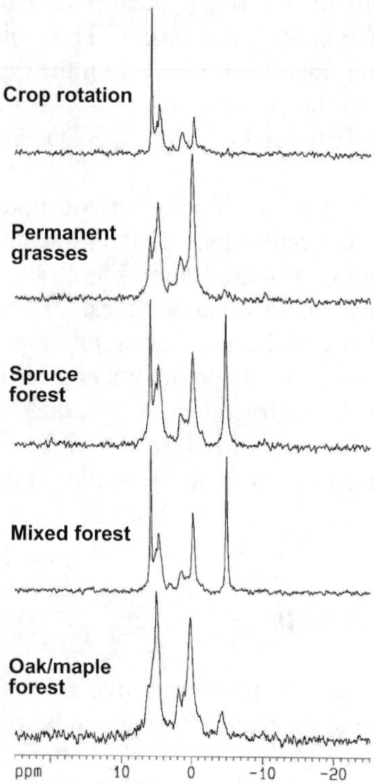

Figure 1. ^{31}P NMR spectra of dialyzed direct 0.1 M NaOH extracts from the Ah horizons of primitive soils.

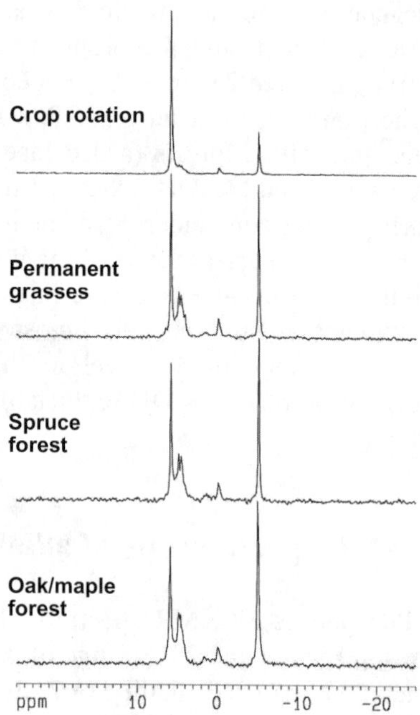

Figure 2. ^{31}P NMR spectra of dialyzed sequential 0.1 M NaOH extracts from the Ah horizons of primitive soils.

The proportions of P_o compounds in the extracts from the soils of grass and tree vegetation were very similar. Monoesters comprised 38–47%, DNA 33–39% and phospholipids and teichoic acids 18–23% of extractable P_o. A somewhat different P_o composition was found in the crop rotation soil, where monoesters P reached 52%, and the proportion of phospholipids and teichoic acid P was higher than DNA P (27 and 21%, respectively). The proportions of P_o compounds extracted after soil pretreatment with HCl were different: monoesters comprised 68–75%, DNA 13–19% and phospholipids and teichoic acids 11–13%. However, similar to the direct alkaline extract, the composition of organic phosphates varied insignificantly in the extracts obtained from different soils.

Table 4. Phosphorus species in dialyzed 0.1 *M* NaOH extracts (1 – direct extract, 2 – extract after pretreatment with 0.1 *M* HCl) from the surface horizon of primitive soils (I – % of P_t in extract, II – mg kg^{-1})

		Inorganic				Organic					
								Diesters			
Treatment	Extract	Ortho-phosphate		Pyro-phosphate		Mono-esters		Phospho-lipids and teichoic acids		DNA	
		I	II	I	II	I	II	I	II	I	II
Crop rotation	1	25	6	1	0	39 (52)*	9	20 (27)	5	16 (21)	4
	2	64	86	15	20	14 (68)	19	3 (13)	4	4 (19)	5
Permanent grasses	1	9	6	3	2	34 (38)	22	20 (23)	13	34 (39)	23
	2	32	33	25	25	32 (74)	33	5 (12)	5	6 (14)	6
Spruce forest	1	16	11	20	13	27 (43)	18	13 (20)	8	24 (38)	16
	2	28	22	28	22	33 (75)	26	5 (12)	4	6 (14)	5
Oak/maple forest	1	7	7	6	7	40 (47)	43	15 (18)	16	30 (35)	32
	2	25	31	34	41	31 (75)	37	5 (12)	6	6 (14)	7
Mixed forest	1	19	11	21	12	27 (46)	15	13 (21)	7	20 (33)	11

* % of P_o in extract

DISCUSSION

Certain differences in the P status were found in the primitive, developing soils of these model ecosystems after 30 years of soil formation on a noncalcareous loam substrate. The soils under forest treatments had the highest concentration of organic C and N, with no pronounced accumulation of P_o except in the oak/maple forest soil. Similar results were obtained for soils of different land-use in Simlipal National Park, India, where evergreen and deciduous forest soils were richest in organic C followed by grasslands and finally cultivated lands, while total and available P levels showed no significant differences (Saikh et al., 1998). However, Chen et al. (2000) and Groenendijk et al. (2002) found that afforestation of grassland soils resulted in elevated mineralization of soil organic C and N.

Organic matter in forest soils usually has low P concentration. The C:P in the soils under trees was high and varied little between forest types in this experiment. The study of the phosphate status of soils under spruce and mixed spruce–birch forests in Sweden also revealed no differences in the content of P_o between two treatments (Saetre et al., 1999). Other projects have found a widening of C:P in the transition from grassland to forest soils, indicating the strong influence of vegetation (Schoenau and Bettany, 1987; Roberts et al., 1989). In forest soils phosphatase activity is often high, and P_o mineralization appears to be an important process for supply of plant phosphorus nutrition (Häussling and Marschner, 1989; Adams, 1992; Dai et al., 1996; Fox and Comerford, 1992; Zou et al., 1992). Condron et al. (1996) suggested that conifers increase mineralization of soil P_o and thereby increase plant available P_i. Studies in New Zealand confirmed that 10- to 30-year-old pine plantations (*Pinus nigra, Pinus ponderosa* and *Pinus sylvestris*) on grassland soils decreased the P_o content of the all studied fractions (extracted by 0.5 M $NaHCO_3$, 0.1 M NaOH, and 0.1 M NaOH after 0.1 M HCl treatment), while inorganic fractions were higher under forest (Ross et al., 1999; Chen et al., 2000). These authors speculated that the lower levels of P_o in the forest soils were due to enhanced microbial and phosphatase activity during earlier stages of forest development when high P uptake from soil by trees may be the driving force for mineralization of soil P_o. High phosphatase activity, decrease of P_o and accumulation of P_i were shown in the rhizosphere of conifers (Häussling and Marschner, 1989; Fox and Comerford, 1992). On the contrary, deforestation and using the territory as a pasture promoted an increase in the content of P_o fractions in soils (Turrion et al., 2000).

Although a high C:P in the organic matter of forest soils may result from the increased P_o mineralization, the fractionation of P in the studied primitive soils showed that the content of inorganic and organic P compounds did not point directly to an increasing mineralization of P_o in soils under forests (including spruce). Inorganic fractions extractable with $NaHCO_3$ and NaOH were lower in the Ah horizon of spruce and mixed forest soils compared with grassland soil, while extractable organic fractions were higher in spruce and oak/maple forest soils.

^{31}P NMR spectroscopy also indicated some differences in the composition of extractable P from different soils. The large concentration of inorganic pyrophosphate in the direct alkaline extract from the soils of spruce and mixed forests may indicate the important role of fungal biomass in these soils, because some fungi are notable sources of pyrophosphate (Makarov et al., 2002b; Makarov et al., 2005). Another specific characteristic of P compounds extracted from the spruce and mixed forest soils was lower proportions and concentrations of relatively unstable phosphate diesters (phospholipids and DNA) that can indicate higher P_o mineralization, although this was not revealed by P fractionation procedure.

The ^{31}P NMR spectra of the alkaline extracts showed more clearly the specific characteristics of the P_o compounds extractable from the primitive soils, which distinguished these primitive, developing soils from the majority of more developed soils. The primitive soils were characterized by a high concentration of inorganic phosphates in the dialyzed alkaline extracts. The presence of P_i in alkaline solutions of preliminarily dialyzed and dried extracts may be explained by several reasons: incomplete dialysis of P_i from the initial alkaline solution; P_o hydrolysis upon the repeated dissolution of extracted material in alkali; or resonance of P_i associated with organic compounds through cation bridges (Guggenberger et al., 1996a; 1996b).

It remains unclear which P_o compounds can be subjected to alkaline hydrolysis to inorganic orthophosphate and pyrophosphate, since the content of P_i in the alkaline solutions was low for the P_o standards typical for organisms and soils (inositol hexaphosphate, DNA, RNA, lecithin, glucose-1-phosphate, glucose-6-phosphate). Moreover, the alkaline solutions of dialyzed and dried NaOH extracts from bacteria and fungi, as well as from tree needles and leaves, contained very little inorganic orthophosphate (Makarov et al., 2002b; Makarov et al., 2005). Therefore, significant amounts of P_i in intensively dialyzed alkaline extract is not likely to be a result of alkaline hydrolysis of mono- and diester P in the solution prepared for NMR spectroscopy.

Many dialyzed extracts from variety of soils have been analyzed by ^{31}P NMR spectroscopy to date (Guggenberger et al., 1996a; 1996b; Turrion et al., 2000; Makarov et al., 2002a; 2004). None of the solutions had as a high proportion of P_i as was found for primitive soils in this experiment. It seems unlikely that P_i in the alkaline extracts of the primitive soils was so specifically protected against dialysis, so the P_i probably originated from the alkaline hydrolysis of specific soil organic phosphates including hydrolysis of P_i bound to humic compounds by Fe and Al bridges. The high proportion of P_i in the dialyzed alkaline extracts may indicate the relatively low resistance of P_o in the primitive soils to alkaline hydrolysis compared to P_o of the mature soils.

A high proportion of diester phosphates appears to be another specific characteristic of the primitive soils. This proportion significantly exceeded the values typical of mature soils both among phosphates directly extracted by NaOH without acid pretreatment (48–62% of P_o is diesters) and converted to the total content of alkali extractable P_o (38–47%). Diesters are more labile P_o compounds then monoesters (Hawkes et al., 1984), and their high proportion can also indicate that P_o in primitive soils is relatively easy to mineralize.

A high amount of monoester P in mature soils (mainly represented by inositol phosphates, with low biological input to soil) results from selective accumulation of more stable organic P compounds. Therefore, the relative accumulation of diester P compounds, which are the main products of biological synthesis (DNA, phospholipids, teichoic acids), is related to the limited time of selective accumulation of stable P_o compounds in the primitive soils.

Diester P is more easily extracted than monoester P that is more strongly fixed by soil components. The extractability differs primarily because of the different density of negative charge in these compounds. Interaction of organic matter with Ca and Mg decreases alkaline solubility of monoesters due to insolubility in alkali of Ca and Mg salts of inositol phosphates (Anderson, 1967). This was the main reason why two phosphate fractions extracted by 0.1 M NaOH directly and after soil pretreatment with 0.1 M HCl differed considerably in diester P proportions. The first fraction is especially rich in diesters. The same result was found for chernozem Ah horizon and for clay fraction of Cambisol Ah horizon flocculated with $CaCl_2$ during the isolation procedure (Makarov et al., 2004). Mahieu et al. (2000) also demonstrated that mobile humic acids extracted without acidic pretreatment from lowland rice soils contained more diester P than Ca-bound humic acids.

CONCLUSION

The plant composition influenced the accumulation of P_o and transformation of P_i at the initial stage of soil formation on this noncalcareous loam. At the first stage of soil formation, organic matter accumulated more rapidly under forest vegetation than under permanent grasses. The accumulation of P_o was less pronounced because of the lower P concentrations in the organic matter of forest soils. Lower concentration of the labile ($NaHCO_3$) inorganic P fraction in the spruce and mixed forest soils may indicate P limitation, if other factors of limiting biological productivity (CO_2, N) increase. Higher concentrations of extractable organic P fractions in forest soils can probably reduce such limitation, but this effect in turn can be reduced by lower proportion and concentration of relatively unstable phosphate diesters (phospholipids and DNA) among organic P species. While the effect of vegetation type on the composition of soil P_o at the initial stage of soil formation is more or less hypothetical, the organic P species of primitive soils are specifically different from those of the mature soils. They are nonresistant to alkaline hydrolysis and contain high proportions of relatively labile diester phosphates that indicate a high chemical and biochemical mineralization potential of P_o and probably testify an important role of organic phosphates in plant P nutrition at the initial stage of soil formation.

ACKNOWLEDGEMENTS

This work was supported by a Research Fellowship awarded to M.I. Makarov by the Alexander von Humboldt Foundation. We thank Professor A.S. Vladychensky and Dr. D.V. Saveliev, providing the soil samples for this study, and Dr. L. Haumaier, providing ^{31}P NMR spectroscopy.

References

Adams J A and Walker T W 1975 Some properties of a chrono-toposequence of soils from granite in New Zealand: 2. Forms and amounts of phosphorus. Geoderma 13, 41–51.

Adams M A 1992 Phosphatase activity and phosphorus fractions in Karri (*Eucalyptus diversicolor* F. Muell.) forest soils. Biol. Fertil. Soils 14. 200–204.

Anderson G 1967 Nucleic acids, derivatives, and organic phosphates. *In:* Soil Biochemistry. Eds. A D McLaren and G H Peterson. pp. 67–89. Edward Arnold, Marcel Dekker, London, New York.

Chen C R, Condron L M, Davis M R and Sherlock R R 2000 Effects of afforestation on phosphorus dynamics and biological properties in a New Zealand grassland soil. Plant Soil 220, 151–163.

Condron L M, Frossard E, Tiessen H, Newman R H and Stewart J W 1990 Chemical nature of organic phosphorus in cultivated and uncultivated soils under different environmental conditions. J. Soil Sci. 41, 41–50.

Condron L M, Davis M R, Newman R H and Cornforth I S 1996 Influence of conifers on the forms of phosphorus in selected New Zealand grassland soils. Biol. Fertil. Soils 21, 37–42.

Dai K H, David M B Vance G F and Krzyszowska A J 1996 Characterization of phosphorus in a spruce–fir Spodosol by phosphorus-31 nuclear magnetic resonance spectroscopy. Soil Sci. Soc. Am. J. 60, 1943–1950.

Fox T R and Comerford N B 1992 Rhizosphere phosphatase activity and phosphatase hydrolyzable organic phosphorus in two forested Spodosols. Soil Biol. Biochem. 24, 579–583.

Groenendijk F M, Condron L M and Rijkse W C 2002 Effects of afforestation on organic carbon, nitrogen and sulfur concentrations in New Zealand hill country soils. Geoderma 108, 91–100.

Guggenberger G, Christensen B T, Rubæk G and Zech W 1996a Land-use and fertilization effects on P forms in two European soils: resin extraction and ^{31}P-NMR analysis. European J. Soil Sci. 47, 605–614.

Guggenberger G, Haumaier L, Thomas R J and Zech W 1996b Assessing the organic phosphorus status of an Oxisol under tropical pastures following native savanna using ^{31}P NMP spectroscopy. Biol. Fertil. Soils 23, 332–339.

Hawkes G E, Powlson D S, Randall E W and Tate K R 1984 A ^{31}P nuclear magnetic resonance study of the phosphorus species in alkali extracts of soil from long-term field experiments. J. Soil. Sci. 35, 35–45.

Häussling M and Marschner H 1989 Organic and inorganic soil phosphates and acid phosphatase activity in the rhizosphere of 80-year-old Norway spruce [*Picea abies* (L.) Karst.] trees. Biol. Fertil. Soils 8, 128–133.

Hedley M J, Stewart J W B and Chauhan B S 1982 Changes in inorganic and organic soil phosphorus fractions induced by cultivation practices and by laboratory incubations. Soil Sci. Soc. Am. J. 46, 970–976.

John M K 1970 Colorimetric determination of phosphorus in soil and plant materials with ascorbic acid. Soil Sci. 109, 214–220.

Magid J, Tiessen H and Condron L M 1996 Dynamics of organic phosphorus in soils under natural and agricultural ecosystems. *In:* Humic Substances in Terrestrial Ecosystems. Ed. A Piccolo. pp. 429–466. Elsevier, Oxford.

Mahieu N, Olk D C and Randall E W 2000 Analysis of phosphorus in two humic acid fractions of intensively cropped lowland rice soils by ^{31}P NMR. European J. Soil Sci. 51, 391–402.

Makarov M I, Malysheva T I, Haumaier L, Alt H G and Zech W 1997 The forms of phosphorus in humic and fulvic acids of a toposequence of alpine soils in the Northern Caucasus. Geoderma 80, 61–73.

Makarov M I, Haumaier L and Zech W 2002a The nature and origins of diester phosphates in soils: a ^{31}P-NMR study. Biol. Fertil. Soils 35, 136–146.

Makarov M I, Haumaier L and Zech W 2002b Nature of soil organic phosphorus: an assessment of peak assignments in the diester region of ^{31}P NMR Spectra. Soil Biol. Biochem. 34, 1467–1477.

Makarov M I, Haumaier L, Zech W and Malysheva T I 2004 Organic phosphorus compounds in particle-size fractions of mountain soils in the northwestern Caucasus. Geoderma 118, 101–114.

Makarov M I, Haumaier L, Zech W, Marfenina O E and Lysak L V 2005 Can ^{31}P NMR spectroscopy be used to indicate the origins of soil organic phosphates? Soil Biol. Biochem. 37, 15–25.

Newman R H and Tate K R 1980 Soil phosphorus characterization by ^{31}P nuclear magnetic resonance. Commun. Soil Sci. Plant Anal. 11, 835–842.

Roberts T L, Bettany J R and Stewart J W B 1989 A hierarchical approach to the study of organic C, N, P, and S in western Canadian soils. Can. J. Soil Sci. 69, 739–749.

Ross D J, Tate K R, Scott N A and Feltham C W 1999 Land-use change: effects on soil carbon, nitrogen and phosphorus pools and fluxes in three adjacent ecosystems. Soil Biol. Biochem. 31, 803–813.

Saetre P, Btandtberg P O, Lundkvist H and Bengtsson J 1999 Soil organisms and carbon, nitrogen and phosphorus mineralisation in Norway spruce and mixed Norway spruce–birch stands. Biol. Fertil. Soils 28, 382–388.

Saikh H, Varadachari C and Ghosh K 1998 Changes in carbon, nitrogen and phosphorus levels due to deforestation and cultivation: A case study in Simlipal National Park, India. Plant Soil 198, 137–145.

Saunders W M H and Williams E G 1955 Observations on the determination of total organic phosphorus in soil. J. Soil Sci. 6, 254–267.

Schoenau J J and Bettany J R 1987 Organic matter leaching as a component of carbon, nitrogen, phosphorus, and sulfur cycles in a forest, grassland, and gleyed soil. Soil Sci. Soc. Am. J. 51, 646–651.

Sharpley A N and Smith S J 1985 Fractionation of inorganic and organic phosphorus in virgin and cultivated soils. Soil Sci. Soc. Am. J. 49, 127–130.

Tate K R and Newman R H 1982 Phosphorus fractions of a climosequence of soils in New Zealand tussock grassland. Soil Biol. Biochem. 14, 191–196.

Tiessen H and Moir J O 1993 Characterization of available P by sequential extraction. *In* Soil Sampling and Methods of Analysis. Ed. M R Carter. pp. 75–86. Soil Science Society of America, Madison.

Turner B L, Mahieu N and Condron L M 2003 Phosphorus-31 nuclear magnetic resonance spectral assignments of phosphorus compounds in soil NaOH–EDTA extracts. Soil Sci. Soc. Am. J. 67, 497–510.

Turrion M B, Glaser B, Solomon D, Ni A and Zech W 2000 Effects of deforestation on phosphorus pools in mountain soils of Alay Range, Khyrgyzia, Biol. Fertil. Soils 31, 134–142.

Walker T W and Syers J K 1976 The fate of phosphorus during pedogenesis. Geoderma 15, 1–19.

Williams B L 1996 Total, organic and extractable P in humus and soil beneath Sitka spruce planted in pure stands and in mixture with Scots pine. Plant Soil 182, 177–183.

Williams J D H and Walker T W 1969 Fractionation of phosphate in a maturity sequence of New Zealand basaltic soil profiles. Soil Sci. 107, 213–219.

Zou X, Binkley D and Doxtader K G 1992 A new method for estimating gross phosphorus mineralization and immobilization rates in soil. Plant Soil 147, 243–250.

CHAPTER 13

SHORT-TERM KINETICS OF SOIL MICROBIAL RESPIRATION – A GENERAL PARAMETER ACROSS SCALES?

Hana Santruckova[1,2], Juliya A. Kurbatova[3], Olga B. Shibistova[4], Miluse Smejkalova[1] and Eva Uhlirova[1,2]

[1]University of South Bohemia, Faculty of Biological Sciences, Branisovska 31, 370 05, Ceske Budejovice, Czech Republic. [2]Institute of Soil Biology AS CR, Na Sadkach 7, 370 05, Ceske Budejovice, Czech Republic. [3]A.N. Severtzov Institute of Ecology and Evolution RAS, Leninski Prospect 33, 117 071 Moscow, Russia. [4]V.N. Sukachev Forest Institute, Akademgorodok 660 036, Kraskoyarsk, Russia. E-mail: hana.santruckova@bf.jcu.cz

ABSTRACT

Microbial parameters derived from the short-term Michaelis-Menten type model are tested and applied on the ecosystem study. Soil dried immediately after sampling and stored at 4°C was moistened to 60% water holding capacity and CO_2 production was measured (GC) after 24 h (respiration response to water supply, V_{DS}) and between the 5th and 6th day of incubation (basal respiration V_{BR}). Then glucose was added into the soil and CO_2 production was measured 16 to 24 h later (maximum respiration, V_{MAX}). Substrate saturation kinetics of respiration was measured after addition of glucose in 6 different concentrations. Soil heterotrophic respiratory potential was expressed as V_{DS}/V_{MAX} ratio; biologically available C (AC_{BR}) and potential flush of the biologically available C (AC_{DS}/AC_{BR}) was estimated using Michaelis-Menten type model. It is shown that the above parameters can get relevant results because they meet the basic assumptions: (i) Fitting of Michaelis-Menten type model is accurate (R^2 = 0.956-0.994), (ii) microbial respiration is substrate limited in natural conditions, (iii) V_{MAX} is relatively invariable for a wide range of C substrate concentrations and microbial populations are not growing. (iv) After moistening of the soil, extra C is released, the amount of which is characteristic for the given soil. Application of the short-term kinetic approach on the upper soil layer of various

D. Binkley and O. Menyailo (eds.),
Tree Species Effects on Soils: Implications for Global Change, 229–246.

ecosystems (Western Canada transect, Central Siberia transect) revealed that except for climate texture, which is closely linked to other physico-chemical parameters of soil, is an important parameter controlling biological activity of soils. The soil profile (from 0 to 100 cm) study performed at two Russian locations (Fyodorovskoye, spruce forest on sedge bogs, and Zotino, pine forest on sands) demonstrates that the deep soil layers contain a high proportion of the biologically available C and are inhabited by microbial populations which are able to efficiently use the available C if the environmental conditions improve. The C accumulated in the peat soil profile is of higher quality than those accumulated in the coarse sand.

INTRODUCTION

The global soil organic C pool (SOC) is considered to be between one-half to two-thirds of the total terrestrial carbon pool (eg. Townsend et al., 1995). This SOC pool provides the longer-term transient sink for much of carbon sequestered by plant biomass and, being continuously decomposed by heterotrophic organisms, it is a source of CO_2 to the atmosphere. Using biologically available SOC, organisms build their biomass, an active part of SOC with a short turnover time and, concurrently, they contribute to the soil CO_2 efflux by respiration activity, an important source of C to the atmosphere. It is thought that heterotrophic microbial respiration contributes in a range of 40% to 60% to the total soil CO_2 efflux (Hanson et al., 2000) but the temporal and spatial variability of this contribution still remains uncertain.

Microbial biomass is frequently used as a measure of heterotrophic respiratory potential of the soil for estimating microbial contributions to the total soil respiration. However, it brings great uncertainties to the estimations, as fluctuation in microbial respiration is, to a certain degree, independent of microbial growth (Santruckova et al., 2004; Tempest and Neijssel; 1978; 1992). Microbial populations reply immediately to any change in the environmental conditions, e.g. substrate availability, water supply or temperature shift, by a change of respiration rate. The change of the environmental conditions, however, must exceed a certain threshold value or persist for a certain period to trigger microbial growth and change in biomass. In addition, there is no positive correlation between microbial biomass and microbial respiration rate (Anderson and Domsch, 1993; Santruckova and Straskraba, 1991). Implicit in the above statements is that the estimation cannot be performed only from the size of microbial populations without respecting microbial physiology as has already been pointed out by Schimel (2001): "we still need more research targeted at understanding when and where microbial community effects, either through variation in physiology or through changes in population size, have large scale impact." Characteristics controlled by microbial population size and physiology and by C availability are needed for estimation of heterotrophic respiratory potential of soil. For these characteristics to be reliable for ecosystem biogeochemical cycles, they should (i) be reproducible and comparable for a wide range of soils and ecosystems, (ii) be simple to measure, (iii) integrate information over time and space, and (iv) reflect sizes of microbial populations and microbial physiology.

This contribution considers that the use of short-term kinetics of microbial respiration is a characteristic, which is controlled by microbial population size, physiology and substrate availability. If a suitable sampling protocol is followed, it is applicable at higher scales providing relevant information about biogeochemical cycles and for ecosystem research.

SITE DESCRIPTION AND METHODS

Soils were sampled in Western Canada, Central European Russia, Central Siberia and Europe, brief descriptions of which are summarized in Table 1. The ecosystems studied represent a wide range of natural (seminatural) ecosystems from tundra and alpine meadows to the different types of forests.

Sampling protocol

Two methods of sampling were used. First, transect sampling was used to study microbial activity of soils in various ecosystems (ecosystem study) along north-south transects (Siberia, Canada) and at Fyodorovskoye and Zotino sites in Russia. Soil cores were collected from each of ten separate locations on a transect generally 0.5 to 1 km in length. After weighing in the field, the material was composited into a bulk sample, which was used for analyses. At all sites except for Fyodorovskoye and Zotino, soil was sampled from a 5 cm layer below a litter, moss or lichen surface. At Fyodorovskoye and Zotino, soil was collected at six depth intervals (0-5, 5-10, 10-15, 15-20, 50-60, 90-100 cm) and the bulk sample for the ecosystem study was composited from the 0 to 5 cm layer as described above. In addition, soil from six to nine representative cores from each site was used for a soil profile study. As the purpose of sampling was to attempt to encompass variability in the ecosystem sampled, no attempt was made to restrict sampling to areas of particular vegetation type of age or having particular disturbance history. Emphasis was placed instead on getting the widest coverage possible from the studied ecosystem. This approach filters out site specific effects and allows general conclusions to be drawn regarding the factors controlling organic C pools (Bird et al., 2001, 2001).

Table 1. Characteristics of soil locations. See references for more details (climate estimated from nearest station).

Locations	Mean annual temperature (°C)	Mean annual precipitation (mm)	Vegetation type
Fyodorovskoye Central European Russia	3.9	711	Southern taiga, spruce forest on sedge bogs (Vygodskaya et al., 2002)
Zotino Central Siberia	-3.9	552	Fire successional pine forest with lichen understorey (Tchebakova et al. 2002; Santruckova et al., 2003; Bird et al. 2002a)
Central Siberia transect	–8.5 to 0.8	450 to 552	Gundra, forest-tundra, mixed taiga, pine forest (Santruckova et al., 2003; Bird et al. 2002a)
Rila Mountains South Europe	3.4	900	Alpine meadows (Santruckova et al., in prep.)
Western Canada transect	–14 to 2.3	120 to 450	Tundra, boreal forest, aspen parkland, cordilleran forest (Bird et al. 2002b)

Second, site sampling was used to study alpine meadows. The basic unit for the sampling was one glacial lake catchment. In each catchment, four to seven 0.25 m^2 pits (50×50 cm) were excavated to the bedrock. From each pit, soil from the litter, organic and mineral layers was taken separately, mixed in the field, and a representative (1-2 kg) sample was taken for analyses. This way of sampling reduced the effect of microscale variability (mm to cm), which can be larger than mesoscale (meters) or macroscale (within and among ecosystems) variability (Morris, 1999; Morris and Boerner, 1999). To test the reliability of the sampling regimes, selected physical and chemical properties from the site sampling were compared with those from grid sampling (50 by 50 m grid, which was established in one catchment using geographic information system (GIS) and soil properties were determined in each node of the grid). No significant difference in the selected properties was found indicating that site sampling is the catchment representative. We found out that the spatial variability can be mainly explained by the thickness of the soil layers but not by the soil properties on a dry weight basis. (Kopacek et al, 2004). Similar findings are shown by Rayment and Jarvis (2000). Soils were always field dried in thin layers and kept at 4°C until analysis.

Methods

Dry soil was moistened to 60% WHC and incubated for 7 days. CO_2 production was measured 24 hours after moistening (respiration response to water supply, V_{DS}) and between the 5th and 6th day of incubation (basal respiration rate, V_{BR}). Then glucose (3.64 mg C g^{-1} dry soil for soils with C_{tot} >10 mg C g^{-1} and 1 mg C g^{-1} dry soil for soils with C_{tot} < 10 mg C g^{-1}) was added into the soil and CO_2 production was measured 16 to 24 hours after glucose amendment. This value represents the maximum respiration rate, V_{MAX}, after input of available C in excess. The substrate saturation kinetic of respiration rate was measured 16 h after glucose addition in concentrations 0, 0.028, 0.056, 0.225, 0.45, 0.9, 1.8, 3.6, 7.27 and 10.9 mg C g^{-1} dry soil, respectively. Glucose was always applied in water solution, drop wise on the soil surface. CO_2 production was measured with an HP 5890 gas chromatograph (Hewlett-Packard, USA), equipped with a TC detector and two parallel steel columns with Porapak Q and/or MS 5A, and operated at 45°C. All measurements were performed in three to four replicates. Microbial biomass (C_{MB}) was measured after 24 hours of soil incubation using the CFE method (Vance et al., 1997). Direct counts_of bacteria were determined in soil stored in 3% formalin at 4°C using fluorescent microscopy and DAPI staining method (Bloem et al., 1992). Total C was measured using an elemental NC 2100 Soil analyser (ThermoQuest, Italia).

Soil heterotrophic respiratory potential was expressed as a ratio V_{DS}/V_{MAX}. The modified Michaelis-Menten type model, which has been used to describe soluble C concentration in water (Wright and Hobbie,1966) and available C concentration in soil (Badalucco and Hopkins1997; Bradley and Fyles, 1995; Sikora and McCoy, 1990) (Figure 1) was used to estimate parameters characterizing the heterotrophic microbial respiration of soil:

$$V_0 = (V_{MAX}\,[AC]/(K_m+[AC])) + V_{BR}, \qquad \text{Eq (1)}$$

where K_m is the half saturation constant and AC initially available C. Constant V_{BR}, which represents basal respiration, shifts the curve up and the intercept of the curve with the x axis to the negative part of the axis. An absolute value of the intercept equals the biologically available C concentration in soil, AC_{BR}. To estimate this intercept and other parameters, the data on respiration rate were fitted to Eq (1) using a non-linear regression procedure (statistical software Prism 4, GraphPad, www.graphpad.com) and the best-fit values of V_{MAX}, K_m and V_{BR} used for calculation of AC_{BR}:

$$[AC_{BR}] = 2(K_m+V_{BR})/V_{MAX} \qquad \text{Eq (2)}$$

Similarly, a flush of microbially available C after water supply to the soil, AC_{DS}, can be expressed as:

$$1/[AC_{DS}]= ((V_{MAX}/V_{DS})-1)/K_m\,, \qquad \text{Eq (3)}$$

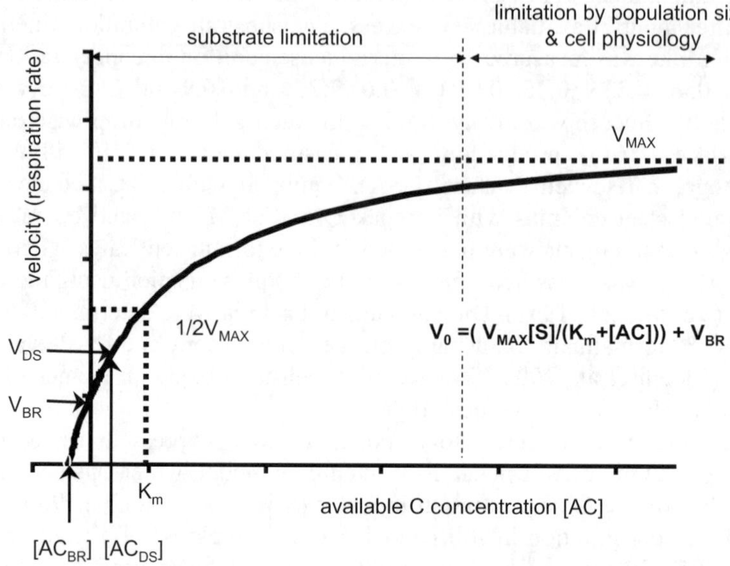

Figure1. The modified equation describing Michaelis-Menten curve with the intercept shifted from zero. The equation approximates soil respiration response to C addition to soil. V_{MAX} is the maximum respiration rate, V_{DS} is respiration rate after moistening of dry soil, V_{BR} is basal respiration rate, K_m half saturation constant, and AC_{BR} and AC_{DS} an amount of C available for basal respiration and for respiration after water supply, respectively.

Combining the two equations (Eqs 2 and 3) a potential flush in C availability after water supply, AC_{DS}/AC_{BR}, was calculated from the respiration rate measurements:

$$[AC_{DS}]/[AC_{BR}] = V_{DS}\,V_{MAX}/(2V_{BR}\,(V_{MAX} - V_{DS})) \qquad \text{Eq (4)}$$

RESULTS AND DISCUSSION

Assessment of the short-term kinetic approach

We propose the parameters derived from the short-term kinetics of microbial respiration, V_{DS}/V_{MAX}, AC_{DS}/AC_{BR} and AC_{BR} are to be used for ecosystem and biogeochemical studies. They fulfill the requirements of simplicity, reproducibility and are controlled by population size and physiology, and C availability, as well. V_{DS}/V_{MAX}, heterotrophic respiratory potential of the soil, describes an increase in C mineralization rate after water supply in relation to maximum mineralization rate. It is a function of microbial population size and physiology, and C availability. V_{MAX} is controlled mainly by the size of microbial populations, growth physiology and

C quality (e.g. Anderson and Domsch, 1978; Panikov 1995), V_{DS} by an amount and quality of available C. In our study, V_{MAX} was induced by glucose addition and represents maximum respiration on sugar-like compounds. If V_{DS}/V_{MAX} approaches unit, the concentration of potentially available C in the soil is close to saturation and available C derives from sugar-like compounds or compounds of similar quality. In this case, high fluctuation of heterotrophic soil respiration can be expected. The lower the V_{DS}/V_{MAX} ratio, the lower the amount or quality of released C and such soils will be distinguished by a low fluctuation of the heterotrophic soil respiration. AC_{DS}/AC_{BR} specifies a maximum flush of an amount of available C after water supply to the soil. We assume that AC_{DS}/AC_{BR} is controlled by the soil physico-chemical and biochemical properties and corresponds to a potential flush in C availability in the soils resulting from drying and re-wetting of soil after rain events following dry periods, melting of soil in spring or mechanical disturbance. Solution from the air-dried samples contains large hydrophilic fractions, the same as those mobilized after soil melting and precipitation after dry periods (Christ and David, 1996). An amount of biologically available C, AC_{BR}, is a distinctive feature to the soil in "steady state" conditions.

The estimation of the suggested parameters should meet several assumptions to get relevant and not skewed results:

(i) Fitting of respiration data to the modified Michaelis-Menten model is accurate.

(ii) Microbial respiration rate is substrate limited in natural conditions.

(iii) V_{MAX} is relatively invariable for a wide range of C substrate concentrations and microbial populations are not growing.

(iv) After moistening of the soil, extra C is released, the amount of which is characteristic for the given soil. The last assumption has been largely confirmed (e.g. Birch, 1958; Fierer and Schimel, 2003; Franzluebbers et al., 2000).

The Michaelis-Menten equation is widely used to model respiration rates (eg. Panikov, 1995; Schimel, 2002). Even though the modified equation with the basal respiration as a variable is used less frequently, existing data exhibit a good accuracy of the fitting (Badalucco and Hopkins, 1997; Bradley and Fyles 1995; Sikora and McCoy, 1990). In our experiments with alpine soils, we compared the goodness of the fit to the Michelis-Menten and modified model. An introduction of V_{BR} to the modified model did not change the goodness of the fit significantly (Table 2, Figure 2). Data fit both models with reasonable accuracy but fitting to the modified model slightly increased the accuracy, particularly when basal respiration was high.

Schimel (2001) showed in his comprehensive study that 1st order kinetics can be used for a description of processes, which follow Michaelis-Menten type kinetics, only if substrate concentrations in the soil are low or moderate but do not vary over an extensively wide range. Thus the concentration of C available in natural conditions should be lower than half of the saturation constant, K_m, and implicitly, basal respiration rate should be lower than

Table 2. The best fit values of V_{MAX} and K_m, coefficient of determination (R^2) and goodness of the fit (sum-of-squares, SS) of the respiration rate data to the simple Michaelis-Menten and to the modified Michaelis-Menten Equation. Respiration rate was measured in organic (A) and mineral (M) layers of alpine meadow soils from various elevations after addition of various concentrations of glucose. For more detail see Figure 2.

Site/ elevation	Michaelis Menten equation				Modified Michaelis Menten equation			
	V_{MAX} μgC g^{-1} h^{-1}	K_m μgC g^{-1}	R^2	SS	V_{MAX} μgC g^{-1} h^{-1}	K_m μgC g^{-1}	R^2	SS
Bu-A/2281	25.8±1.0	1218±178	0.975	30.8	25.2±1.0	1800±211	0.993	8.82
Bu-M	1.14±0.1	30.6±10.8	0.782	0.22	0.87±0.04	65.1±12.1	0.980	0.02
Ka-A/2391	10.4±0.9	580±190	0.847	19.3	9.3±0.5	1287±232	0.984	2.03
Ka-M	1.91±0.1	30.3±6.9	0.931	0.25	1.63±0.1	42.2±7.1	0.981	0.07
Le-A/2709	25.2±0.7	494±52	0.985	16.9	24.4±0.7	582±66	0.990	10.7
Le-M	5.04±0.2	101±19	0.955	2.25	4.9±0.3	109±27	0.956	2.19

$V_{MAX}/2$ and V_{BR}/V_{MAX} ratio should be lower than 0.5. If ratio V_{BR}/V_{MAX} approaches one, the process becomes substrate saturated, independent on substrate concentration and 1st order kinetics cannot be used to describe time dependency of the process. We examined V_{BR}/V_{MAX} ratio in upper soil layers of various ecosystems at different temperatures and in soil profiles of the Russian sites. V_{BR}/V_{MAX} in upper soil layers from various Central Siberian and Canadian ecosystems along the north-south transect increased with temperature but never exceeded a value of 0.25 (Table 3). It did not exert an evident latitudinal effect but it was higher in coarse textured soils (data are not given). A change of V_{BR}/V_{MAX} down to the soil profile was measured at Fyodorovskoye (southern taiga) and Zotino (pine forest) sites in Russia (Table 4). V_{BR}/V_{MAX} ratio in soil was generally low with no apparent dependency on soil depth. The above findings add to those of Schimel (2001) demonstrate that respiration follows 1st order kinetics across a wide range of ecosystems and total C content in soil.

If the process follows 1st order kinetics, V_{MAX} should be invariable for a certain range of C concentrations until inhibited by C surplus. The range of C concentrations inducing V_{MAX} was estimated using saturation curves that were determined in the soils from alpine meadow ecosystems (Figure 2). V_{MAX} remained relatively constant for a wide range of C inputs, and decrease of V_{MAX} as a result of C surplus was observed when C was added in concentration about 10% of C_{tot}. The half saturation constant, K_m, ranged from 0.04 mg C $g^{-1}h^{-1}$ (C_{tot} 26 mg C g^{-1}) to 1.8 mg C $g^{-1}h^{-1}$ (C_{tot} 85 mg C g^{-1}). The data are far from exhaustive but represent the range of soils with different C content and availability. The concentration of glucose, which induced V_{MAX}, was lower in C poor than in rich soils but no correlation between C_{tot} and glucose concentration was found. V_{MAX} was invariable across the range of

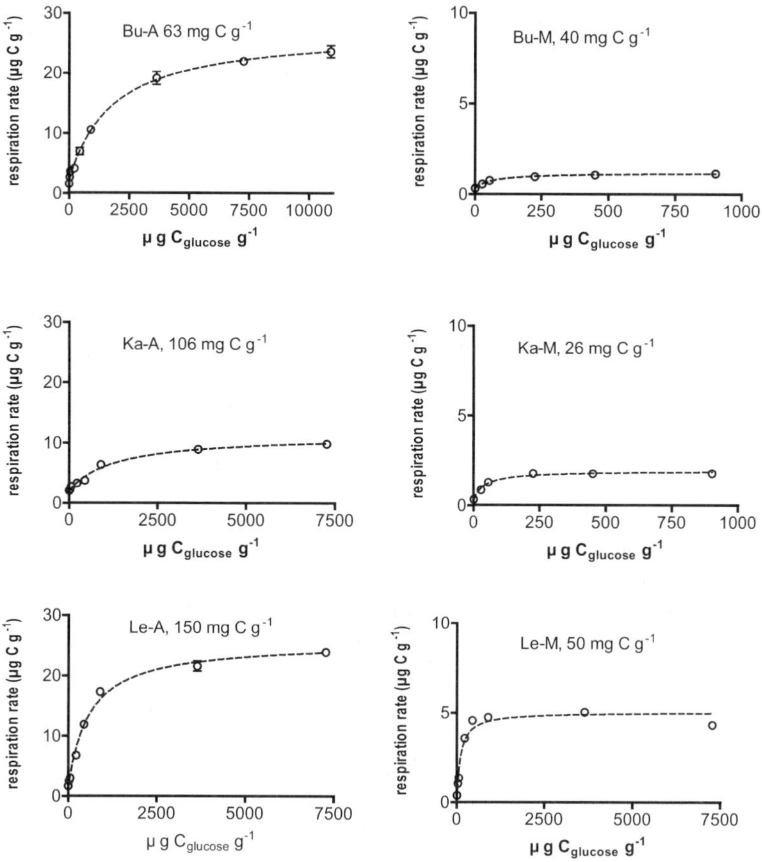

Figure 2. The data of substrate saturation of microbial respiration fitted to modified Michaelis-Menten equation. The alpine meadow soils from Rila Mnts. were used, each point is mean from four replicates with s.d. bars, the coefficient of variation did not exceed 10%. C_{tot} is given in the legend of each graph, for more detailed characteristics of the soils see Table 6.

concentrations indicating low sensitivity of V_{MAX} measurements to errors caused by an addition of unsuitable concentration of C substrate. Our findings correspond to the results from agriculture soils (Anderson and Domsch, 1985). K_m referred to for various soils is in a range from 0.040 to 1.7 mg C $g^{-1}h^{-1}$, which is very close to our observations (Anderson and Domsch, 1986; Anderson and Gray, 1990; Coody et al., 1986; Sikora and McCoy, 1990). The short-term kinetic concept is based on the initial response of microbial populations before they start growing. Before growth is induced, microorganisms increase respiration rate that is invariable and independent on an amount of C substrate over a wide range of the C substrate concentrations. Our measurements of respiration rate in the period between 16 to 24 hours was based on the preliminary investigation of microbial growth on glucose at

Table.3. Basal/maximum respiration rate (V_{BR} /V_{MAX}) in the surface layers of soils from various ecosystem along north-south transect in Central Siberia and Western Canada and its change with temperature. See Santruckova et al. (2003) and Bird et al. (2002) for the detailed description of the ecosystem and soil properties.

Temperature	5°C	10°C	15°C	20°C
Central Siberia (n=7)	0.10±0.06	0.11±0.05	0.11±0.04	0.16±0.03
Western Canada (n=5)	0.10±0.06	0.09±0.03	0.13±0.01	0.19±0.01

20°C. The lag phase lasted for 20 hours at least, which is in agreement with Panikov (1995) and Colores et al. (1996).

Application of short-term kinetic approach to ecosystem research

We determined the heterotrophic respiratory potential of soils (V_{DS}/V_{MAX}) and a flush in the amount of available C (AC_{DS}/AC_{BR}) in upper soil layers from the different world ecosystems (Western Canada transect, Russian forests, alpine meadows), and soil profiles of the Fyodorovskoye and Zotino sites. In addition, the soils from the Canada transect were used to determine a temperature dependency of the parameters. V_{DS}/V_{MAX} and AC_{DS}/AC_{BR} varied with site, temperature and soil depth. In seven out of nine ecosystems, AC_{DS}/AC_{BR} increased with water supply less than five times and V_{DS}/V_{MAX} being lower than 0.5 increased with AC_{DS}/AC_{BR} (Figure 3). It implies that upper soil layers contains a low amount of biologically available C, which remains undecomposed. Larger fluctuation is to be induced by an input of fresh organic compounds. V_{DS}/V_{MAX} ratio in pine forest (Zotino) and alpine meadows (Rila Mnts.) soils was lower than in the other soils corresponding to lower quality of SOC at these sites (Santruckova et al., 2003; unpubl. results). In two soils from the Canadian transect (YKN and BUF) AC_{DS}/AC_{BR} increased more than 5 times indicating higher amount of biologically available C and weak bounding of organic C in these soils. The soils from the YKN and BUF sites were distinguished by having high proportion of organic C in a coarse fraction > 500 μm and low C/N ratio (Bird et al.; 2002). C in the coarse fraction is obviously more available than C bound on a fine fraction (Hopkins et al., 1993; Schulten and Leinweber, 2000). The AC_{DS}/AC_{BR} in the YKN soil was significantly lower than in the BUF soil, which coincides with

Table. 4. Basal/maximum respiration rate (V_{BR} /V_{MAX}) in soil profile of spruce forest on sedge bog (Fyodorovskoye, n=6) and of pine forest (Zotino, n=5)

Soil depth (cm)	0-5	5-10	10-15	15-20	50-60	90-100
Fyodorovskoye	0.13±0.05	0.08±0.02	0.06±0.03	0.08±0.04	0.2±0.08	0.15±0.08
Zotino	0.05±0.001	nd	0.11±0.05	nd	0.06±0.04	0.05±0.03

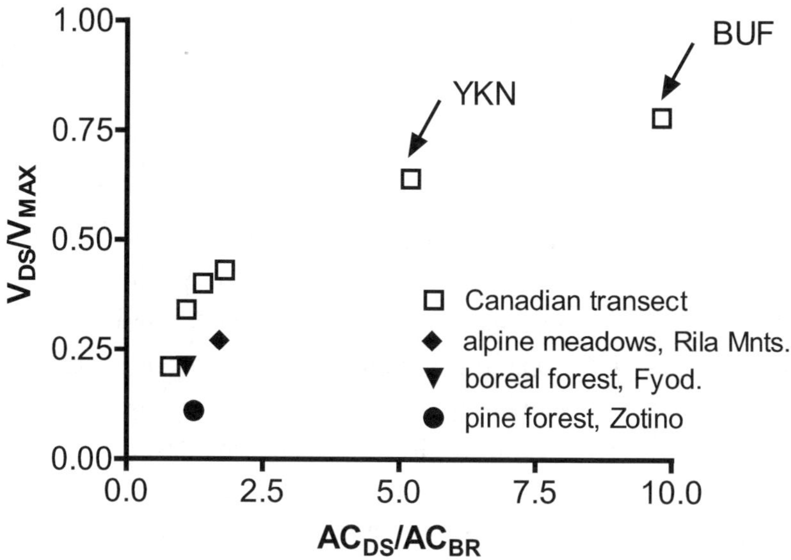

Figure 3. Relationship between V_{DS}/V_{MAX} and AC_{DS}/AC_{BR} as determined in soils from various world ecosystems. The soils analysed were sampled along the North-South transect in Canada, in Central Siberia, in Central European Russia and in alpine meows from Rila Mnts (Europe). The arrows show coarse soils at YKN and BUF sites.

lower proportion of the coarse fraction in the YKN than BUF soil VDS/VMAX and ACDS/ACBR in the soils from the Canadian transect were obviously higer at a high temperature (Table 5). It corresponds to an increasing solubility of organic compounds and increasing ability of microbial populations to produce extracellular enzymes and mineralize more complex organic material (Andrews et al. 2000; Christ and David, 1996; Mackey 1984). No dependence on latitude was found. Soil texture is an important factor controlling biological activity, in part because of its close link with features such as bedrock type, nutrient status, water holding capacity, illuviation and bioturbation rates, root penetration resistance and the availability of oxygen to support microbial mineralization. Soil C inventory in the coarse textured soils distinguished by weak bounding of SOC, high proportion of the biologically available C and, consequently, high turnover rate. It is consistent with the conclusions of Bird et al (2002a, 2002b), who worked in the Central Siberia and Western Canada transect and found that the observed differences in SOC inventories are largely the result of variation in soil texture.

In the soil profiles of the Fyodorovskoye soils, V_{DS}/V_{MAX} increased linearly with AC_{DS}/AC_{BR} in all samples, except for two (Figure 4). AC_{DS}/AC_{BR} did not increase more than five times and was not related to C_{tot}

Table 5. A temperature dependency of a heterotrophic respiratory potential (V_{DS}/V_{MAX}) and a flush of available C (AC_{DS}/AC_{BR}) in soils from the ecosystems along the Western Canada transect. For more detailed description of the ecosystem and soil properties see Bird et al. 2002.

Site (Latitude/°N)	C_{tot} mg g^{-1}	Temperature (°C) 5	10	15	20	25
				V_{DS}/V_{MAX}		
WIN (68)	50	0.03±0.001	0.07±0.005	0.16±0.01	0.21±0.01	0.28±0.02
RAY (65)	108	0.37±0.02	0.35±0.02	0.36±0.03	0.34±0.02	0.48±0.02
YKN(63)	130	0.57±0.03	0.53±0.02	0.60±0.05	0.64±0.05	0.67±0.05
BUF (61)	121	0.78±0.03	0.75±0.04	0.76±0.03	0.78±0.04	0.86±0.04
EDM(54)	172	0.35±0.01	0.21±0.01	0.27±0.01	0.40±0.03	0.51±0.03
CYP (50)	204	0.10±0.004	0.19±0.01	0.18±0.02	0.43±0.02	0.49±0.03
				AC_{DS}/AC_{BR}		
WIN (68)	50	0.6±0.02	0.8±0.05	0.8±0.03	0.8±0.02	1.1±0.05
RAY (65)	108	1.5±0.07	1.2±0.04	1.4±0.16	1.1±0.03	1.5±0.06
YKN(63)	130	5.5±0.38	4.6±0.54	5.8±0.32	5.2±0.02	4.6±0.11
BUF (61)	121	11.1±0.96	11.0±0.37	10.9±0.57	9.8±0.73	13.3±0.87
EDM(54)	172	1.4±0.09	0.8±0.06	1.0±0.03	1.4±0.07	1.7±0.01
CYP (50)	204	0.6±0.03	1.2±0.94	0.8±0.02	1.8±0.04	1.2±0.006

or direct counts of bacteria (Figure 5). The two samples that displayed much higher V_{DS}/V_{MAX} and AC_{DS}/AC_{BR} were from bottom layers of the peat bog locations with a peat layer deeper than 1m. High values of both characteristics show an accumulation of easily available C, which is in agreement with the limitation of microbial activity on polysaccharide rich *Sphagnum* by low temperature, high acidity, deficiency of oxygen and nutrients other than C (Breemen and Finzi, 1998). Comparison of individual cores and layers revealed that AC_{DS}/AC_{BR} and V_{DS}/V_{BR} increased down to the profile being higher in the peat bog than mineral profiles (Figure 5). In Zotino soils both characteristics were comparable with Fyodorovskoye only in the upper 30 cm layers. In deeper layers, an enormous flush of the biologically available C, AC_{DS}/AC_{BR}, exceeding a value of 25 was detected but the heterotrophic respiratory potential, V_{DS}/V_{MAX}, was comparable with the upper layers and did not increase with the increase AC_{DS}/AC_{BR} (Figure 4). Zotino soils are sandy soils with a high proportion of coarse particles from 1 to 2 mm. The high AC_{DS}/AC_{BR} corresponds to a weak bounding of organic C on the coarse particles (Schulten and Leinweber, 2000). It can be expected that a variation in the amount of biologically available C resulting from fluctuations of the environmental factors will be of high importance at this site, which is consistent with a high increase in soil respiration after a rain event (Lloyd and

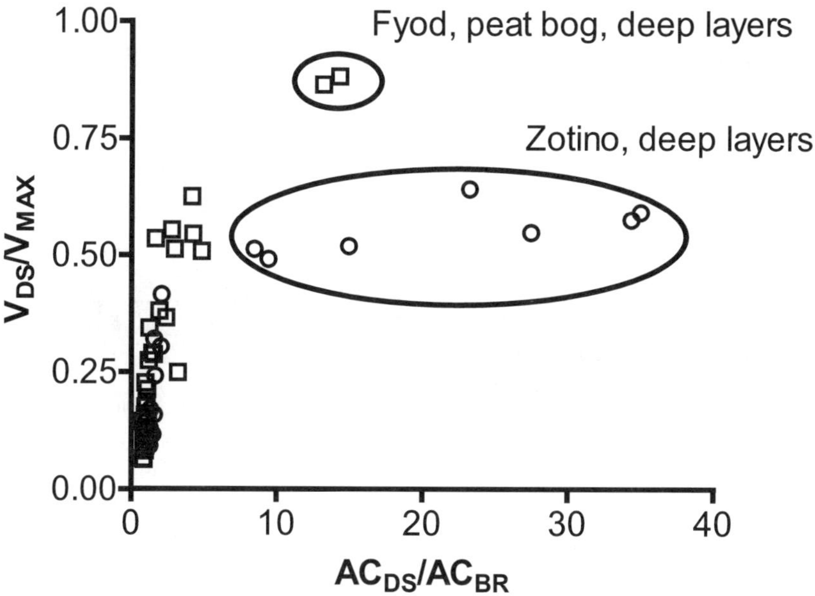

Figure 4. Relationship between V_{DS}/V_{MAX} and AC_{DS}/AC_{BR} as determined in soil profiles of Central Siberia (Zotino, open circles) and in Central European Russia (Fyodorovskoe, open squares) sites.

Kelliher, personal comm.). V_{DS}/V_{MAX} of about a value of 0.5 indicates either a lower quality of the biologically available C or low microbial biomass. It is consistent with a low heterotrophic soil respiration as estimated from eddy covariance data at Zotino site (Lloyd et al., 2002) and our previous conclusions (Santruckova et al., 2003). We found a low mineralisation potential of the Zotino soils and a substantial part of the consumed C in heterotrophic metabolism to be built into microbial products and transformed to SOC rather than mineralized. Low microbial biomass and activity, and the dominance of pine and lichen litter of low quality has already been found at this site (Ross et al., 1999; Santruckova et al., 2003; Shibistova et al., 2002). The profile study demonstrates that the deep soil layers, regardless of low C content and microbial biomass, contains a high proportion of the biologically available C and are inhabited by microbial populations which are able to efficiently use the available C if the environmental conditions improve. The C accumulated in the peat soil profile is of higher quality than those accumulated in the coarse sand. The available C in deep profiles can be leached into ground water in the wet period but also transported into the upper layers in dry periods, being an additional C source for microbes.

An amount of the biologically available C and its quality is a key determinant of many microbial processes in soil. It affects heterotrophic

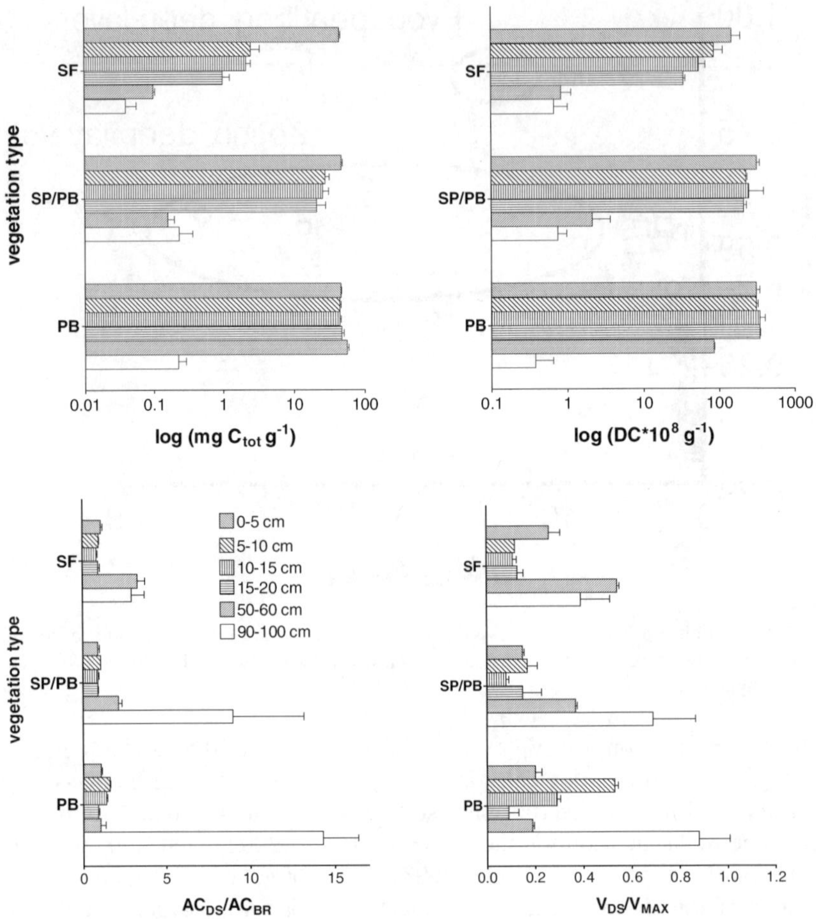

Figure 5. Total carbon content (C_{tot}), direct counts (DC), V_{DS}/Vmax and AC_{DS}/AC_{BR} in soil profile of different vegetation types at Fyodorovskoye site. From each vegetation type (SF-sphagnum-vaccinium spruce forest, PB - peat bog with sphagnum vaccinium spruce-pine stand, SF/PB - transient type between SF and PB) means from three sampling points and s.d. are given.

respiration rates in the soil in "steady state" conditions and its fluctuation determines a range of fluctuation of the heterotrophic respiration rate. AC_{BR} represents the biologically available C and the microbial community's ability to utilize it. If microbial utilization is limited by unfavorable soil conditions AC_{BR} remains in the soil. The higher AC_{BR} the higher the limitation of microbial C transformation either by a small size of microbial populations or by the environmental conditions. AC_{BR} was measured in soils from the alpine meadows. A poor, if any, correlation between AC_{BR} and C_{MB} or V_{BR} was detected, and, as expected, an amount of available C did not correlate with C_{tot} (Table 6). A ratio AC_{BR} to C_{tot} or AC_{BR} to C_{MB} was usually higher in mineral

Table 6. An amount of microbially available C, (AC_{BR}), total C (C_{tot}), microbial biomass (C_{MB}), ratio of AC_{BR} to C_{tot} (AC_{BR}/C_{tot}) and microbial biomass (AC_{BR}/C_{MB}), bazal respiration rate (V_{BR}) and specific respiration rate of microbial biomass (V_{BR}/C_{MB}) in the alpine meadow soils from the Rila Mnts. The parameters were determined in the organic (A) and mineral (M) layer, respectively, from the surface to the bedrock.

Site-layer/ elevation (m)	AC_{BR} µgC g^{-1}	C_{tot} mg C g^{-1}	C_{MB} µgC g^{-1}	AC_{BR}/C_{tot} %	AC_{BR}/C_{MB} %	V_{BR} µgC g^{-1} h^{-1}	V_{BR}/C_{MB} mgC gC^{-1} h^{-1}
Bu-A/2281	145.3	62.7	831	0.23	17.5	1.604	1.9
Bu-M	151.5	39.8	193	0.38	78.8	0.493	2.6
Ka-A/2391	275.4	106.2	2209	0.26	12.5	2.091	0.9
Ka-M	52.2	26.4	137	0.20	38.0	0.324	2.4
Le-A/2709	47.8	150.3	1917	0.03	2.5	1.671	0.9
Le-M	44.5	50.1	174	0.09	25.6	0.352	2.0

than organic horizons indicating a higher limitation of the microbial C transformation in the deeper soil layers as compared to the upper organic layers. A higher specific respiration activity of microbial biomass in the mineral than in organic soil supports it. The higher specific respiration rate of a microbial community could also indicate a higher metabolic activity of the community, but we do not assume that this is the case. Microbial communities were not growing in the time of the measurements and specific respiration rates higher than 1 mg C g $C^{-1}h^{-1}$ is too high for populations not growing (Anderson and Domsch, 1985). Even the presented data set is rather limited, yet it demonstrates that AC_{BR} measurements can contribute to better understanding of a rate of heterotrophic microbial respiration.

Our observations show that:

(i) The short-term kinetics of microbial respiration is a relevant parameter across scaling but great attention must be paid to the sampling protocol to eliminate the effect of lower scale factors to upper scale determination.

(ii) The texture except from the climatic conditions is an important factor controlling biological activity of soil on the ecosystem scale.

(iii) The increase of the biologically available C in the deeper soil layers of the soil profile resulting from the change of the environmental conditions is comparable with the upper soil layers. Microbial populations surviving in the deeper layers are smaller in the size but comparable in the efficiency of C use with the populations in the upper layers.

ACKNOWLEDGEMENTS

This contribution could not have been completed without the kind assistance of our Russian colleagues from Sukachev's Forest Institute, Krasnoyarsk and from Severtzov's Institute of Evolution and Ecology Problem RAS, Moscow. The national project MSM 123100004 and EU project TCOS-Siberia (EVK2-2002-00555) supported it. We thank Keith R. Edwards for correcting the language.

REFERENCES

Anderson T H and Domsch K H 1978 A physiological method for the quantitative measurement of microbial biomass and soils. Soil Biol. Biochem. 10, 215-221.

Anderson T H and Domsch K H 1985 Maintenance carbon requirements of actively-metabolizing microbial populations under in situ conditions. Soil Biol. Biochem. 17(2), 197-203.

Anderson T H and Domsch K H 1986 Carbon assimilation and microbial activity in soil, Zeitschrift fur Pflanzenernahrung und Bodenkunde. 149, 457 - 468.

Anderson T H and Domsch K H 1993 The metabolic quotient for CO_2 (qCO_2) as a specific activity parameter to assess the effects of environmental conditions, such as pH, on the microbial biomass of forest soils. Soil Biol. Biochem. 25, 393-395.

Anderson T H and Gray T R G 1990 Soil microbial carbon uptake characteristics in relation to soil management. FEMS Microbiol Ecol. 74, 11-20.

Andrews J A, Matamala R, Westover K M and Schlesinger W H 2000 Temperature effects on the diversity of soil heterotrophs and the δ^{13} of soil-respired CO_2. Soil Biol. Biochem. 32, 699-706.

Badalucco L and Hopkins D W 1997 Available carbon in soil determined from substrate utilization kinetics: comparison of substrates and soil amendments. J. Microbiol. Meth. 30, 43-47.

Bird M I, Santruckova H, Arneth A, Grigoriev S, Gleixner G, Kalaschnikov Y N, Lloyd J and Schulze E D 2002a Soil carbon inventories and carbon - 13 on a latitude transect in Siberia. Tellus 54B, 631-641.

Bird M, Santruckova H, Lloyd J and Veenendaal E 2001 Global soil organic carbon pool. *In* Global Biogeochemical Cycles in the Climate System. Eds. E D Schulze, M Heimann, S Harrison, E Holland, J Lloyd, I C Prentice and D Schimel. pp. 350. Academic Press, San Diego.

Bird M, Santruckova H, Lloyd J and Lawson E 2002b The isotopic composition of soil organic carbon on a north - south transect in western Canada. Eur. J. Soil Sci. 53, 393-403.

Birch H F 1958 The effect of soil drying on humus decomposition and nitrogen availability. Plant Soil X, 9-31.

Bradley R L and Fyles J W 1995 A kinetic parameter describing soil available carbon and its relationship to rate increase in C mineralization. Soil Biol. Biochem. 27, 167-172.

Christ M J and David M B 1996 Temperature and moisture effects on the production of dissolved organic carbon in a spodosol. Soil Biol. Biochem. 28, 1191-1199.

Coody P N, Sommers L E and Nelson D W 1986 Kinetics of glucose uptake by soil mikroorganisms. Soil Biol. Biochem. 18, 283 – 289.

Colores G M, Schmidt S K and Fisk M C 1996 Estimating the biomass of microbial functional groups using rates of growth-related soil respiration. Soil Biol. Biochem. 12, 1569-1577.

Franzluebbers A, Haney R, Honeycutt C, Schomberg H and Hons F 2000 Flush of carbon dioxide following rewetting of dried soil relates to active organic pools. Soil Sci. Soc. Am. J. 64, 613-623.

Hanson P J, Edwards N T, Garten C T and Andrews J A 2000 Separating root and soil microbial contributions to soil respiration: a review of methods and observations. Biogeochemistry 48, 115-146.

Kieft T L, Soroker E and Firestone M K 1987 Microbial biomass response to a rapid increase in water potential when dry soil is rewetted. Soil Biol. Biochem. 19, 119-126.

Kopacek J, Kana J, Santruckova H, Picek T and Stuchlik E 2004 Chemical and biochemical characteristics of alpine soils in the Tatra Mountains and their correlation with lake water quality. Water Air Soil Poll. 135, 307-327.

Lloyd J, Shibistova O, Zolotoukhine D, Kolle O, Arneth A, Wirth K, Styles J M, Tchebakova N M and Schulze D E 2002 Seasonal and annual variations in the photosynthetic productivity and carbon balance of a central Siberian pine forest. Tellus 54B, 590-610.

Mackey B M 1984 Lethal and sublethal effects of refrigeration, freezing and freeze-drying on mico-organisms. *In* The Revival of Insured Microbes. Eds. M H E Andrew and A D Russell. pp. 45-75. Acad. Press, London.

Morris S J and Boerner R E J 1999 Spatial distribution of fungal and bacterial biomass in southern Ohio hardwood forest soils: scale dependency and landscape patterns. Soil Biol. Biochem. 31, 887-902.

Morris S J 1999, Spatial distribution of fungal and bacterial biomass in southern Ohio hardwood forest soils: fine scale variability and microscale patterns. Soil Biol. Biochem. 31, 1375-1386.

Panikov N S 1995 Microbial Growth Kinetics. Chapman&Hall, London. 378 pp.

Rayment M B and Jarvis P G 2000 Temporal and spatial variation of soil CO_2 efflux in a Canadian boreal forest, Soil Biol. Biochem. 32, 35-45.

Ross D J, Kelliher F M and Tate K R 1999 Microbial processes in relation to carbon, nitrogen and temperature regimes in litter and a sandy mineral soil from a central Siberian Pinus sylvestris L. forest. Soil Biol. Biochem. 31, 757-767.

Santruckova H and Straskraba M 1991 On the relationship between specific respiration activity and microbial biomass in soils. Soil Biol. Biochem. 23, 525-532.

Santruckova H, Bird M I, Kalaschnikov Y N, Grund M, Elhottova D, Simek M, Grigoryev S, Gleixner G, Arneth A, Schulze E D and Lloyd J 2003 Microbial characteristics of soils on latitudinal transect in Siberia. Global Change Biol. 9, 1106-1117.

Santruckova H, Picek T, Tykva R, Šimek M, Pavlů B 2004 Short-term partitioning of ^{14}C-[U]-glucose in soil microbial pool in different aeration status. Biol. Fertil. Soils. (accepted)

Schimel J 2001 Biogeochemical models: implicit versus explicit microbiology. *In* Global Biogeochemical Cycles in the Climate System. Eds. E D Schulze, M Heimann, S Harrison, E Holland, J Lloyd, I C Prentice and D Schimel. pp. 350. Academic Press, San Diego.

Schulten H R and Leinweber P 2000 New insight into organic-mineral particles: composition, properties and models of molecular structure.

Shibistova O, Lloyd J, Evgrafova S, Savushkina N, Zrazhewskaya G, Arneth A, Knohl A, Kolle O and Schulze E D 2002 Seasonal and spatial variability in soil CO_2 efflux rates for a central Siberian Pinus sylvestris forest. Tellus 54B, 631-641.

Sikora L J and McCoy J L 1990 Attempts to determine available carbon in soils. Biol. Fertil. Soils 9, 19-24.

Tempest D W and Neijssel O M 1978 Eco-physiological aspects of microbial growth in aerobic nutrient-limited environments. Adv. Microb. Ecol. 2, 105-153.

Tempest D W and Neijssel O M 1992 Physiological and energetic aspects of bacterial metabolite overproduction. FEMS Microbiol. Lett. 100, 169-176.

Tchebakova N M, Kolle O, Zolotoukhine D, Arneth A, Styles J M, Vygodskaya N N, Schulze E D, Shibistova O and Lloyd 2002 Interannual and seasonal variations of energy and water vapour fluxes above a Pinus sylvestris forest in the Siberian middle taiga. Tellus 54B, 537-551.

Towndsend A R, Vitousek P M and Trumbore S E 1995 Soil organic matter dynamics along gradients in temperature and land-use on the island of Hawaii. Ecology 76, 721-733.

Van Breemen N and Finzi A C 1998 Plant soil interactions: ecological aspects and evolutionary implications. Biogeochemistry 42, 1-19.

Vygodskaya N N, Schulze E D, Tchebakova N M, Karpachevskii L O, Kozlov D, Sidorov K N, Panfyrov M I, Abrazko M A, Shaposhnikov E S, Solnzeva O N, Minaeva T Y, Jeltuchin A S, Wirth C and Pugachevskii 2002 Climatic control of stand thinning in unmanaged spruce forests of the southern taiga in European Russia. Tellus 54B, 443-461.

Wright R T and Hobbie J E 1966 Use of glucose and acetate by bacteria and algae in aquatic ecosystems. Ecology 47, 447-464.

Chapter 14

THE INFLUENCE OF STAND DENSITY ON GROWTH OF THREE CONIFER SPECIES

R. S. Sobachkin, D. S. Sobachkin, and A. I. Buzykin
V.N. Sukachev Institute of Forest, Russian Academy of Sciences, Siberian Branch, Academgorodok, Krasnoyarsk, Russia, 660036
Fax: +7(3912) 43-36-86
e-mail: leshii@ksc.krasn.ru

ABSTRACT

We evaluated the influence of stand density on the growth and development of monoculture plantations of Scots pine (*Pinus sylvestris*), larch (*Larix sibirica*) and Norway spruce (*Picea abies; = Picea obovata* Ledeb.). The planting density ranged from 500 to 128,000 trees/ha. After 22 years, tree height was not affected by density, though the height of pine and larch (8 m) was greater than spruce (6 m). Diameter declined exponentially with increasing density, but the slope was too low to offset the contribution of more trees to the stand total basal area and stem volume. Basal area and stem volume per hectare were greatest at the most extreme density for all three species, refuting ideas of "stagnation" at high density.

INTRODUCTION

The rates of wood growth and biomass accumulation in forests depends in part on the density of trees per unit area. The initial planting density in monocultures is particularly important in relation to the effects of tree density on the development of competing understory vegetation. Many research projects in Russia have examined the influence of stand density on tree size, productivity, and biomass (Kayrukshtis and Yuodvalkis, 1976; Kondrat`ev,

D. Binkley and O. Menyailo (eds.),
Tree Species Effects on Soils: Implications for Global Change, 247–255.

1959; Martynov, 1974; Pschenichnikova, 1978; Ryabokon, 1979; Timofeev, 1957; Usoltsev, 1988; Hozumi, 1980), though the range of densities tested in these studies did not deal with extremely low or high densities. The influence of density could also be examined in natural, non-plantation forests (Usoltsev, 1994), but confounding factors that underlie the differences in density across stands limit the interpretability of non-plantation studies.

Very little information is available from Siberia on the effects of stand density on growth and stand structure. In this chapter we present the results of the first experiment in Siberia that examined a full range of densities for three species: Scots pine (*Pinus sylvestris*), larch (*Larix sibirica*) and Norway spruce (*Picea abies; = Picea obovata* Ledeb.). Species were compared and the dependences between inventory parameters and stand density are revealed.

MATERIALS AND METHODS

The experiment was established in 1982 by the department of forestry of V. N. Sukachev Institute of Forest under the supervision of A. I. Buzykin, in the southern taiga subzone (Bolshaya Murta Forest Enterprise, Krasnoyarsk region). The gray forest soils appeared uniform across the 5-ha site, and the prior landuse was pasture. Eighteen planting densities were used, ranging from 500 to 128,000 trees/ha. The size of plots varied across this 256-fold range in density; all plots had at least 500 seedlings, and plot sizes ranged from 1 ha (for the lowest density) to 0.007 ha for the highest. Two-yr.-old seedlings were planted according to the square scheme (Table 1). The plots for each species were grouped in a tight arrangement without gaps, forming an integrated nested set of one-species plantations. The plantation blocks with different species were separated by five-meter gaps.

Routine measurements of the plots included tree diameter (by 1-cm classes at 1.3 m height; Pobedinskiy 1966, Moiseev 1971) for 25-30 representative trees/plot. The total tree height, and height to base of live crown, were measured on 2 representative trees for each 1-cm size class. Representative trees were selected to represent healthy trees in the population, with full crowns and no mechanical damage. The trees in the forest border were not included in the general measurement of height and were measured separately. The effects of tree species and stand density on inventory parameters were determined by two-way analysis of variance, and relationships among inventory parameters were studied with Spearman rank order correlations (STATISTICA (5.0 for Windows, Stat Soft, 1997). Although the experimental treatments were not replicated within species, the

Table 1. Treatment levels of density and spacing

Density levels	Initial density trees/ha	Initial density trees/km^2	Initial spacing (m)
1	500	50,000	4.47
2	750	75,000	3.65
3	1,000	100,000	3.16
4	1,500	150,000	2.58
5	2,000	200,000	2.24
6	3,000	300,000	1.83
7	4,000	400,000	1.58
8	6,000	600,000	1.29
9	8,000	800,000	1.12
10	10,000	1,000,000	1,00
11	12,000	1,200,000	0.91
12	16,000	1,600,000	0.79
13	24,000	2,400,000	0.64
14	32,000	3,200,000	0.56
15	48,000	4,800,000	0.46
16	64,000	6,400,000	0.40
17	96,000	9,600,000	0.32
18	128,000	12,800,000	0.28

use of regression to test for patterns across the 18 density levels allows for a powerful analysis of the effect of density.

RESULTS AND DISCUSSION

After 18 years, the span of mortality across all species and densities was about 4-fold (from 20% to 80%), far smaller than the 256-fold range in initial density (Figures 1-3). The density of pine plots had declined by about 25% (in density treatments 6 to 11, from 3,000 to 12,000 initial trees/ha), or up to 75% in the highest density (treatment 18, 128,000 trees/ha). Mortality was greater for larch, ranging mostly from 40 to 80%, and for spruce (30 to 70%). Early mortality was probably driven by competition with grass and herb competitors, and later mortality by competition among trees (natural thinning or self-thinning).

Pine and larch trees averaged about 8 m tall, compared with 6 m for spruce. Tree height did not vary among density treatments ($P>0.25$), but diameter responded very strongly to density treatments. Diameters declined exponentially with increasing density (Figure 4), and the ANOVA revealed strong effects of density, species, and species x density interaction ($P<0.001$).

Figure 1. The spruce plot with an initial planting density of 500 trees/ha declined to 260 trees/ha at age 20 years, probably as a result of competition with non-tree vegetation (left). The spruce plot with an initial density of 8,000 trees/ha declined to 4,100 trees/ha, and tree-to-tree competition was more important.

The diameters were relatively insensitive to density when density exceeded about 5000 trees/ha. Correlation coefficients between density and diameter were -0.993 (n=17) for pine, -0.925 (n=18) for larch, and -0.863 (n=18) for spruce. The distribution of diameters within plots was much narrower for high-density plots than for lower density plots (Figure 5).

Figure 2. The pine plot with an initial density of 500 trees/ha declined to 239 trees/ha at 20 years (left), and the larch plot with 1500 trees/ha initially declined to 740 trees/ha (right).

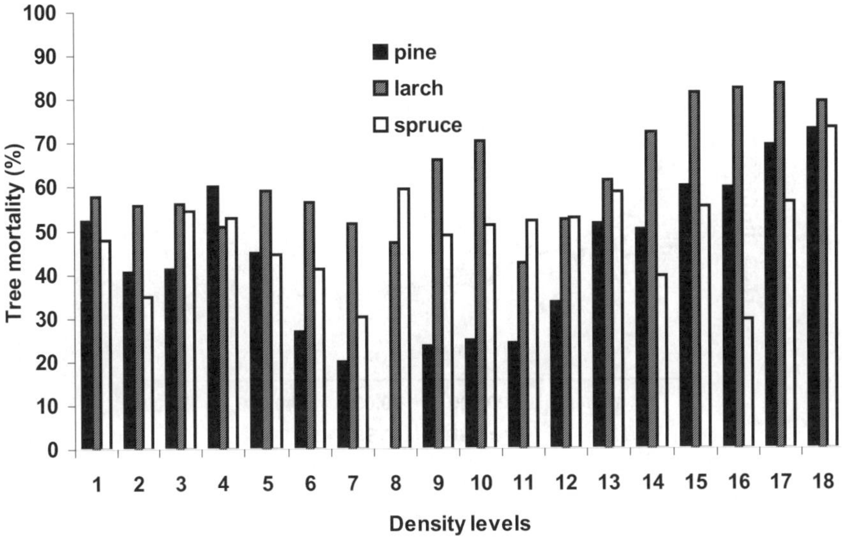

Figure 3. Trees mortality in relation to stand density treatment (see Table 1 for treatment densities).

Given the lack of density effect on tree height, and a strong effect on diameter, the relationship between height and diameter shifted dramatically with density. Trees in high-density plots had much less taper, reaching the same height as trees at lower density despite smaller diameters (Figure 4; Vysotskiy, 1962, Eytingen, 1918). The greater taper in widely spaced trees is consistent with the expectation that tree taper is a function of the wind environment experienced by a tree; wider spacings expose larger canopies to greater wind velocities, and trees respond by adjusting taper to provide for uniform resistance to wind stress along the stem (see for example West et al. 1989).

The height of the base of the live crown increased with density increases up to about 8,000 to 10,000 trees/ha, beyond which there was little response (Figure 4). As density and diameter correlated strongly, the height of the base of the live crown also correlated strongly with diameter (pine r=-0.894, larch r=-0.841, spruce r=-0.725). The shade-tolerant spruce had much lower heights to the base of the live crown, though this may relate to the shorter stature of spruce rather than to shade tolerance of spruce needles; the depth of live crown (total height minus height of base of live crown) was about 4 m for all species.

Stand basal area and total stem volume increased with increasing density; at 10,000 trees/ha, basal area and stem volume were just 60-80% of the maximum values at the highest densities (Figure 6). Some silviculturists expect increments should "stagnate" at high densities (see for example Smith

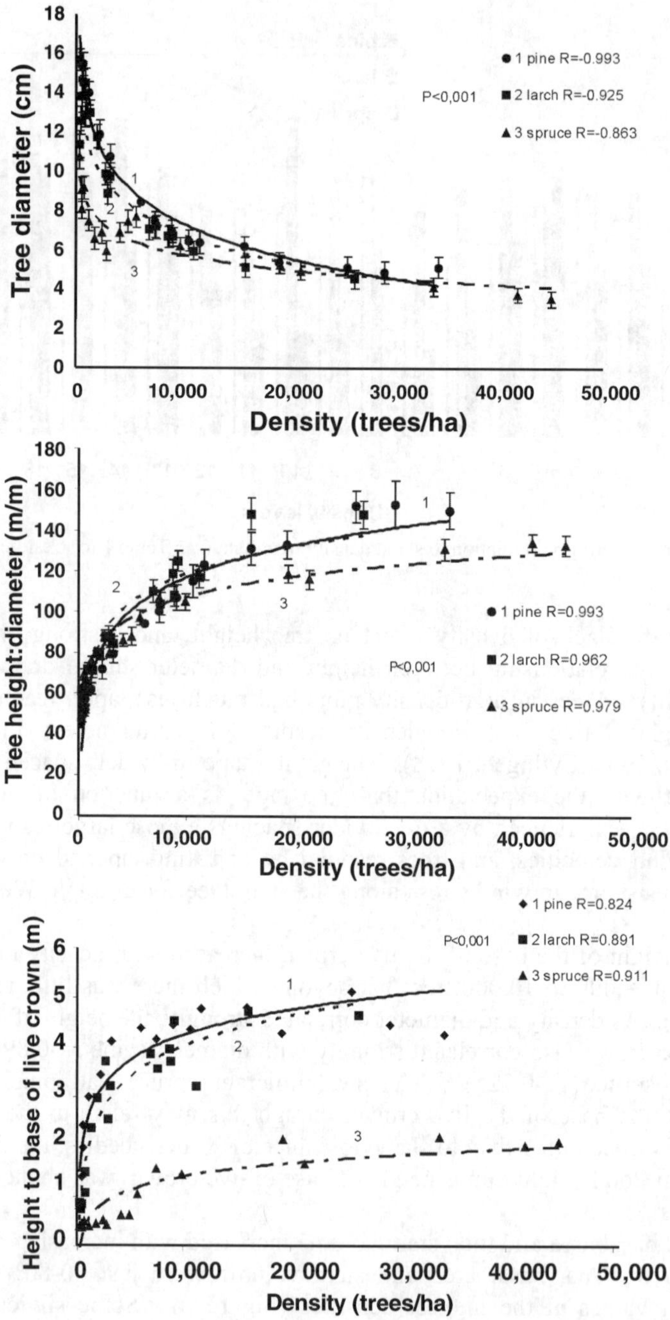

Figure 4. Tree diameter, height:diameter, and height of the base of the live crown as a function of density.

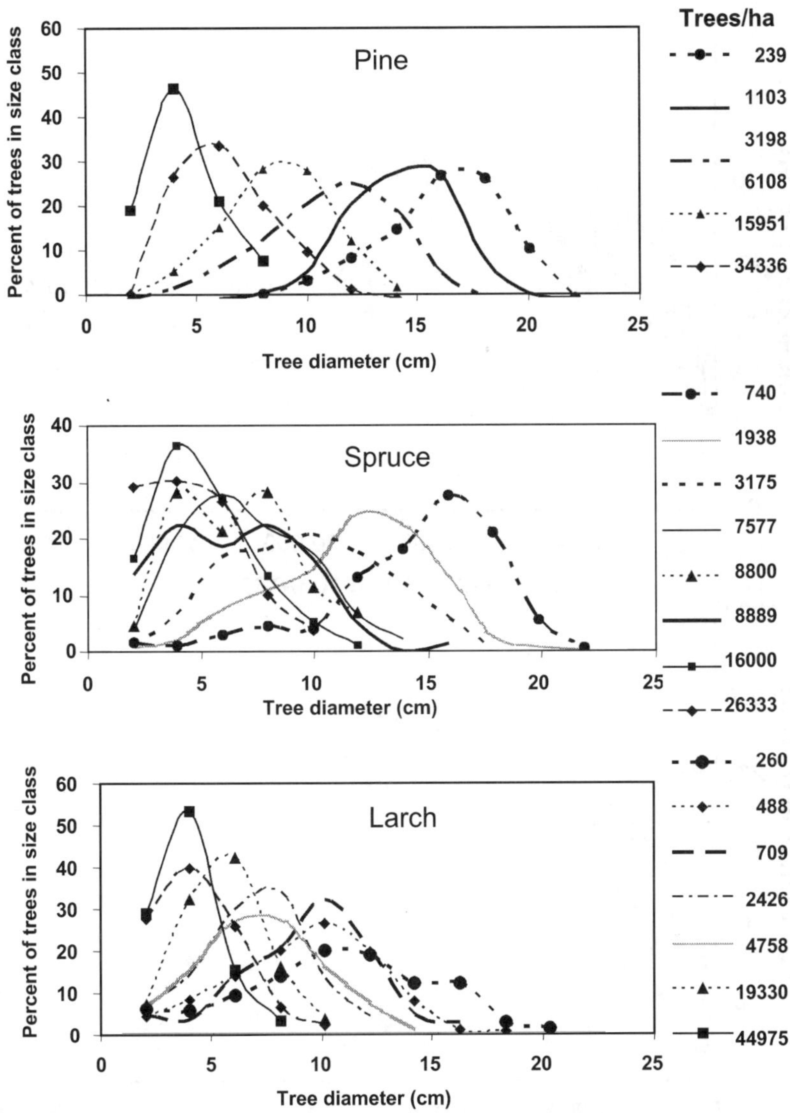

Figure 5. Distribution of tree sizes by 2-cm diameter steps (numbers are stand density).

et al. 1997), but our results disprove this idea. High-density plots had more than 10 times the basal area and volume of the lowest density treatments. The lower yield of low-density stands reflected in part the greater competitive pressure of grasses and herbs. In Siberia, intensive control of competing vegetation is uncommon, so this experiment's design did not attempt to

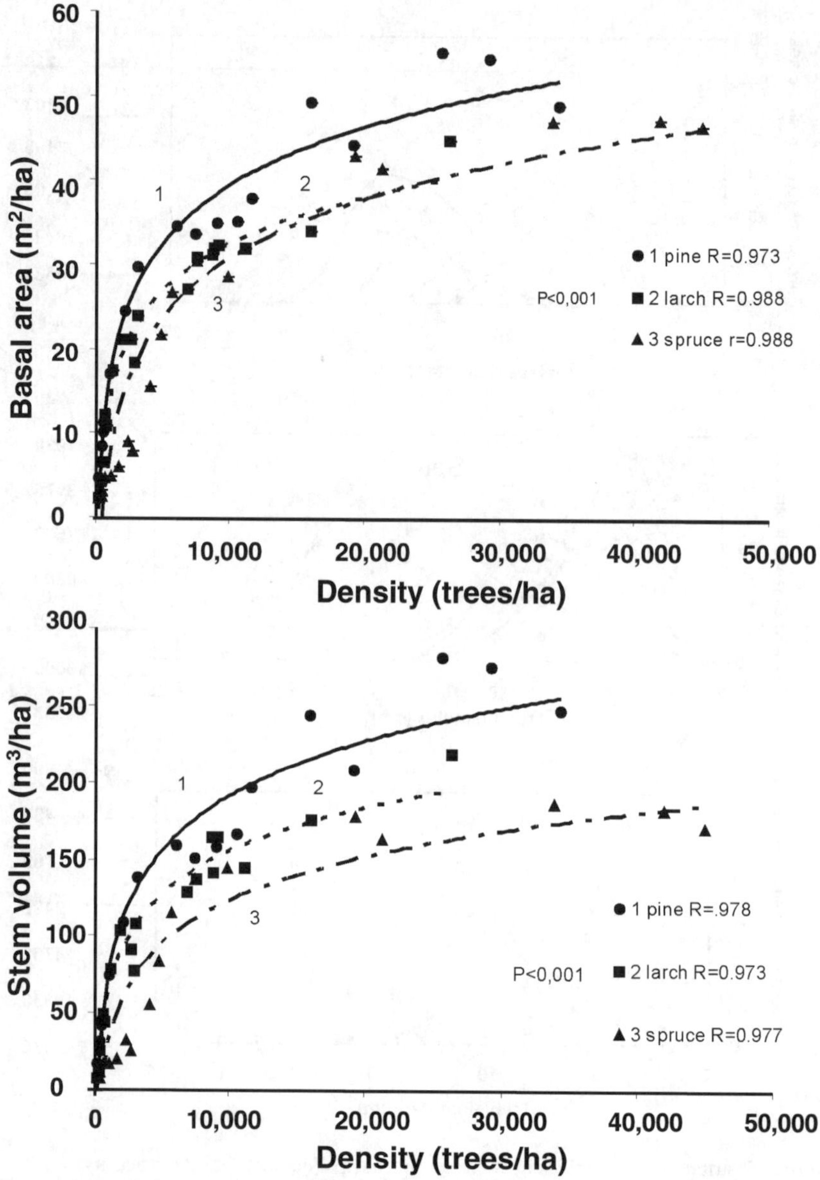

Figure 6. Basal area and total stem volume as a function of density.

separate the effects of tree spacing alone from the interactive effects of tree spacing and understory competition.

Overall, tree density strongly influenced all parameters for all species, except for height. The lack of sensitivity of height to the density of trees reinforces the classical idea that comparisons of site productivity can be based on the heights of dominant trees, as this index is relatively insensitive to the density of trees in a stand.

ACKNOWLEDGEMENTS

This research was supported by the grant from the Russian Foundation of Fundamental Investigations (N 04-04-49279).

REFERENCES

Buzykin A I Pschenichnikova L S 1999 Density impacts on pine plantation morphological ctructure and productivity. Lesovedenie. 3, 38-43. (In Russian)

Eytingen G R 1918 Influence of stand density on plantings growth. Petrograd. 38 pp. (In Russian)

Hozumi K 1980 Ecological and mathematical considerations on self-thinning in even-aged pure stands. Bot. Mag. Tokyo. 93, 149–166.

Kayrukschtis L A and Uodvalkis A I 1976 The phenomenon of intracpecies competition change on mutual tolerance of the individuals in spruce coenosis *In* Contemporary researches of productivity and wood cutting. pp. 47-64. Kaunas. (In Russian)

Kondrat`ev P S 1959 The new data about pine growth at different density levels. Proc. TSHA. 2 (27), 141-165. (In Russian)

Martynov A N 1974 Density of coniferous stands and its importance. CBNTIleschos, Moskow, Russia. 60 pp. (In Russian)

Moiseev V S 1971 Inventory of young forests. LTA, Leningrad. 342 pp. (In Russian)

Pobedinskij A V 1966 Investigation of reforestation processes (Methodical instructions). Nauka, Moskow. 64 pp. (In Russian)

Pshenichnikova L S 1978 Productivity of pine young trees at different density levels *In* Forest Productivity factors. pp. 36-52. Nauka, Novosibirsk. (In Russian)

Ryabokon` A P 1979 Definition of a pine stand density biological optimum. Lesovedenie. 3, 16-23. (In Russian)

Smith D M, Larson B C, Kelty M J and Ashton P M S 1997. The Practice of Silviculture: Applied Ecology. 9th Edition. Wiley and Sons, Inc. 537 pp.

Timofeev V P 1957 Forest density and layering as a condition of their productivity. *In* Science achievement in a wood facilitiesUSSR for 40 years. pp. 109-142. Proceedings of the symposium. Goslesbumisdat, Moskow, Leningrad. (In Russian)

Usoltsev V A 1988 Growth and structure of stand phytomass. Nauka, Novosibirsk. 254 pp. (In Russian)

Usoltsev V A 1994 A partition of edaphic and ecosystem components of stand productivity on the density experiment data In Ural forests and its economy. Proceedings of Russian scientific conference. Ekaterinburg. 77-85 pp. (In Russian)

West P W, Jackett D R and Sykes S J 1989 Stresses in, and the shape of, tree stems in forest monoculture. J. Theor. Biol. 140, 327-43

Vysotskiy K N 1962 Rules of a structure compound stands. Lesnaja promischlennost`. 5, 177. (In Russian).

Chapter 15

THE SIBERIAN AFFORESTATON EXPERIMENT: HISTORY, METHODOLOGY, AND PROBLEMS

L.S. Shugalei
Sukachev Institute of Forest, Siberian Division, Russian Academy of Sciences, Akademgorodok, Krasoyarsk, 660036, Russia
e-mail:biosoil@forest.akadem.ru institute@forest.akadem.ru

INTRODUCTION

Many of the environmental issues of the 21st Century will drive changes in our forests and their soils, and some of the changes in the forests and soils may feedback to alter the size and impact of environmental changes. Increasing concentrations of atmospheric CO_2 may affect tree growth directly, and any changes in climate could alter the geographic location of vegetation zones, and the frequency of major disturbances. Scientists at the soil science laboratory at the V.N. Sukachev Institute of Forest and Wood SB of the USSR Academy of Sciences have studied all the complexity of forest and soil interaction and interdependence. The influence of tree species on soils in natural forests is difficult to isolate from the wide array of other factors that influence soil over variable time periods.

In 1968 Professor N.V. Orlovsky decided to establish a long-term forest experiment to explore the effects of the major forest species of Siberia. The effects of the species would be determined by planting them on the same substrate under the same climatic regimes (Orlovsky, 1968). The study site was 1.7 km^2 located on the vast terrace of the Kacha river, a western tributary of the Yenisei river (56° 13′ N, 92° 19′ E). The soil on the site was originally a grey soil developed on chestnut-brown clay (Anonymous 1997, Table 1); agricultural activity on the site resulted in the incorportation of the original O horizon into a partially homogenized anthropogenic epipedon. Despite a long agricultural utilization, the agro-grey soils have remained lythogenic and palynological features of profile differentiation by elluvial-illuvial type: illuviation of silt, humus, segregation of organogenic – mineral compounds to ortsteins.

D. Binkley and O. Menyailo (eds.),
Tree Species Effects on Soils: Implications for Global Change, 257–268.

In preparation for the planting of the common garden experiment, the top 20-cm layer of the soil was removed from the whole site area in 1969 and 1970, and piled into two large mounds at the edge of the site. The 20-60 cm depth soil was plowed, packed, and then the top layer was spread evenly across the site. The success of the soil homogenization treatment was variable; the actual depth of topsoil distributed across the site averaged 30 $\pm$ 1 cm, and covered about 60% of the area. The uniformity of the soil was tested by planting a crop of wheat before the trees. Unfortunately, the wheat crop showed dramatic variation across the site, and the variation in yield correlated with depth of the applied layer of topsoil ($r^2 = 0.64$). Despite the lack of uniformity, the pattern in wheat yield and topsoil depth allowed the experiment to be planted in an area with average soil fertility (Shugaley et al. 1984). The plantation soil (and the soil under a 40-year-old birch stand nearby) was sampled for textures, micro- and macroaggregate structure, concentrations of organic matter, N, extractable cations, pH (water and KCl), titratable acidity, and bulk density were determined by standard methods (Arinushkina, 1970, Khaziev, 1976, Shugaley et al. 1984, Zvyagincev 1991).

Six 2400 m^2 plots were established, with each plot planted to monocultures in 1971-1972: arolla pine (*Pinus sibirica*), Scots pine (*Pinus silvestris*), Siberian larch (*Larix sibirica*), Norway spruce (*Picea abies;* = *Picea obovata* Ledeb.), aspen (*Populus tremula*) and birch (*Betula fruticosa*). The density of planting was unusually high (0.5 x 0.5 m) to hasten crown closure and the development of species-driven effects (Figure 1).

Over the course of the experiment, we focused on quantifying the effect of the tree species on: (1) water and temperature regimes, (2) the development of microbial and soil animal coenoses (communities) and (3) the biological activity and cycling of substances. We used these characterizations over a period of 25 years to answer a series of questions about the impacts of the six tree species on soils.

HOW DID THE SOILS CHANGE AS A RESULT OF THE INFLUENCE OF TREES?

Soil horizon formation has been important and variable among species. For example, the AY1 horizon was 4 cm thick under conifers and aspen, compared with 2 cm thick under birch. The transition line to the AY2 horizon was uneven. The layer embraced by modern soil formation was characterized by the more dark color, rich inclusions of organic mass and bulk density of 1.05 to 1.09 kg/L under arolla pine, Siberian larch, pine, aspen and 1.140 to

Table 1. Chemical and physical properties of grey and agrogrey soils

Soils	Horizon	Depth (cm)	Organic matter (g kg^{-1})	Exctractable cations (mg/kg)			-----pH------		Base saturation (%)	$\frac{SiO_2}{R_2O}$	Clay (%)	Aggregates >0.05 mm, (%)	Water stable aggregates <0.25 mm, (%)	Bulk Density (Mg/m^3)
				Ca^{2+}	Mg^{2+}	H^+	H_2O	KCl						
Grey	AY1	0-5	105.8	4380	780	57	6.1	5.2	83	6.6	85	58	64	1.05
	AY2	5-10	53.4	3640	720	57	6.1	5.2	81	7.0	85	47	64	1.09
	AEL	10-20	45.6	3640	650	65	6.0	4.8	78	6.8	94	58	66	1.19
		20-30	28.4	3480	700	64	5.8	4.7	78	5.6	54	47	64	1.28
	BT1	30-40	17.7	3640	756	60	5.8	4.5	80	4.7	64	52	61	1.30
		40-50	11.2	3800	756	59	5.7	4.3	84	4.2	69	39	42	1.36
	BT2	50-80	4.9	4100	744	58	5.8	4.3	82	4.6	80	39	59	1.40
	BT3	80-100	3.2	4240	780	54	6.0	4.3	84	4.8	88	39	47	1.42
Agro -grey	PY	0-5	37.9	3780	696	41	6.4	5.1	86	5.2	78	53	65	1.29
		5-10	37.9	3780	696	44	6.4	5.1	85	5.6	78	53	65	1.29
		10-20	35.1	3880	840	46	6.4	5.0	85	5.2	72	53	66	1.31
	AEL	20-30	26.1	3760	852	46	6.2	4.7	85	5.1	69	56	67	1.33
		30-40	18.2	3800	804	46	6.1	4.6	85	4.8	68	44	67	1.33
	BT1	40-50	13.1	3840	756	43	5.8	4.4	85	4.8	76	42	66	1.37
		50-80	7.8	3940	864	43	5.8	4.4	86	4.8	86	39	71	1.38
	BT2	80-100	3.2	4020	852	44	6.1	4.5	86	5.1	90	40	62	1.40

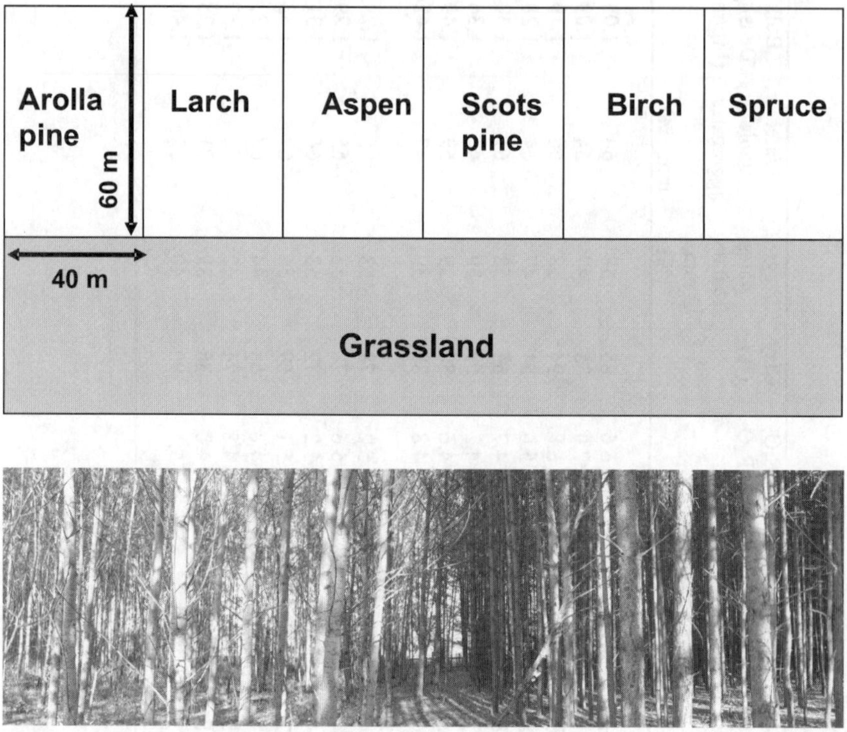

Figure 1. Layout of the Siberian Afforestation Experiment (upper), and the boundary between aspen and Scots pine plots 32 years after planting (lower).

1.16 kg/L under spruce and birch. The loose soil structure was characterized by the powder-like small grains, cloddy small grains or fragile, foliated- small grains (under larch).

The main root mass of trees and grasses (85- 98%) under spruce, arolla pine and Scotch pine was concentrated in the 0-30 cm depth; larch, birch and aspen concentrated 63%, 77%, and 81% of their root mass in this depth. The maximum depth of rooting for all species was 100 to 110 cm. Rich mushroom mycelium was found under spruce at the depth 72 cm. Its abundance and thickness under other tree species was much less: 40 cm under arolla pine, 22 cm under pine, and 12 cm under larch. Mushroom mycelium under aspen and birch was poor and its main mass is focused in the top 10 cm. The largest organic matter increases in the 0-5 cm depth developed under larch and aspen (Figures 2, 4). Organic matter also increased in the 5-10 cm depth under arolla pine, larch, pine and aspen, but the change under spruce and birch at this depth was not reliable. On a relative basis, the accumulation of soil organic matter was greatest in the 0-10 cm depth under larch and aspen (18 and 23%), followed in decreasing order by arolla pine, Scots pine, spruce and birch. The patterns in N accumulation in this upper mineral soil layer were similar to the accumulation of organic matter (Figure 3).

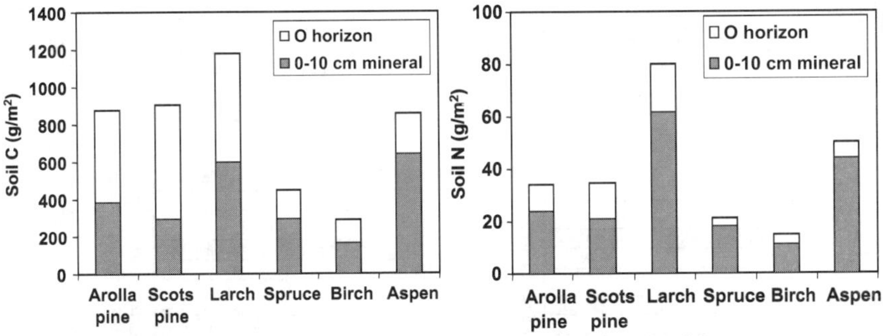

Figure 2. Carbon and nitrogen accumulation under six tree species after 25 years.

We characterized the soil organic matter pool by measuring the concentration of fulvic and humic acids. The ratio of humic to fulvic acids decreased from larch (1.5) to Scots pine (1.4) to aspen (1.2) to arolla pine and spruce (1.1) to birch (1.0). All of these represent a relative increase in humic acids, as the ratio of humic to fulvic acids in the agro-grey soil under cropland was 0.9. The degree of humification of organic matter under larch, pine and aspen trees was 55-59%, compared with 51-53% under arolla pine, spruce, and birch, and 49% in the agro-grey cropland soils it makes 49%. The newly formed humus had a wider C:N ratio (15-16) than the prior cropland soil.

The content of extractable base cations did not vary much among species. Soil pH in water declined by about 0.5 units in 25 years across all species, matched by increases in exchangeable and titratable (hydrolytic) acidity; any differences among species appeared minor (Figure 4).

The impacts of larch, pine and birch trees may be especially strong in this experiment, as the southern taiga of the Kemchug Hills provide very suitable conditions for these species, and they also grew extremely well at this site. The growth of these species would fall within the Class I bonitet ("bonitet" is a quantitative assessment of soil productivity obtained at the compatible input level (Karmanov, 1980); bonitet class I has the highest productivity), indicating the trees have good access to water and nutrients. High foliar concentrations of nutrients at ages 6 and 25 years also indicate adequate nutrient supplies. The lower productivity of arolla pine and spruce likely results from physiological factors (such as competition between trees at high density for light or water) rather than limitations in soil nutrient supply (Shugalei and Popova, 1999). The plots in the afforestation experiment differed substantially in the accumulation of carbon and nitrogen in the soil, as well as the distribution between the O horizon and the 0-10 cm mineral soil (Figure 2). At 25 years of age, the greatest differences were between birch and larch; the soil in the birch plot had 890 g/m^2 less C, and 65 g/m^2 less N. If these soils were identical when the plantations were established, the rate of divergence would have been quite large, about 35 g/m^2 for C and 2.6 g/m^2 for N. The O horizon comprised about 25% of the C under aspen, compared with 67% under Scots pine. The C:N ratio was lower in the mineral soil than

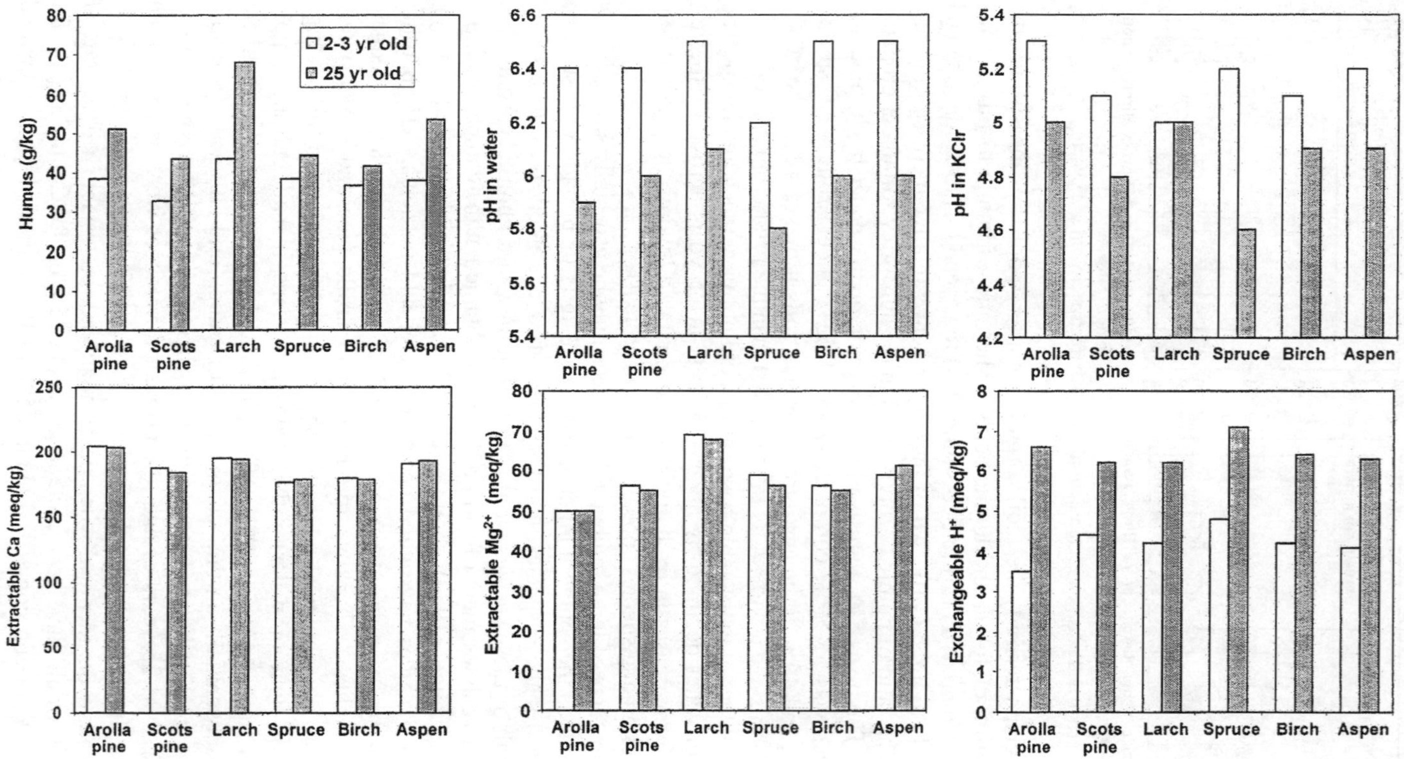

Figure 3. Species effects on concentrations of humus, soil pH, and exchangeable cations in 0-10 cm depth mineral soil. Increases in humus concentrations were associated with acidification of the soil, with only minor changes in extractable cations.

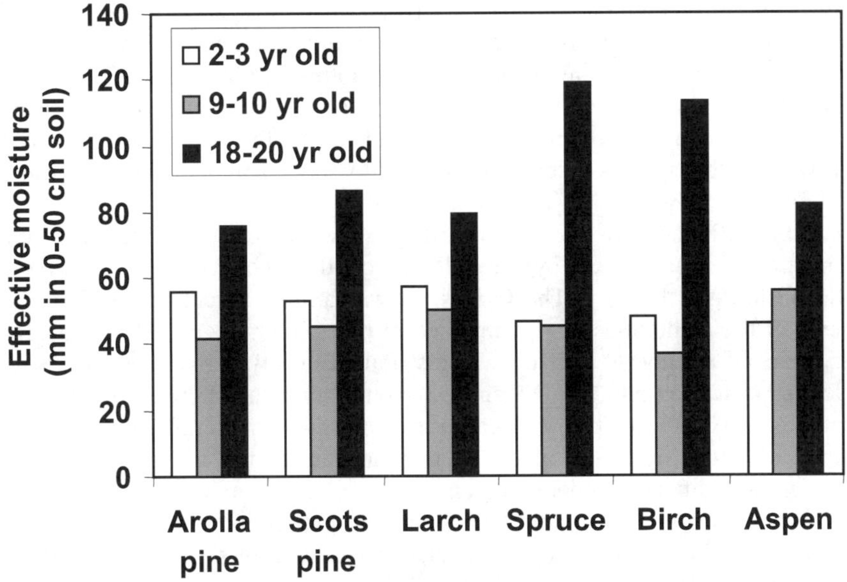

Figure 4. Pool of effective moisture in 0-0.5 m layer of grey soil under six tree species.

in the O horizon, so the percentage of N in the O horizon ranged from 12% under aspen to 40% under Scots pine. Along with the increases in soil C, the soils acidified by about 0.5 units (in water) under all species (Figure 3).

WHICH ELEMENTARY SOIL FORMING PROCESSES CREATED THESE CHANGES?

The water regime of the soil at this site is characterized by a thorough wetting in early spring, and sometimes in late fall (Beskorovainaya, Vedrova et al., 1997). The development of closed canopies in the plots has resulted in varying levels of interception loss and transpiration by the trees. Through age 10, differences in hydrology among species appeared to be minimal (Figure 4). By 18 years, the available soil water appeared to be about 30% greater under spruce and birch; at the same time, the available soil water appeared to have increased by more than 50% for the other species. Differences in available water storage could result from greater interception loss (evaporation of precipitation from the canopy, without entering the soil or trees), or from differences in transpiration.

The soils generally freeze in late October to early November, with freezing temperatures penetrating to 80 to 150 cm. Thawing begins after snow melt, in late April to early May. For the first 10 years of the experiment, the temperature regime across the plots was variable, with cooler summer temperatures (by 1 to 3 °C) under the developing canopies in the upper 20 cm of mineral soil. By the 18-20 years of age, the pattern of soil temperature within plots had become more uniform and cooler (by 2 to 4 °C) in the summer, approaching the soil temperature regime in nearby forests. Overall, the development of soil moisture and temperature regimes were representative of the regimes found in forests of the southern taiga, with growth of the species likely constrained by cool soil and summer water deficits.

The influence of the forest canopy on mineral soil moisture and temperature was joined by the effects of the accumulating O horizon (Karpachevsky, 1999). The O horizon was almost absent in the first few years of the plantations (and comprised of grass litter; Figure 5), and by 18-25 years of age the O horizon was distributed evenly across the stand with clear stratification into O1, O2, and O3 subhorizons (the O3 was absent under spruce). The O horizon was comprised mostly of material from trees, as input of litter from understory plants accounted for only 3% of total aboveground litterfall under conifers, and 3-6% of that under aspen and birch.

Decomposition of organic compounds in conifer litters runs 1.5-2 times slower than in litters of deciduous stands (see chapters in this volume by E.Vedrova, L.Mukhortova). Early in the development of the plantation, the decomposition of O horizons was highest where grass vegetation was most dominant (such as the aspen plot; Figure 6). By 1995, the differences among species were much lower. The overall hydrolytic activity (catalase) remained higher for autumn-deciduous species (larch, aspen, and birch) than for evergreen conifers (Figure 6). The overall biological activity of the O horizon increased over time for all species.

The availability of N in the O horizon can be examined by characterizing the inorganic extractable pool (readily available to plants), the pool that is somewhat easily hydrolized (coming available over longer periods), and the pool that is more difficult to hydrolize (very unavailable). More than 90% of the N occurred in the hard-to-hydrolize pool for all species (Figure 7). The easily hydrolyzed pool differed by about two-fold among species, with the lowest concentration under spruce, and highest under aspen. Given that the spruce O horizon had twice the mass of the aspen O horizon (Figure 5), these concentration differences would even out in terms of mass of easily hydrolyzed N /m^2.

Overall, we expect that the pool of soil organic matter in these experimental plots will reach the levels found in natural forests (never plowed) within 50 to 70 years for larch and aspen, and 110 to 150 years under arolla pine, Scots pine, and spruce. The birch plot would likely take 260 years.

We conclude that all experimental plots demonstrated strong activity as carbon sinks following afforestation under conditions in the southern taiga, but the rate of carbon accumulation depended strongly on species. The differences among species related to the qualitative and quantitative structure of the litterfall inputs, and these impacts on the soil biota. The data obtained

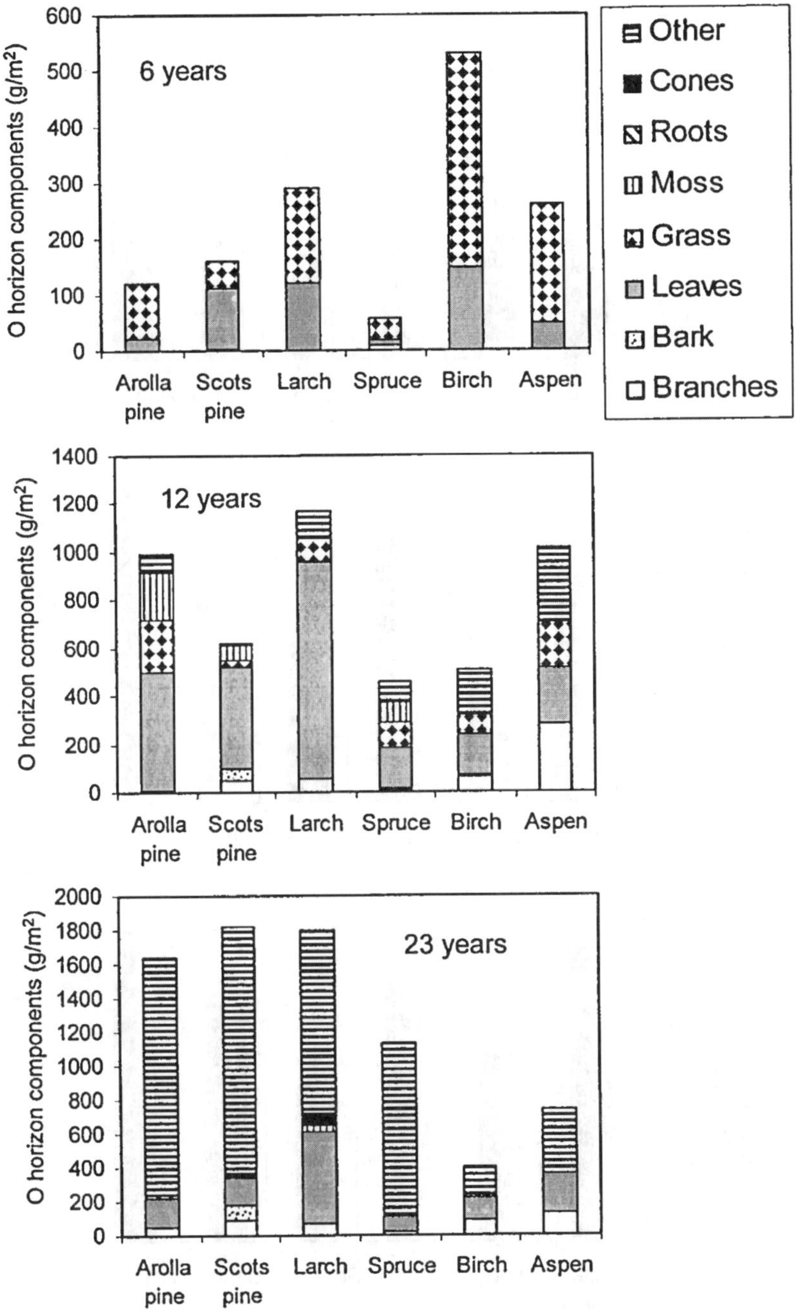

*Figure 5.*Accumulation of O horizon components at Dynamics of litter stocks under six tree species.

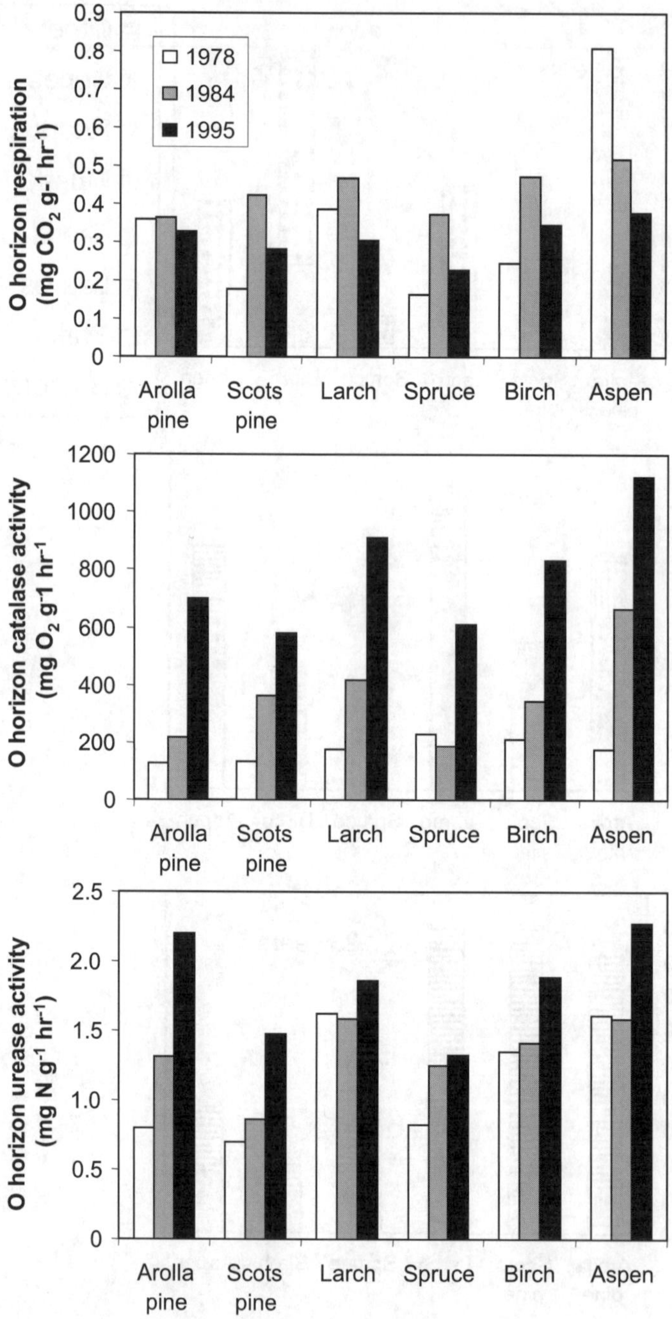

Figure 6. Dynamics of biochemical activity of forest litters.

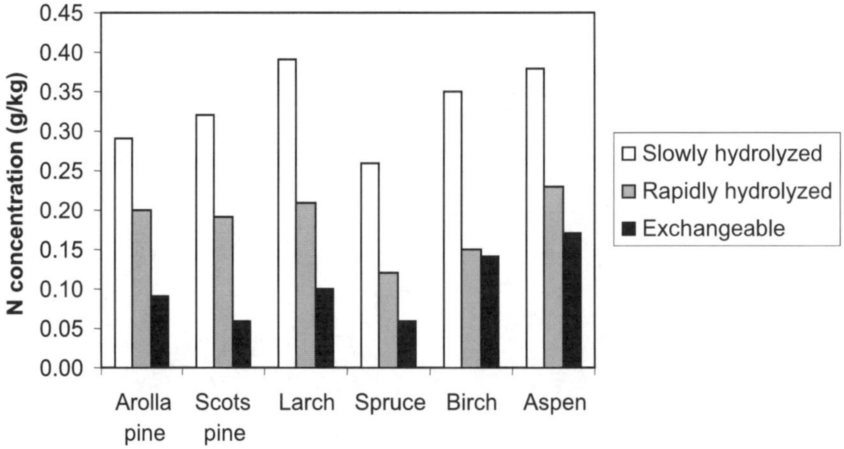

Figure 7. Concentrations of N in exchangeable, rapidly hydrolyzed, and slowly hydrolyzed pools in the O horizon.

from this experimental plantation may be used for building mathematical models to predict soil response to changes in both climate and species composition.

We speculate that longer-term development of these plots would lead to self-thinning in the spruce, arolla pine, larch and pine stands, as well as the replacement of the aspen and birch stands by coniferous species that are typical of the southern taiga (and climate) of the Kemchug Hills. These changes would complicate the evaluation of the impact of individual tree species on soil properties, and novel techniques need to be developed.

REFERENCES

Anonymous 1997 Classification of Russian soils. V.V.Dokuchaev Soil Institute. Moscow. 236 pp. [In Russian]

Arinushkina E V 1970 Manual in chemical soil analysis. Publication House of Moscow University, Moscow. 487 pp. [In Russian]

Bezkorovainaya I N Vedrova E F Popova E P Spiridonova L V Shugalei L S and Yashikhin G I 1997 Development of artificial forest ecosystems. Siberian Ecological Journal. 4, 393-403. [In Russian]

Karmanov I I 1980 Soil Fertility in the USSR (Natural Regularities and Quantitative Assessment). Kolos, Moscow. 224 pp. [In Russian]

Karpachevsky L O 1999 New approaches to forest soil studying. Soil Sci. (Russia). 1, 152-160. [In Russian]

Khaziev F K 1976 Enzymatic soil activity. Nauka. Moscow. 176 pp. [In Russian]

Orlovsky N V 1968 Afterword. *In* Soil and Forest. pp. 593-598. Book Publ. House, Krasnoyarsk. [In Russian]

Shugalei L S and Popova E P 1999 Regime of culture nutrition on old arable soils. Lesnoye khozyaistvo (Forestry). 4, 34- 37. [In Russian]

Shugaley L S Semichkina M G Jashikhin G I and Dmitrienki V K 1984 Development modelling of artificial forest ecosystems. Nauka, Siberian Branch of RAS, Novosibirsk. 151 pp.

Zvyagincev D G 1991 Methods of soil microbiology and biochemistry. Publication.House of Moscow University, Moscow. 303 pp. [In Russian]

Chapter 16

PRODUCTIVITY OF SIX TREE SPECIES PLANTATIONS FOR THREE DECADES IN THE SIBERIAN AFFORESTATION EXPERIMENT

V.V. Kuzmichev, L.S. Pshenichnikova and V.A. Tretyakova
Sukachev Institute of Forest, Siberian Division, Russian Academy of Sciences, Akademgorodok, Krasoyarsk, 660036, Russia Akademgorodok, Krasoyarsk, 660036, Russia

ABSTRACT

Measurements were made of height, diameter and number of trees in the Siberian Afforestation Experiment for monoculture plots with six tree species (*Picea abies (=Picea obovata), Pinus sylvestris, Pinus sibirica, Larix sibirica, Populus tremula, Betula pendula*), established in the central part of the Krasnoyarsk region. Yield tables were developed for these plantations, and in three of the plantations the tables were developed for management with and without thinning. Almost 150 felled trees were measured to evaluate aboveground biomass fractions and determine biomass dynamics through stand development. Local uncertainties in needle mass dynamics were revealed for Scots and arolla pine stands. Carbon storage was evaluated for live trees and in biomass of dead trees and fall of branches and needles. Over the 35 years of stand development, these species accumulated 100 to 300 m^3/ha of wood, with a likely carbon content of about 25 to 75 Mg/ha. The trees also added a large amount of C to the soil in fallen leaves, branches, and dead roots. In this region, Norway spruce and arolla pine form the longest-lived forests, and may have higher productivities than the other species later in stand development.

INTRODUCTION

Prediction of future development in forests depends on information gained from long-term experiments that examine the influence of stand

D. Binkley and O. Menyailo (eds.),
Tree Species Effects on Soils: Implications for Global Change, 269–279.

density and species composition. The early stages of development of stands may be the most important, as many later developments depend strongly on the conditions and trajectories early in the life of the stand. The development of research on forests under Siberian conditions began in the 1920s, but the available information was generally not sufficient to develop stand growth and yield tables. Classic work includes the papers of V.N. Gabeyev (1971) and V.V. Ogievsky (1966), dealing with the growth of pine and larch growth up to 60 years of age. These assessments were based on temporary sample plots and growth analysis of individual trees, with stand densities of 5,000 to 1,500 trees/ha.

The Siberian afforestation experiment was established in 1968 by the researchers of the forest soil science laboratory to study the impacts of tree species on soil processes (Anonymous 1984). The experiment was located on the vast terrace of the river Kacha, the western tributary of the Yenisei River. Seedlings of 6 species (2 to 3 years old) were planted in 1971-1972 in 2,400 m^2 monoculture plots at a density of 40,000 stems/ha, following homogenization of the 0-20 cm soil (Shugalei, this volume). The species were Scots pine (*Pinus sylvestris)*, Norway spruce (*Picea abies;* = *Picea obovata* Ledeb.), arolla pine (*Pinus sibirica)*, larch (*Larix sibirica)*, aspen (*Populus tremula)* and birch (*Betula pendula*). The genotype for birch was poor, so the development of this plot may not be representative of genotypes more suited for this site. The initial planting density was extremely high, at 40,000 trees/ha (4 trees/m^2).

Figure 1. Boundary between aspen and Scots pine stand at 32 years after planting.

Annual observations of tree growth began in 1975 (Semechkina 1984, Semechkina and Kostoustova 1985), and have been continued by L.S. Pshenichnikova (unpublished data) and E.F. Vedrova et al. (2000), V.V. Kuzmichev and V.A. Tretyakova (Figure 1). In this chapter we summarize the information obtained from these repeated observations, based on both permanent and temporary sample plots. The primary parameters measured were tree height and diameter, as well as crown dimensions. A full inventory of all trees was made in 1988 and 2002. The heights of all trees were measured in 1975-1984, but only 15-25 trees (across the range of stem diameters) were measured in 2002. Crown diameter was measured in two directions, along with height of the base of the live crown.

Eighteen 0.01 ha subplots were thinned at age 11 or 15 in the Scots pine, arolla pine, spruce, and larch plots; data are not available for the thinned subplot of larch. The growth in the birch and aspen plot was lower than for the other species, and no subplot was thinned. The frequency distributions for diameter were determined in all plots, and the mean diameter calculated based on mean basal area. The average height was determined by equations relating height to measured diameter. We fit the pattern of number of trees/ha with a Weibull function through year 35:

$$N= B1*(1\text{-}exp(-1*(A/B2)^{B3})), \qquad (1)$$

where B1 is an index characterizing stand density two years after planting; B2 is a coefficient of scale variation of age range; B3 is an index of tree thinning rate; A is an age; and N is a number of trees per hectare.

The pattern of height in relation to age was fit with a Weibull function and Koller's equation:

$$H= b1* (a^{b2})*exp\ (-b3*a) \qquad (2)$$

An exponential function was fit to characterized tree stand growth in diameter:

$$D=b1*(a-a_{1.3})^{b2} \qquad (3)$$

Wood volume per area unit was assessed by results of stem-section measurements on model trees. Local tables of volumes of small sized trees were also used (Zhilkin et al. 1977, Isaev and Korovin 1997, Usol'tsev 2003). Growing stock was equalized by Koller's function depending on age (2). The sampling of model trees included 73 Scots pine, arolla pine and larch at 8 years of age (Semechkina 1984). Another 58 Scots pine and arolla pine trees were sampled at age 15 or 16 years, along with 64 Scots pine and arolla pine, larch, aspen and birch trees at 21 years. Volume estimates were converted to mass at age 27 (Vedrova et al. 2000, Usol'tsev 2001) based on specific density. The mass of branches and leaves was equalized for Scots pine and arolla pine by the following equation:

$$P = (b1 * \exp(b2 * x) / (b3 + b4 * \exp(b5 * x))) + b6 * x \text{ age} \quad (4)$$

The volume loss from mortality was based on the ratio between the volume of mean-sized trees. The dead fallen branches and fallen needle (leaves) were assessed taking into account their life duration.

RESULTS AND DISCUSSION

Observations on separate 0.01 ha subplots showed great variation in survival, tree size, and growth. Two control subplots of arolla pine stand (#122 and #134, Figure 2) differed by more than 50% in density, though height and diameter were more similar, providing about a 20% difference in stem volume/ha. Variation was similar in the plots of other species.

Mortality was very high in these high-density stands (Tables 1-4), including drought-related mortality within the first few years that killed 86% of the aspen, 51% of the birch and 42% of the larch. Initial mortality was much lower in spruce (9%), Scots pine (22%) and arolla pine (24%). Although the plots developed from different initial stand densities, the differences among species became clear as self-thinning developed later in stand development (Figure 3). Self-thinning mortality was most rapid in the larch plot, eventually catching up with the density of the aspen and birch plots that experienced high initial mortality. The self-thinning trajectories appeared similar for the group of deciduous trees (aspen, birch, larch), and for the group of evergreen trees (spruce, Scots pine, and arolla pine).

The high early survival of spruce was similar to the results obtained by Kairyukshtis and Yuodval'kis (1976), where survival after 1 year was 86% following planting of 50,000 trees/ha. Similarly, the moderate early survival in our stands of Scots pine and arolla pine were similar to the 66% survival reported by Zhilkin et al. (1977) for high density plantations (40,000 trees/ha) of these species.

Height growth differed substantially among species (Tables 1-4), decreasing from aspen > larch > Scots pine > birch > spruce > arolla pine. The slowest growing species (spruce and arolla pine) also have the longest typical life span (about 300 years) and maximum height (35-40 m).

The pattern of height growth was consistent for the 3 tallest species (aspen, larch, and Scots pine), but the ranking for diameter shifted with arolla pine > birch > spruce (Tables 1-4). The pattern of stem volume accumulation included the effects of stand density as well as tree size, and Scots pine had more than 50% greater stem volume than the other species (300 m^3/ha).

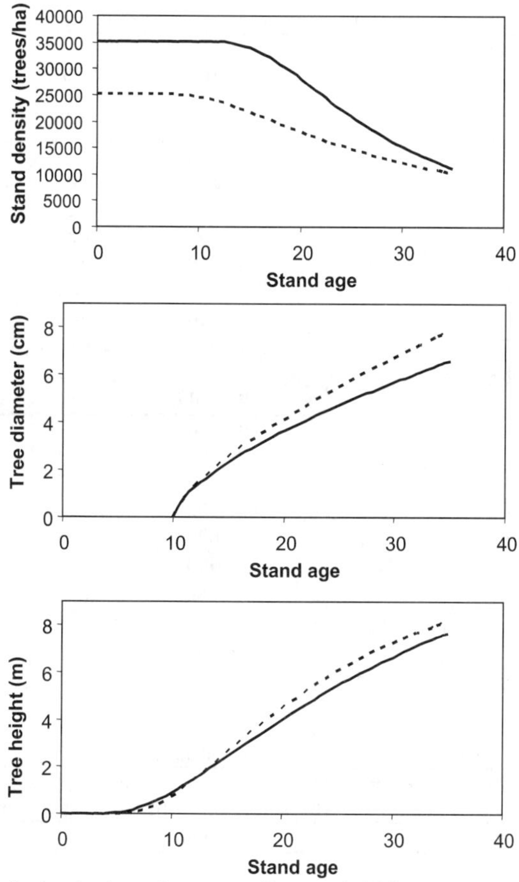

Figure 2. Changes in density (upper), average diameter (middle) and average height of arolla pine trees in two 0.01 subplots of the 0.24 ha arolla pine plot (#122 dashed line, #134 solid line).

Spruce and aspen had about 190 m^3/ha, followed by arolla pine and larch (165 m^3/ha) and birch (100 m^3/ha).

Thinning of these high-density plantations by removing half the rows of trees in Scots pine, arolla pine and spruce plots increased stem diameter growth but not height (Tables 1-3). After 20 years, the accumulated stem volume in thinned plots matched that of the high-density control plots, as the control plots experience higher self-thinning mortality (Table 5).

The mass of branches increased to an asymptote for both Scots pine and arolla pine, with more branch mass for Scots pine (Figure 4). The mass of leaves was similar for the two pines, based on an estimated curve with stand age.

Table 1. Dynamics of inventory indices in Scots pine stands.

Age (years)	Height (m)	Diameter (cm)	Density (trees/ha)	Basal area (m^2/ha)	Volume (inside bark, m^3/ha)
			No thinning		
5	0.3		31,400		1
8	1.1	0	31,100	0	7
10	1.9	1.7	30,200	7.2	18
11	2.4	2.2	29,500	11.3	26
15	4.5	3.6	25,200	26.2	77
20	7.0	5.0	19,500	38.2	159
25	9.1	6.1	15,100	44.5	228
30	10.7	7.1	11,900	47.3	275
35	11.8	8.0	9,600	48.3	302
			With thinning		
5	0.3		30,300		0
8	1.2	0.0	29,700	0.0	1
10	2.0	1.7	29,400	6.7	7
11	2.5	2.2	29,200	11.1	18
11 after	2.5	2.2	14,600	5.6	26
15	4.6	4.7	11,300	19.3	13
20	7.2	6.8	8,900	31.8	54
25	9.5	8.3	7,400	39.7	126
30	11.4	9.4	6,400	44.5	199
35	12.8	10.4	5,600	47.2	260

Table 2. Dynamics of inventory indices in arolla pine stands.

Age (years)	Height (m)	Diameter (cm)	Density (trees/ha)	Basal area (m^2/ha)	Volume (inside bark, m^3/ha)
			No thinning		
5	0.1		35,200		0
10	0.9	0	35,200	0	2
15	2.4	2.3	33,900	14.2	27
20	4.0	3.6	28,100	29.0	77
25	5.4	4.7	20,900	36.7	122
30	6.6	5.7	15,200	38.9	151
35	7.7	6.6	11,200	38.2	167
			With thinning		
5	0.1		35,200		0
10	0.8	0	35,200	0	2
15 before thinning	2.7	2.3	33,900	14.1	27
15 after thinning	2.7	2.3	14,300	5.9	12
20	4.6	4.7	11,700	20.2	59
25	6.0	6.1	10,200	29.5	106
30	7.0	7.0	9,200	35.7	145
35	7.9	7.7	8,400	39.6	175

Table 3. Dynamics of inventory indices in spruce stands

Age (years)	Height (m)	Diameter (cm)	Density (trees/ha)	Basal area (m^2/ha)	Volume (inside bark, m^3/ha)
			No thinning		
5	0.2		35,800		7
10	1.0	0	35,800	0	23
15	2.3	2.2	35,000	12.9	46
20	3.7	3.3	31,200	26.1	74
25	5.1	4.1	25,900	33.8	108
30	6.2	4.8	20,900	37.0	147
35	7.0	5.3	16,700	37.5	191
			With thinning		
5	0.2		35,800		7
10	0.9	0	35,800	0	23
15 before thinning	2.2	2.2	35,000	12.9	46
15 after thinning	2.2	2.2	15,500	5.7	27
20	4.0	4.0	10,500	13.2	51
25	5.9	5.2	8,800	18.5	87
30	7.6	6.2	8,200	24.4	135
35	8.9	7.1	7,900	30.8	198

Table 4. Dynamics of inventory indices in larch, aspen and birch stands

Age (years)	Height (m)	Diameter (cm)	Density (trees/ha)	Basal area (m^2/ha)	Volume (inside bark, m^3/ha)
			Larch		
5	0.6	0	22,900	0	2.9
10	2.1	1.8	20,500	5.4	18.2
15	4.4	3.7	13,400	14.2	46.2
20	6.9	5.5	8,300	20.1	80.6
25	9.4	7.4	5,400	23.4	114.5
30	11.5	9.3	3,700	25.2	142.8
35	13.2	11.2	2,700	26.2	162.9
			Aspen		
5	0.5		4,900		2
7	1.2	0	4,900	0	6
10	2.8	2.0	4,900	1.5	14
15	6.4	4.7	4,400	7.6	38
20	10.0	7.2	3,800	15.1	72
25	13.2	9.5	3,100	22.3	112
30	15.7	11.8	2,600	28.5	154
35	17.5	14.0	2,200	34.0	194
			Birch		
5	1.3	0	19,000	0	15
10	2.7	1.9	18,800	5.3	62
15	4.2	3.0	16,100	11.3	109
20	5.6	3.9	12,300	14.7	134
25	7.0	4.7	9,200	16.0	136
30	8.3	5.5	6,900	16.2	123
35	9.6	6.2	5,300	15.9	102

The shape of the curve was assumed from the literature (Usoltsev 2003), combined with one or two direct sampling. The mass of leaves:branches showed a very dynamic pattern of 2-fold increases and declines; this indicates either a very dynamic situation or that the pattern of leaf mass with age from the literature may not apply here. The effects of thinning on mass of branches and leaves lasted only a few years.

We estimated the cumulative mass of fallen leaves and branches for 35 years of stand development (Table 6). The mass of fallen leaves was 4 to 10 times greater than the mass of fallen branches. The greatest litter inputs were found for birch and larch, and the least for arolla pine and spruce.

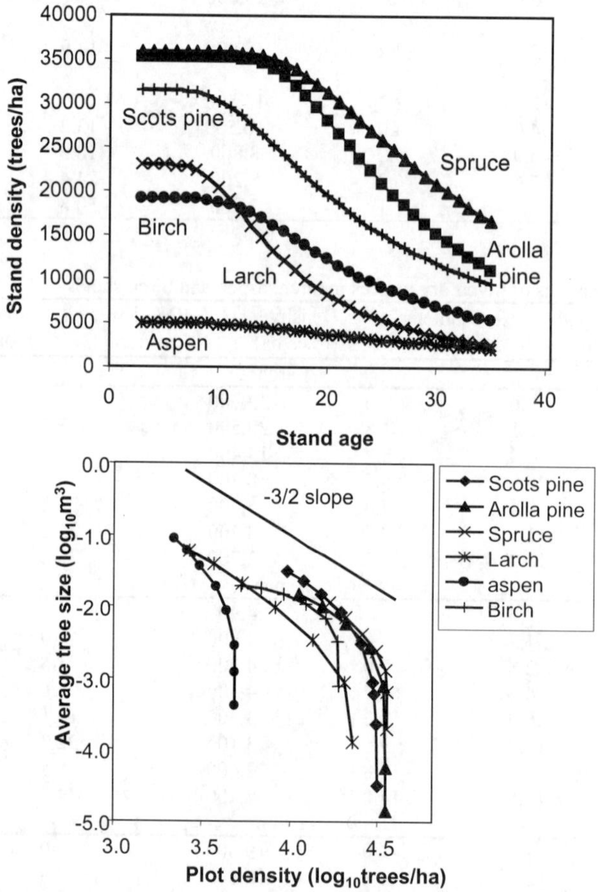

Figure 3. Self-thinning in the plots; trees/ha with age (upper), and self-thinning curves (lower).

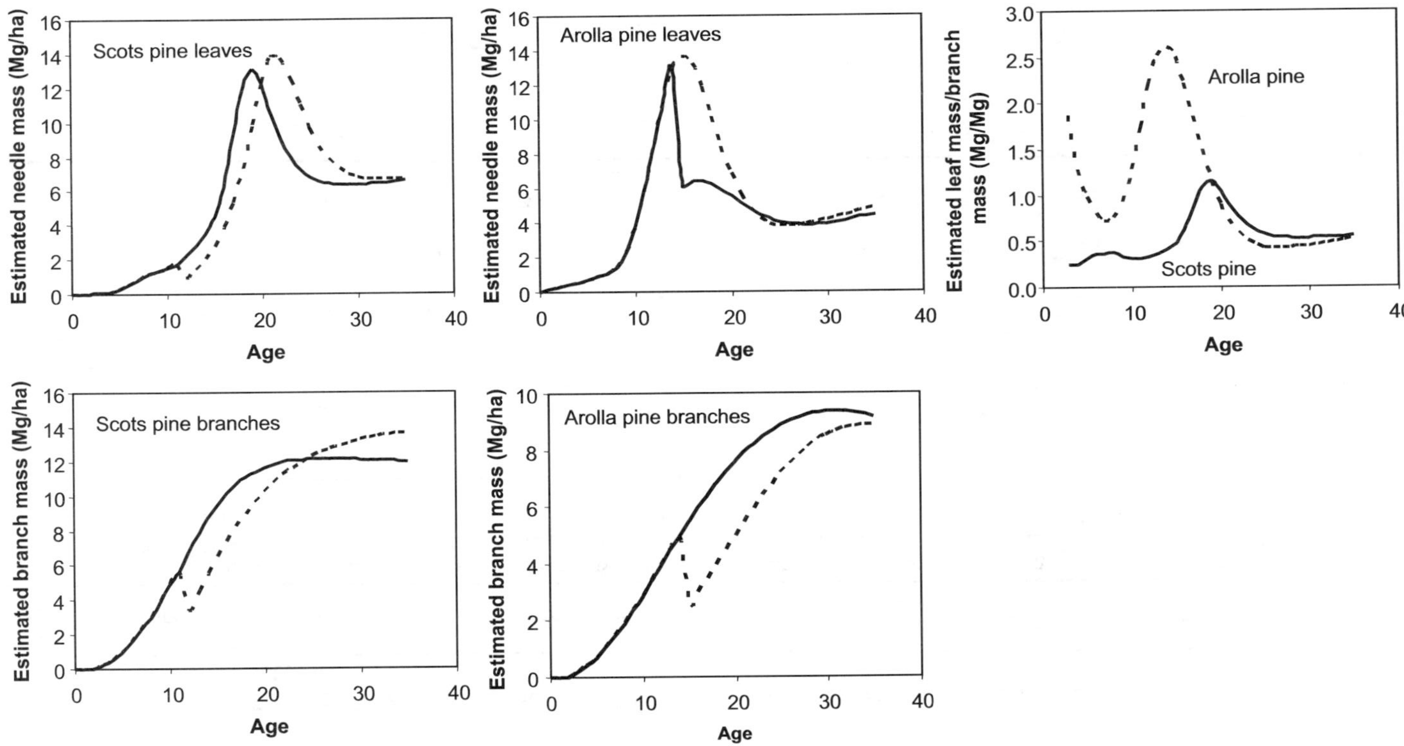

Figure 4. Estimated development of leaf (upper) and branch (lower) mass in the Scots pine (left) and Arolla pine plot (center). The ratio of leaf mass:branch mass (right) indicates a very dynamic relationship necessary for the peak in leaf mass.

Table 5. Tree number and mortality between measurement periods for unthinned subplots.

Age, years	Scots pine		Norway spruce		Arolla pine		Larch		Aspen		Birch	
	Mortality (trees/ha)	Volume loss (m^3/ha)	Mortality (trees/ha)	Volume loss (m^3/ha)	Mortality (trees/ha)	Volume loss (m^3/ha)	Mortality (trees/ha)	Volume loss (m^3/ha)	Mortality (trees/ha)	Volume loss (m^3/ha)	Mortality (trees/ha)	Volume loss (m^3/ha)
8	1.2	0.1	0	0	0	0	2.5	0.1	0.1	0.1	0.2	0.1
13	5.1	0.2	0.8	0.1	1.3	0.1	7.1	0.5	0.4	0.2	2.7	0.6
18	5.6	1.2	3.8	0.2	5.8	0.2	5.0	1.5	0.7	0.5	3.8	2.0
23	4.4	3.8	5.4	0.4	7.2	1.0	2.9	2.8	0.6	1.2	3.1	3.2
28	3.2	8.1	5.0	0.6	5.7	2.2	1.7	4.2	0.5	2.1	2.3	3.6
33	2.3	13.6	4.1	0.8	4.0	3.0	1.0	5.5	0.4	3.2	1.6	3.4

Table 6. Cumulative mass of fallen branches and leaves after 35 years (Mg/ha)

Species	Fallen branches		Fallen needle (leaves)	
	Control	Thinned	Control	Thinned
Scots pine	11.6	13.2	45.3	46.5
Arolla pine	6.8	7.0	48.2	40.8
Norway spruce	6.5	6.5	47.4	46.9
Larch	6.4		81.3	
Aspen	5.7		60.5	
Birch	10.2		82.4	

REFERENCES

Anonymous 1984 Modeling of artificial forest ecosystem development. Siberian Branch of the USSR Academy of Sciences. Nauka, Novosibirsk. 152 p.

Gabeyev V N 1971 Impact of stand growing density and cleaning cuts on productivity of pure pine plantations. *In* Efficiency of using forest resources and their regeneration in West Siberia. pp.220-229. Novosibirsk.

Isaev A S and Korovin G N 1997 Carbon deposition in forests of Russia. *In* Carbon in ecosystems. Ed. V N Sukachev. pp. 57-98. Readings in memory of the academician. Moscow.

Kairyukshtis L A and Yuodval'kis A I 1976 Phenomenon of changing intraspecific competition to mutual tolerance of trees in spruce phytocenoses. *In* Modern studies of forest productivity and cuts. pp. 47-57. Kaunas.

Ogievsky V V 1966 Forest stands of West Siberia. Nauka, Moscow. 187 pp.

Semechkina M G 1984 Growth and productivity of artificial forest phytocenoses. *In* Productivity of forest phytocenoses. pp. 112-120. Krasnoyarsk.

Semechkina M G and Kostoustova L A 1985 Assessment of pine stand productivity. *In* Productivity and structure of forest associations. pp. 116-122. Krasnoyarsk.

Usoltsev V A 2001 Forest phytomass of the northern Eurasia. Database and geography. The Russian Academy of Sciences, the Urals Branch Botanical Garden, Yekaterinburg. 707 pp.

Usoltsev V A 2003 Forest phytomass of the northern Eurasia. Peak productivity and geography. The Russian Academy of Sciences, the Urals Branch Botanical Garden, Yekaterinburg. 406 pp.

Vedrova E F, Spiridonova L V and Stakanov V D 2000 Carbon cycling in young tree stands of the main forest forming tree species. J. Lesovedenie (Silviculture) 3, 40-48.

Zhilkin B D, Lakhtanova L I and Richter I E 1977 Biological productivity and turnover of substances in pine stands of different density. Moscow. Botany 19, 58-63.

REFERENCES

Chapter 17

BIOCHEMISTRY OF CARBON AND NITROGEN IN THE SIBERIAN AFFORESTATION EXPERIMENT

E.F.Vedrova
V.N.Sukachev Institute of Forest, Siberian Branch of the Russian Academy of Science, 660036 Krasnoyarsk, Akademgorodok, Russia, Fax: (3912) 433686; E-mail: Institute@forest.akadem.ru

ABSTRACT

We evaluated parameters (stock and flux rates) carbon, nitrogen and ash element (Ca, Mg, Si, Al, Fe, P, and Mn) cycling in the plant-soil system. The study was performed in arolla pine (*Pinus sibirica*), Scots pine (*Pinus sylvestris*), Siberian larch (*Larix sibirica*), Norway spruce (*Picea abies (=Pinus obovata))*, aspen (*Populus tremula*) and birch (*Betula fruticosa)* plantations established on Kemchug upland (56°13′N, 92°19′E), Chulym-Yenisei denudational plain, in 1968-1972.

Young ecosystems (age class II) that have developed under relatively similar environmental conditions (heat, moisture, and available nutrients) have thus accumulated different amounts of organic and mineral matter in biomass and detritus, mainly due to differences in major woody species characteristics. All ecosystems are a net atmospheric carbon sink, with a pooling rate of 1.81-4.43 t C ha^{-1} yr^{-1}. The sink (NEP) is made up by C accumulation in stand biomass production, vegetation detritus, and soil humus and accounts for 46-76% of the total C allocated in annual increment. The latter two sequestrators contribute not more than 22-27% to the total C accumulation; the main component is net C flux to biomass.

The rate of N-min flux from soil pool to NPP is in fact in balance with that of soil N-min accumulation in the arolla pine, larch, spruce, and birch plots. In addition, the ecosystems are characterized by a transitional cycling regime evident through non-balanced fluxes controlling organic and mineral matter stored in "plant-soil" system blocks.

D. Binkley and O. Menyailo (eds.),
Tree Species Effects on Soils: Implications for Global Change, 281–292.

INTRODUCTION

Forests are very relevant to issues about increasing atmospheric CO_2 concentrations and potential changes in global climate. Developing forests accumulate carbon (C) in live biomass, and in undecayed dead biomass. The initial fixation of C by forest canopies depends in part on the cycling and availability of nitrogen (N), and N supply may also influence rates of decomposition release of C. This study provides detailed C and N for 6 forest species planted on a common site in central Siberia. The tree species were arolla pine *(Pinus sibirica)*, Scots pine *(Pinus sylvestris)*, larch (*Larix sibirica)*, spruce (*Picea abies (=Picea obovata)*, aspen *(Populus tremula)*, and birch (*Betula fruticosa)*, and the plots were assessed when the plantations were 25-30 years old.

METHODS

The Siberian afforestation experiment is described in detail in other chapters in this volume, particularly Shugalei. They were established on Kemchug upland (56° 13′ N, 92° 19′ E), Chulym-Yenisei denudational plain, following removal, mixing, and reapplication of topsoil.

Our ecosystem approach to quantifying the budgets of C and N involved measuring 1) above- and belowground stand biomass by components, 2) the O horizon of the soil, and 3) dead organic matter in the 0-20 cm mineral soil (dead roots, other root detritus, and soluble soil humus). We quantified the mass of these components, as well as the fluxes between them. The mass and element content of trees was extrapolated from samples of two model trees within each 2-cm diameter class in each plot (Utkin and Rozgdestvenskiy, 1988). O horizon (forest floor) material was sampled using a template of 0.063 m^2, with 10 replicate samples within each plot (each tree species was planted in just one plot). The O horizon samples were divided into O_L, O_F, and O_H layers (=Oi, Oe, and Oa). Slightly decomposed materials within the O horizon were sorted into leaves, bark, branches, and cones. Both living and dead roots were removed from the forest floor samples. Each fraction was weighed, and its moisture content determined by oven-drying.

The mass of roots was determined by weighing samples removed from the O horizon, and from 20-cm cubes of mineral soil. Roots were washed (over a 0.25 mm mesh), and live roots (light-colored core and bark (for conifers), flexible and solid) were separated from dead roots (dark-colored, with wrinkled bark, fragile). The remaining dead residues (pilled-off bark, root cases, mycelium, rotted wood dust, and charcoal particles) were fractionally classified as “other root detritus”. The samples were oven-dried and weighed for moisture content. The total dead root material in the O horizon and 0-20 cm mineral soil layer, plus "other root detritus" fraction, indicated the amount of 0.5-10mm root detritus.

The annual increment of coarse roots was estimated as 50% of the increments of stems plus branches (Kazimirov et al. 1977). The mortality of coarse roots was calculated assuming that above- and below-ground plant parts die at the same rate relative to their amount stored. The net rates of fine root rates of increment, mortality, and accumulation in "other root detritus" fraction were determined in field by Persson's (1980) method.

Litterfall was sampled twice annually for 4 years, using 15 to 20 traps (0.0625 m^2) in each plot. Samples were separated into components of leaves, <10mm-diameter branches, bark, cones, and other material. A representative composite sample was chemically analyzed at the end of the year. The annual rate of decomposition release was calculated based on 4-years of litterfall collection, matched with annual measurements of O horizon mass, using these equations (same equation for C and for N):

$$\text{Loss of C} = C_o + C_{\text{litterfall}} - C_t.$$

Where C_o is the C (or N) content of the O horizon at the beginning of the year, $C_{\text{litterfall}}$ is the year's addition of C, and C_t is the C content of the O horizon at the end of the year (Vedrova 1997; Vedrova et al 2002).

The release of water-soluble products from materials in the O horizon was determined by analyzing chemical composition of solutions collected in plastic lysimeters (Vedrova, 1995). The release from root detritus was determined by laboratory simulation with natural conditions: the material was decomposed for 27 months in the darkness at a temperature close to that of soil and was periodically treated with distilled water on the 300 mm normal summer season precipitation basis. Solutions were filtered and analyzed for pH and C concentration.

The annual sampling of O horizon pools, litterfall inputs, and solution outputs allowed us to estimate the decomposition constant (k) of plant litter residues, and to determine the relationship between the two main processes of litter processing, mineralization and humification (Vedrova 1997).

The decomposition of the coarse dead roots was determined by following the decomposition of material fractions placed 10 to 20 cm into the mineral soil in each plantation over a 3 year period. The release of inorganic N from the pool of labile organic N (exctractable with 0.1 *M* NaOH) was estimated based on the decline in the size of this pool through the growing season.

The C concentration of vegetation samples was determined by Anstet's method (oxidation by H_2SO_4+CrO_3) as modified by Ponomareva-Nikolaeva (1959), and N concentration by Kjeldahl digestion and titration. The vegetation samples were analyzed for ash elements after heating samples to 450° C (Rodin et al 1968). Soil C was determined by Turin's method (oxidation by $K_2Cr_2O_7$).

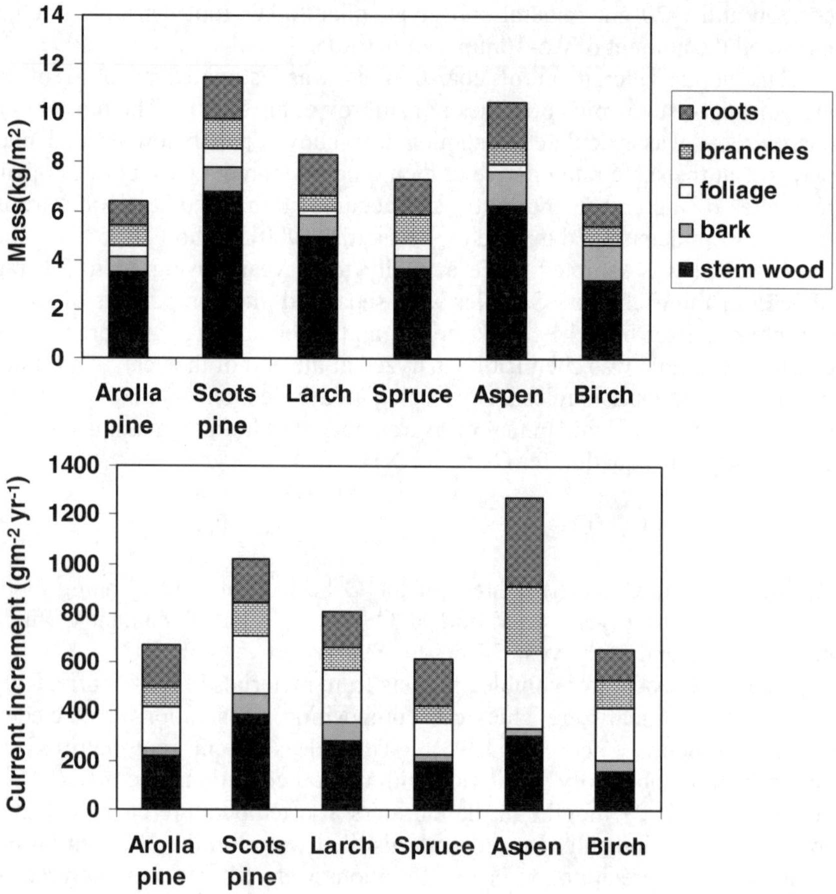

Figure 1. Structure of biomass stand (kg m^{-2}) and current increment (g m^{-2} yr^{-1}), oven-dry matter.

RESULTS

The accumulation of living biomass differed by up to 80%, with the lowest value for birch and highest for Scots pine (Figure 1). The average value for the 4 conifer species was the same as the average for the two hardwood species. More than half of the biomass accumulated in stems, compared with about 3% to 8% occurring in leaves. The concentrations of N differed across species, leading to lower differences in N mass between species than for biomass or C (Figure 2). Birch accumulated about 230 g of biomass per g of N accumulated, whereas larch accumulated about 400 g of biomass per g of N accumulated. Most of the N content of trees was found aboveground (75-90%) for all species, and the leaves contained about half of

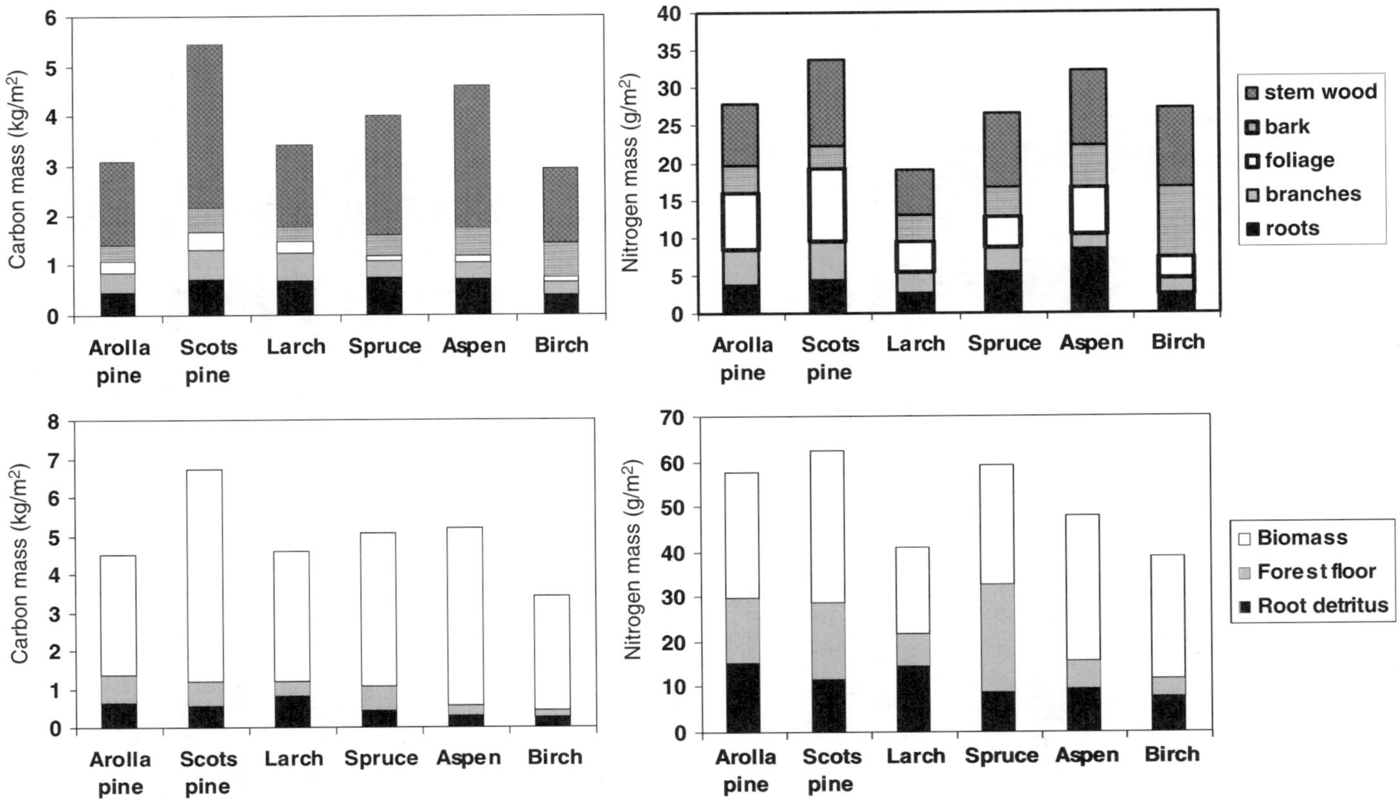

Figure 2. Carbon and nitrogen accumulated in trees tissues, and in forest floor.

the aboveground N despite their minor contribution to total mass. The O horizons differed substantially among species, with conifers accumulating larger masses (by 1.5 to 4.9 fold) than the hardwood species. The ratios of recent litter to more decomposed litter ($Ol:O_f+O_h$) ranged from 1:20 (under larch) to 1:2 under slowly decomposing spruce. The amount of carbon stored in root detritus was higher under evergreen conifers than larch or hardwoods (Figure 2). The C content of vegetation detritus ranged from 1.2 to 1.4 kg C m^{-2} under the conifers, much greater than the 0.4 to 0.5 kg C m^{-2} under the hardwoods.

Carbon fluxes

Net primary productivity of the two hardwood species exceeded the range among the 4 conifer species, with birch NPP of 330 g m^{-2} yr^{-1} and aspen NPP of 580 g m^{-2} yr^{-1} (Table 1, Figure 3). Belowground allocation of NPP accounted for only 15% to 25% of NPP, which is lower than commonly reported for forests. The proportion of fine roots in the total root production ranged from 5% (aspen) to 52% (arolla pine).

Table 1. Rate of carbon fluxes, g C m^{-2} yr^{-1}.

Carbon fluxes	Arolla pine	Scots pine	Larch	Spruce	Aspen	Birch
1. Net Primary Production (NPP)	410.8	528.4	443.3	342.5	579.0	329.1
1a. Aboveground	250.6	407.0	314.5	199.3	387.8	247.0
1b. Roots	77.0	77.4	64.4	84.6	153.1	51.9
1.c. Other root detritus (0-20 cm)	83.2	44.0	64.4	58.7	38.1	30.2
2. Detritus production	272.2	230.0	229.8	124.2	245.1	167.1
2a. Aboveground litterfall	157.3	163.5	137.3	47.0	178.5	122.7
2b. Root death	31.7	22.5	28.1	18.5	28.5	14.2
2c. Other detritus	83.2	44.0	64.4	58.7	38.1	30.2
3. Humus formation	62.6	38.3	26.7	23.9	41.2	20.2
3a. O horizon	48.5	25.3	18.9	8.7	31.3	14.6
3b. Root detritus	14.1	13.0	7.8	15.2	9.9	5.6
4. Mineralisation (decomposition)	221.1	148.2	152.6	124.7	136.4	147.9
4a. O horizon	162.4	132.8	138.6	85.3	117.8	107.0
4b. Root detritus	30.1	6.1	7.3	35.3	13.4	29.7
4c. Humus	28.6	9.3	6.7	3.9	5.2	11.2
5. Net Ecosystem Production (NEP)	189.7	380.2	290.7	217.8	442.6	181.2
5a. Live biomass (1a + 1b – 2a – 2c)	138.6	298.4	213.5	218.3	333.9	162.0
5b. Detritus (2 – 4 – 5c)	17.0	52.8	57.2	- 20.5	71.7	10.1
5c. Humus (0-20 cm) (3 - 4c)	34.0	29.0	20.0	20.0	37.0	9.0

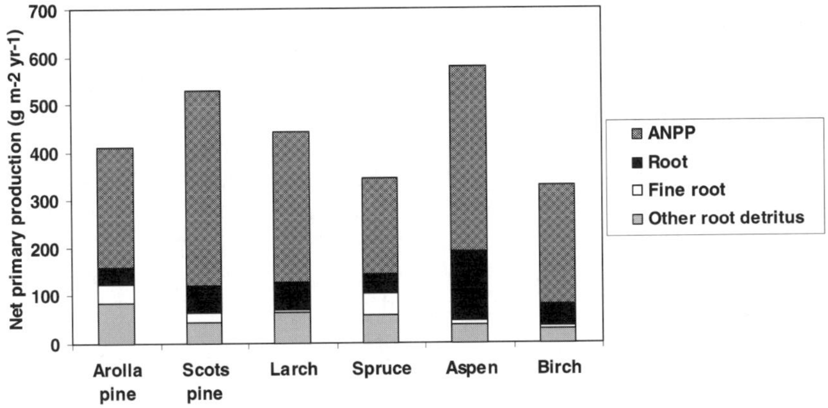

Figure 3. Net primary production (g m^{-2} yr^{-1}).

In addition to the increment in above- and belowground biomass, C accumulates in the soil in newly developed fractions of "other root detritus" consisting mainly of mycorrhiza-forming fungus mycelium. The rate of this C accumulation was higher in the conifers than in the hardwood plots: 83, 44, 64, 59, 38, and 30 g C m^{-2} yr^{-1} under arolla pine, Scots pine, larch, spruce, aspen, and birch, respectively. Overall, the NPP of the conifers (g C m^{-2} yr^{-1}) ranked Scots pine (528) > larch (443) > arolla pine (411) > spruce (342); NPP of aspen (579) was far greater than that of birch (329).

Aboveground litterfall ranged from 120 to 180 g C m^{-2} yr^{-1} among the plots, except for the very low rate of 47 g C m^{-2} yr^{-1} under spruce. Leaves comprised the majority of aboveground litterfall in all plots, though the contribution of branches was substantial (9-16% of litterfall) under larch, aspen and birch. The net increment in live biomass varied by more than 2-fold, from just 140 g C m^{-2} yr^{-1} for the arolla pine plot to 334 g C m^{-2} yr^{-1} for aspen.

The differences in aboveground litterfall among plots were smaller than the differences in the mass of the O horizon (forest floor), indicating greater differences among plots in rates of decomposition (Figure 2). Decomposition rates decreased with increasing C:N ($r^2 = 0.6$; Vedrova and Mindeeva, 1998).

About 65-80% of the detrital organic matter is oxidized to CO_2 during decomposition, with 2-11% leaching as water-soluble C into the soil, and 11-22% accruing as humus (Table 1). Newly-formed soil humus substances are partially re-utilized by microorganisms contributing thereby to the ecosystem C efflux to the atmosphere (Grishina 1986; Aristovskaya 1980). In our study, 84 and 90% of the total newly- formed humus carbon is sequestrated, respectively, under spruce and aspen, 54% under arolla pine, 76% under Scots pine, 75% under larch, and 45% under birch. Annual C-humus accumulation in 0-20cm soil layer is 0.6% of the total soil humus under aspen, arolla pine and Scots pine, 0.3% under larch, 0.35% under spruce, and 0.2% under birch.

Tikhomirov's (1977) experimental data suggested that decomposition of forest floor C on a loamy substrate be accompanied by sequestrating 48-74% of the total annual carbon transported to loam. Grishina (1986) stated that 88% of the humic acids were immobilized in a loam soil.

In most of the plots, C efflux to the atmosphere did not exceed 28-36% of NPP (Table 1). The arolla pine plot had the largest C efflux (221 g C m^{-2} yr^{-1}), amounting to 54% of the rate of C accumulation in the current tree increment C. The net rate of C accumulation (NEP) in the plots ranged from 190 to 443 g C m^{-2} yr^{-1}, with 74-100% of the accumulation occurring in live biomass, and up to 26% accumulating in detritus and humus.

Nitrogen fluxes

The N content of the current biomass production affects both the concentration of N in tissues, and the mass of tissues produced. Leaves have higher N concentrations than other tissues in all species; the N concentrations of current-year leaves averaged 20.3 g/kg in aspen, 18.0 mg/g for larch, 14.6 mg/g for arolla pine, 13.0 g/kg in birch, 11.2 mg N/g for Scots pine, and 8.0 g/kg for spruce. Nitrogen concentrations in the current increment branches ranges 2.5 to 6.8 g/kg, while the current root increment has the lowest N concentration (1.8-4.9 g/kg).

The current biomass production of trees required much more N for aspen (9.3 g/m^2; lines 1a + 1b in Table 2) than for the other species (larch 5.8 g/m^2, Scots pine 4.8 g/m^2, arolla pine 4.6 g/m^2, 4.3 g/m^2 birch, and 2.2 g/m^2 spruce). Between 47 and 69% of this N requirement was allocated to production of leaves. The production of "other root detritus" (primarily fungal mycelium) required from 0.9 to 2.1 g N m^{-2} yr^{-1} among the species. Combining the N requirement for growing trees and fungal mycelia, the N requirement for NPP ranged from a low of 3.1 g N m^{-2} yr^{-1} for spruce (line 1, Table 2) to 10.5 g N m^{-2} yr^{-1} for aspen. Between 44 and 70% of the N content of current NPP was returned to the soil each year in litterfall, and mortality of roots and fungal mycelia (line 2, Table 2), and 20-25% accumulates in wood.

The release of C and N from decomposing recent litter ranged from 20 to 50% across species (Table 3). Release from the forest floor (O horizon) was slower than the decomposition of recent litter, ranging from 17 to 44%. Root decomposition was much slower, with just 8 to 22% of the C and N released annually. The annual release of N from decomposing recent litter, roots, and the forest floor ranged from 2.9 to 5.7 g m^{-2} yr^{-1} (line 4, Table 2). The N content of the forest floors appeared to be declining, with a net release (negative immobilization) of 0.4 to 2.4 g m^{-2} yr^{-1} (line 5a, Table 2). Conversely, the pool of root detritus served as a sink for N, accumulating between 0.2 to 1.8 g m^{-2} yr^{-1} for all species (line 5b, Table 2) except birch (where there was a net reduction in this pool of 0.5 g N m^{-2} yr^{-1}).

Table 2. Rate of nitrogen fluxes in the major Siberian woody species plantations, g N m^{-2} $yr.^{-1}$

Nitrogen fluxes	Arolla pine	Scots pine	Larch	Spruce	Aspen	Birch
1. Net Primary Production (NPP)	6.89	5.77	6.77	3.10	10.51	5.31
1a. Aboveground	4.02	4.29	5.32	1.89	7.52	4.01
1b. Roots	0.62	0.48	0.49	0.33	1.77	0.30
1c. Fungal mycelium (0-20 cm)	2.25	1.00	0.96	0.88	1.22	1.00
2. Input into detritus	4.32	3.21	4.40	1.76	4.57	3.76
2a. Aboveground litterfall	1.62	1.92	2.98	0.65	3.08	2.48
2b. Root mortality	0.45	0.29	0.46	0.23	0.27	0.28
2c. Fungal mycelium mortality	2.25	1.00	0.96	0.88	1.22	1.00
3. Net biomass accumulation (1-2)	2.56	2.56	2.37	1.34	5.94	1.55
4. Mineralization release	4.04	2.96	5.68	2.86	4.50	4.30
4a. Aboveground litterfall	0.65	0.48	1.19	0.16	1.48	0.98
4b. Forest floor	2.47	2.09	4.21	1.82	2.43	1.53
4c. Root mortality	0.06	0.02	0.05	0.03	0.06	0.04
4d. Root mortality from prior years	0.24	0.06	0.02	0.01	0.01	0.03
4e. Other root detritus	0.62	0.31	0.21	0.74	0.51	1.72
4f. Water-soluble N-release	0.58	0.55	0.25	0.54	1.03	0.53
5. Immobilization in detritus	0.28	0.25	- 1.28	- 1.10	0.07	-0.54
5a. Forest floor	-1.50	-0.65	-2.42	-1.33	-0.83	-0.04
5b. Root detritus	+1.78	+0.90	+1.14	+0.23	+0.90	-0.50
6. Soil NH_4^+ and NO_3^- (0-20 cm.), g N/m^2 (3 yr grow. season avg.)	10.71	6.53	9.16	4.83	7.24	6.50
7. NPP requirement of N from the mineral soil (1 – 4)	2.85	2.81	1.09	0.24	6.01	1.01
8. Transfer of N_{min} from N_{mobile} soil humus	3.09	2.18	1.42	1.71	3.12	2.27
9. Soil input N_{min} (4 + 8)	7.13	5.14	7.10	4.57	7.62	6.57
10. Tree and soil net immobilization (3+5)	2.84	2.81	1.09	0.24	6.01	1.01
11. Balance (9-10), or net loss	4.29	2.33	6.01	4.33	1.61	5.56

The available mineral nitrogen pool is 10.71 and 6.53 g m^{-2} in the 0-200 mm soil layer under arolla pine and Scots pine, respectively (Bezkorovaynaya 1993; Shugaley and Popova 1999). Across all species, the annual addition of N to the soil in the form of litterfall, root and fungal mycelium death (line 2, Table 2) equals 92 to 94% of the rate of N release from the detrital pools (line 4, Table 2), with 6 to 8% accumulating in the forest floor plus decaying roots (line 5, Table 2).

In addition to the N release from recent detritus, the available pool of soil N receives input from the decomposition of the labile N pool in the mineral soil (line 8, Table 2), which ranged from 1.4 g N m^{-2} yr^{-1} for larch to 3.4 g N

Table 3. Specific rate of decomposition for mass, carbon, and nitrogen (mg g^{-1} yr^{-1}).

Material	Pool	Arolla pine	Scots pine	Larch	Spruce	Aspen	Birch
Recent litter	C	552	500	335	267	435	395
	N	400	250	200	250	385	440
Forest Floor	C	274	216	254	250	435	473
	N	172	124	176	251	385	395
Fresh root detritus	mass	131	47	98	106	223	1128
Older root detritus	mass	99	60	34	83	72	84
Other root detritus	mass	41	27	24	53	49	116

m^{-2} yr^{-1} to 3.1 g N m^{-2} yr^{-1} for aspen. The release of N from labile organic pools in the mineral soil contributed 20-43% of the N available annually for tree use, net immobilization into detritus, and leached below 20 cm in the mineral soil.

The apparent leaching losses are very large from all plots, ranging from 1.6 to 6.0 g N m^{-2} yr^{-1} (line 11, Table 2). The value of these estimates depends on the accuracy of the estimates for the N content of NPP, N release from decomposition of recent detritus, and the annual release of N from labile organic pools in the mineral soil. Direct measurements of the N content of soil leachate would be needed to validate the complete N budget.The estimated rates of turnover for C, N, and various other elements differed substantially among the species plots (Figure 4). Overall, aspen and birch showed the shortest turnover times. The turnover time for N tended to be slightly shorter than the time for C.

CONCLUSION

The experimental plots with the major forest-forming species in Siberia showed transitional biogeochemical regimes, with non-steady state fluxes between the vegetation and soil. All plots showed a net accumulation of C (positive NEP), ranging from 181 to 443 g C m^{-2} yr^{-1}, and the accumulation in biomass accounted for about ¾ of the C accumulation for all species. The annual rate of C release from the 0-20 cm mineral soil was very slow, ranging from 0.1 to 0.5% annually. The rates of N use in NPP and the rate of production of available N in the soil were roughly balanced for the arolla pine, larch, spruce, and birch ecosystems.

The plots differed substantially in the soil nitrogen flux used for net N sequestration by stand biomass, and that used for N immobilization by detritus. This flux has the lowest rate in the spruce stand and the highest rate in the aspen stand (respectively, 5% and 79% of the annual N return to the

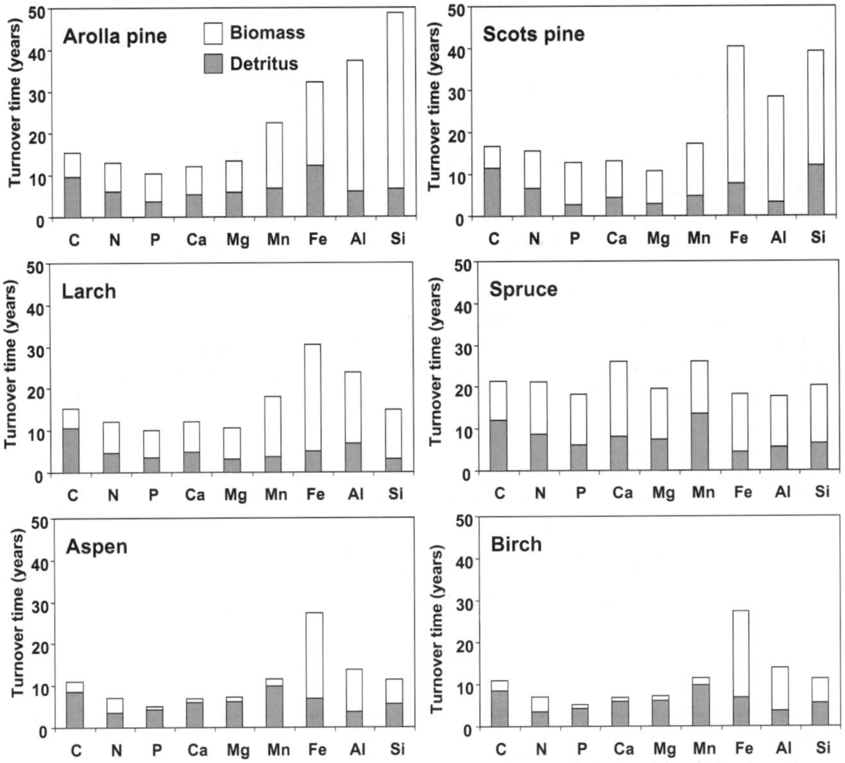

Figure 4. Whole ecosystem turnover time for biomass stand ($t_{biomass}$) and detritus ($t_{detritus}$) elements. Calculated as biomass element content divided by the rate uptake by NPP; detritus elements storage divided by litterfall + root fall+ fungus mycelium death.

soil pool or 8% and 57% of NPP consumption). The pools of N in larch and spruce detritus were remarkably unstable relative to the other species, and the annual N input from litterfall did not balance the larger rate of N loss through decomposition.

ACKNOWLEDGEMENTS

Grants # 99-04-96024, 02-04-49459 from the Russian Fund of Fundamental Research and grants from Krasnoyarsk regional Fund of Science supported this study.

REFERENCES

Aristovskaya T V 1980 Microbiology of the soilformation processes. pp. 12-55. Leningrad, Nauka.

Bezkorovaynaya I N 1993 Soil mezofauna formation and its role for litter transformation in tree plantations. Ph.D.Dissertation. Krasnoyarsk, manuscript. 205 pp.

Grishina L A 1986 Humusformation and humus state of soil. Moscow, MGU. 244 pp.

Kazimirov N I, Volkov A D, Zyabchenko C C, Ivanchikov A A and Morozova P M 1977 Exchange of substance and energy in Scotch pine forests of the European North. Leningrad, Nauka. 304 pp.

Kobak K I 1988 Biotic components of carbon cycles. Leningrad, Hydrometeoizdat. 248 pp.

Kurz W A, Apps M I, Webb T M and McNamee P J 1992 The carbon budget of the Canadian forest sector: Phase I. Northwest Region. Inform. Report Nor-X-326. Canada. 93 pp.

Nabuurs G J and Mohren G M J 1993 Carbon in Dutch forest ecosystems. Netherlands J. Agricult. Sci. 41, 309-326.

Persson H. 1980. Fine-root production, mortality and decomposition in forest ecosystems. Vegetatio 41, 101-109.

Ponomareva V V and Nikolaeva T A 1959 Methods for studying organic matter in bogs. *In* Modern Soil Processes in the Forest Zone of European USSR. pp. 170-173. Moscow Academy Science, Moscow, USSR.

Rodin L E, Remesov N P and Bazilevich N I 1968 Methodical instructions for the study of dynamics and biological turnover in phytocoenosis. Leningrad, Nauka. 143 pp.

Tikhomirova N A 1977 Study of plant residues decomposition in the meadow and steppe ecosystems. Ph.D. Thesis. Novosibirsk. 18 pp.

Titlyanova A A and Tesarõvá 1991 Regimes of the biological turnover. Novosibirsk. Nauka. 149 pp.

Shugaley and Popova 1999 Mineral nutrition of the tree plantation on old arable soils. Lesnoe hozyaystvo. 4, 36-37.

Utkin A I and Rozgdestvenskiy C G 1988 Comparison of methods for estimation of phytomass and production of aspen stands different aged. *In* Analysis of the forest stands production structure. pp. 10-28. Moscow, Nauka.

Vedrova E F 1995 Transformation on plant residues in 25-year-old plantations of main forest-forming species in Siberia. Lesovedenie. 4, 13-21.

Vedrova E F 1997 Organic matter decomposition in forest litters. Eurasian Soil Sci+ 2. 181-188.

Vedrova E F, Spiridonova L V and Stakanov V D 2000 Carbon cycle in young forests of main forest-forming species in Siberia. Lesovedenie. 3, 40-48.

Vedrova E F and Mindeeva T N 1998 Intensity of carbon dioxide production in decomposing forest litters. Lesovedenie. 1, 30-41.

Chapter 18

TREE SPECIES EFFECTS ON POTENTIAL PRODUCTION AND CONSUMPTION OF CARBON DIOXIDE, METHANE, AND NITROUS OXIDE: THE SIBERIAN AFFORESTATION EXPERIMENT

Oleg V. Menyailo[1] and Bruce A. Hungate[2]
[1]*Institute of Forest SB RAS, Krasnoyarsk 660036, Russia*
[2]*Department of Biological Sciences and Merriam Powell Center for Environmental Research, Northern Arizona University, Flagstaff AZ 86001, USA*
Corresponding address:Institute of Forest SB RAS, Krasnoyarsk 660036, Russia e-mail: menyailo@hotmail.com

ABSTRACT

Changes in tree species composition could affect how forests produce and consume greenhouse gases, because the soil microorganisms that carry out these biogeochemical transformations are often sensitive to plant characteristics. We examined the effects of thirty years of stand development under six tree species in Siberian forests (Scots pine, spruce, arolla pine, larch, aspen and birch) on potential rates of soil CO_2 production, N_2O-reduction and N_2O production during denitrification, and CH_4 oxidation. Because many of these activities relate to soil N turnover, we also measured net nitrification and N mineralization. Overall, the effects of tree species were more pronounced on N_2O and CH_4 fluxes than on CO_2 production. Tree species altered substrate-induced respiration (SIR) and basal respiration, but the differences were not as large as those observed for N transformations. Tree species caused similar effects on denitrification potential, net N mineralization, and net nitrification, but effects on N_2O reduction were idiosyncratic, resulting in a decoupling of N_2O production and reduction. CH_4 oxidation was affected by tree species, but these effects depended on soil moisture: increasing soil moisture enhanced CH_4 oxidation under some tree species but decreased it under others. If global warming causes deciduous

D. Binkley and O. Menyailo (eds.),
Tree Species Effects on Soils: Implications for Global Change, 293–305.

species to replace coniferous species, our results suggest that Siberian forests would support soil microbial communities with enhanced potential to consume CH_4 but also to produce more N_2O. Future predictions of CH_4 uptake and N_2O efflux in boreal and temperate forests need to consider changes in tree species composition together with changes in soil moisture regimes.

INTRODUCTION

Natural biological processes are critical regulators of exchanges of greenhouse gases between the biosphere and atmosphere, particularly for carbon dioxide (CO_2), methane (CH_4), and nitrous oxide (N_2O). The processes mediating these exchanges change in direct response to altered temperature, precipitation, and other aspects of global change. However, the indirect effects of global change on exchanges of greenhouse gases are not well understood, particularly the impacts of altered distributions of plant species, a widely predicted (and likely ongoing) consequence of the Earth's changing climate. Because plant species have strong and idiosyncratic influences on ecosystem properties that regulate fluxes of greenhouse gases, their effects on these fluxes are likely to be large and individualistic, in some cases amplifying and in others counteracting the direct influences of global climate change. Furthermore, afforestation in northern latitudes is increasingly considered to be a viable way to increase the terrestrial carbon sink, helping to mitigate global climate change. The effects of such efforts on biosphere-atmosphere exchanges of N_2O and CH_4, however, have not been considered. Because CH_4 has 200 times and N_2O 290 times the climate forcing potential of CO_2, tree species that increase CO_2 uptake but enhance emissions of these more potent greenhouse gases could actually do more harm than good, creating a net positive feedback to global warming (Figure 1).

The effects of tree species are thus important to know because global change will alter species composition of forests, changing ecosystem processes and greenhouse gas fluxes. The direction and magnitude of these effects are unknown. The areal extent of artificial afforestation (conversion of grassland to forestry) is also likely to increase substantially in the coming decades. To maximize the sink strength of these landscapes for greenhouse gases, the effects of individual tree species needs to be clarified across a broad range of sites and environmental conditions.

Artificial afforestation experiments provide an opportunity to identify explicitly the influences of individual tree species on soil microbial processes related to production and consumption of greenhouse gases. These "common garden" experiments are initially homogenous in microbial, chemical, and physical properties, and trees are of the same age (Wedin and Tilman, 1990, Binkley, 1994, Menyailo et al., 2002a). Only a few artificial afforestation

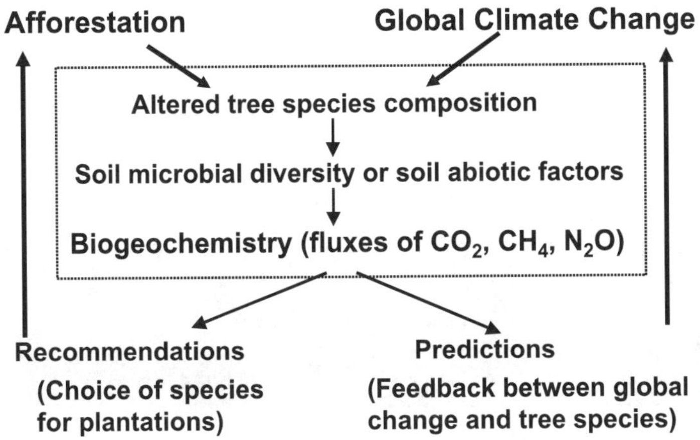

Figure 1. Conceptual scheme showing feedback between climate change and tree species composition in forests. The box in the center shows the hypothesized influences of tree species on the fluxes: either through changes in soil abiotic factors or through changes in microbial community structure and composition.

experiments have examined microbiological processes, and results have been inconsistent. In some cases, no distinct species effect was found after 23-24 years of stand growth (Priha and Smolander, 1997; Scott 1998). By contrast, in another experiment tree seedlings caused changes in C transformation activities in soil after only 4 months of growth (Priha et al., 1999b). The effects of tree species in the Siberian Afforestation Experiment (see other chapters in this volume) on microbial transformations of CO_2, CH_4 and N_2O were presented in a number of publications (Menyailo and Huwe, 1999, Menyailo et al., 2002b, Menyailo and Hungate 2003), here we synthesize the most important results.

THE SIBERIAN AFFORESTATION EXPERIMENT

The research plots (Figure 2) are located 50 km Northwest of Krasnoyarsk and were established by the Laboratory of Soil Science of the Institute of Forests, Siberian Branch of the Russian Academy of Sciences (Menyailo et al., 2002a). The upper 0-50 cm of soil of a 1.5-ha area were removed, mechanically homogenized to minimize vertical and spatial heterogeneity of chemical, physical and biological properties, and subsequently returned to the site prior to experimental planting. In 1971-1972, 2-3 y old seedlings of Norway spruce (*Picea abies, = Picea obovata*), birch (*Betula pendula*), Scots pine (*Pinus sylvestris*), aspen (*Populus tremula*), larch (*Larix sibirica*) and arolla pine (*Pinus cembra*) were planted into monoculture plots, each occupying 2400 m^2. An 600 m^2 area was left for grassland as a control, and

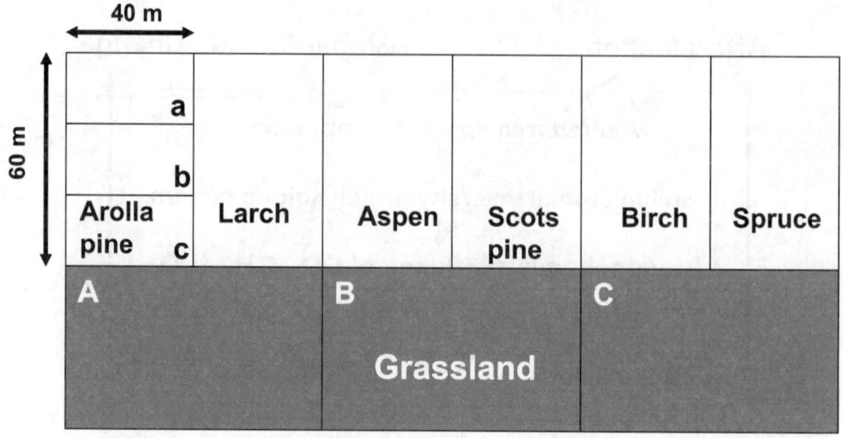

Figure 2. Layout of the Siberian afforestation experiment, organized by laboratory of soil science of the Institute of Forest SB RAS in 1971-1972 y under idea of Prof. N.V. Orlovsky.

the soil under grass was not mechanically homogenized. The region is characterized by continental climatic conditions with average rainfall 500 mm $year^{-1}$, average daily summer temperature of 20 ^{o}C (at noon), depth to permafrost 70-170 cm and soil temperature to 20 cm depth in winter -4° to -14°, in summer 10° to 12°. The soil is the gray forest type according to the Russian Soil Classification System and Greyzem according to FAO (FAO, 1990). Litterfall is asynchronous among the six study species (Mukhortova, this volume), so soil samples were collected in August to avoid the influence of fresh litter. In August 2001, each plot was sub-divided into three parts: A, B and C (as shown in Figure 2). From each sub-plot, two trees were randomly chosen and four soil samples were taken at 50 cm apart of the stem of each tree. In the grassland plot, three sub-plots (each of 2 m^2) were chosen along the forest plantation; at each sub-plot six soil samples were taken from the 0-10 cm depth. Soil samples from each sub-plot were mixed. The total number of soil samples was 21: six species plus grassland by three subplots.

STUDY OF MICROBIAL ACTIVITIES

Denitrification potential

Five grams of soil were placed in glass flasks (25 ml) and pre-incubated at 28°C for 3 days to initiate microbial activity. Thereafter, 5 ml of distilled water with KNO_3 and glucose were added to each sample. The resulting concentrations of nitrate and glucose were 100 μg NO_3-N g^{-1} soil and 100 μg glucose-C g^{-1} soil. The flasks were closed with air-tight rubber stoppers and fixed with clamps. Anaerobic conditions were induced by exchanging the gas phase with helium for 15 min. Acetylene (2.5 ml of C_2H_2,10% v/v) was then added to inhibit N_2O-reductase, and prevent catalysis of N_2O to N_2. All samples were incubated for 24 hours, then a sample of headspace gas was taken from each flask for determination of N_2O using gas chromatography (Shimadzu 14A, N_2 carrier gas, equipped with an electron capture detector (ECD ^{63}Ni), Menyailo and Huwe, 1999). The sample volume was 1 ml. The results were expressed as mg N_2O-N kg^{-1} dry soil day^{-1}.

Potential N_2O reduction

From each of the 63 soil samples, 4 sub-samples were used to measure potential N_2O reduction. As above, five grams of soil were placed in glass flasks (25 ml) and pre-incubated at 28 °C for 3 days to initiate microbial activity. Thereafter, 5 ml of distilled water with glucose were added to each sub-sample. The resulting concentration of glucose was 100 mg C kg^{-1} soil. The flasks were closed with air-tight rubber stoppers and fixed with clamps. Anaerobic conditions were induced by exchanging the gas phase with helium for 15 min. After removing O_2 from the flasks, 1 ml of N_2O was added to two sub-samples as a final electron acceptor. To another two sub-samples, 1 ml of N_2O and 2.5 ml of C_2H_2 (10% v/v) were added. The last series of sub-samples were necessary to estimate abiotic N_2O consumption by soil (e.g., dissolution in water). One ml of the headspace from each flask was sampled at 0 h, 24 h and 48 h, in order to analyze for N_2O concentration as described above. Biotic N_2O consumption rates were calculated as the difference between changes in N_2O concentration between C_2H_2 treated and untreated samples. For the first incubation day, the following formula was used:

$$N_2O \text{ reduction rate} = (B_0 - B_{24}) - (A_0 - A_{24})$$

where B_0 is the initial concentration of N_2O in flask without C_2H_2, B_{24} is the concentration of N_2O in this flask after 24 h incubation, A_0 is the initial concentration of N_2O in flask with C_2H_2 and A_{24} is the concentration of N_2O in this flask after 24 h of incubation. For the second incubation day, the analogous formula was used substituting concentrations of N_2O at 24 h as initial and at 48 h instead of 24 h. As the rate of N_2O consumption was linear during two days of incubation, mean values were calculated and expressed as mg N_2O-N kg^{-1}day^{-1}.

Net N-mineralization and net nitrification

Approximately 15 g were extracted with 1M KCl (1:5) to determine the initial concentration of ammonium (NH_4^+-N) and nitrate (NO_3^-N) with a flow injection analyzer (Lachat). Additionally, 15 g from each soil sample were placed in plastic flasks (150 ml), moistened to 60% of water-holding capacity, sealed with stoppers and incubated at 28 °C for 30 days. To avoid anaerobic conditions, the flasks were opened every 3 days for 5 min. On day 30, soil samples were analyzed for NH_4^+-N and NO_3^--N as described above.

Net mineralization rate was calculated as the difference in [NO_3^- + NH_4^+] before and after the incubation and was expressed as mg (NO_3^- + NH_4^+)-N kg^{-1} 30 $days^{-1}$. Net nitrification was calculated as the difference in [NO_3^-] before and after the incubation and was expressed as mg NO_3^--N kg^{-1} 30 $days^{-1}$.

Basal and substrate-induced respiration

To study basal respiration, 5 g of soil were placed in a 25 ml-flask. Water was added to achieve 60% of water-holding capacity. The flasks were closed with rubber stoppers, fixed with clamps and incubated at 28 °C for 3 days. Afterwards, flasks were open for 30 min to allow aeration, sealed and incubated at the same temperature for 24 h. A gas sample of the headspace (1 ml) from each flask was taken for analysis of CO_2 concentrations using gas chromatography as described above. For substrate-induced respiration, five grams of soil were placed in 25 ml-flasks and moistened with distilled water. The flasks were closed with rubber stoppers, fixed with clamps and pre-incubated at 28 °C for 3 days. Thereafter, flasks were open and a water solution with glucose as a C-source was added to obtain 60% water holding capacity and 100 mg glucose-C kg^{-1} soil. Soil samples were then incubated for 24 hours and 1 ml of the headspace air of each flask was sampled and analyzed for CO_2 as described above. The results were expressed as g CO_2-C $kg^{-1}day^{-1}$.

Methane oxidation

We incubated three replicate soil samples from each of the six tree species with two amounts of soil water content (60 and 90% of water-holding capacity). In each replicate incubation, 10 g of soil were placed in 250-ml flasks and CH_4 was added to a concentration of 10 ml L^{-1}; incubations were conducted at room temperature (approx. 25 °C).

The decline in CH_4 concentration was measured during 6 d of incubation. Each day, 1 ml of the headspace was sampled by syringe and injected to 20 ml filled with He and sealed glass flask to store 1-4 h before concentration measurements by SRI gas chromatograph with FID and Porapak Q column (2m), injection volume was 5 ml. The rate of CH_4 oxidation was calculated using linear regression and expressed as nmol CH_4 g dry $soil^{-1}$ h^{-1}.

RESULTS AND DISCUSSION

N_2O production and consumption via denitrification

Denitrification potential (N_2O production) showed high variation and ranged from 5.8 to 15 mg N_2O-N kg^{-1} day^{-1} (Figure 3). Rates differed significantly among species ($P<0.001$), with highest N_2O production under arolla pine and larch. Production of N_2O under arolla pine and larch was significantly higher than beneath spruce, Scots pine and birch, where the lowest N_2O production was measured. Aspen had a lower N_2O production than larch (P=0.025) and higher than spruce (P=0.017), and in the same range as under Scots pine, arolla pine and birch.

Potential N_2O reduction varied from 4.6 to 13.7 mg N_2O-N kg^{-1} day^{-1} and values were comparable to those observed for N_2O production (Figure 3), indicating that neither NO_3^- nor N_2O was a limiting substrate in our incubation experiments. N_2O reduction was significantly affected by trees species ($P<0.001$). As for N_2O production, the highest rate of N_2O reduction was found under arolla pine and larch, significantly higher than under all other species ($P<0.035$). In contrast to N_2O production, low values of N_2O reduction were found under deciduous species aspen and birch, significantly lower than beneath spruce, arolla pine and larch.

Soil emissions of N_2O depend on the rates of both N_2O production (due mainly to denitrification) and consumption (N_2O-reduction), so the ratio of these processes is indicative of the potential ratio of N_2O/N_2 in the denitrification end products. The ratio of N_2O production-to-N_2O reduction may indicate conditions of species and environment that will produce the greatest rates of N_2O production. We found higher ratios of these two steps of denitrification under deciduous species than under conifers. We propose that higher N_2O fluxes should be observed under birch and aspen than under conifers, in accord with Butterbach-Bahl et al. (1997, also Butterbach-Bahl and Kiese, this volume). Pastor and Post (1988) found that changes in temperature and precipitation resulting from an increase of atmospheric CO_2 concentrations caused a northward migration of the hardwood-conifer forest border in North America. Such migration of the hardwood forests is likely to also occur in Russian Siberia. Based on our results, replacing conifers by deciduous species may double the ratio of N_2O/N_2 as end products of denitrification. Potential denitrification in Siberia was even higher than in the tropical forest soils (Menyailo et al., 2002b), the main terrestrial N_2O source (Matson et al., 1990), indicating that the microbial community in temperate forest zone can potentially produce higher amounts of N_2O. By changing environmental conditions due to global changes, atmospheric N deposition or forest fires, which increase soil inorganic N, temperate forests will likely play an increasingly important role in the global N_2O budget.

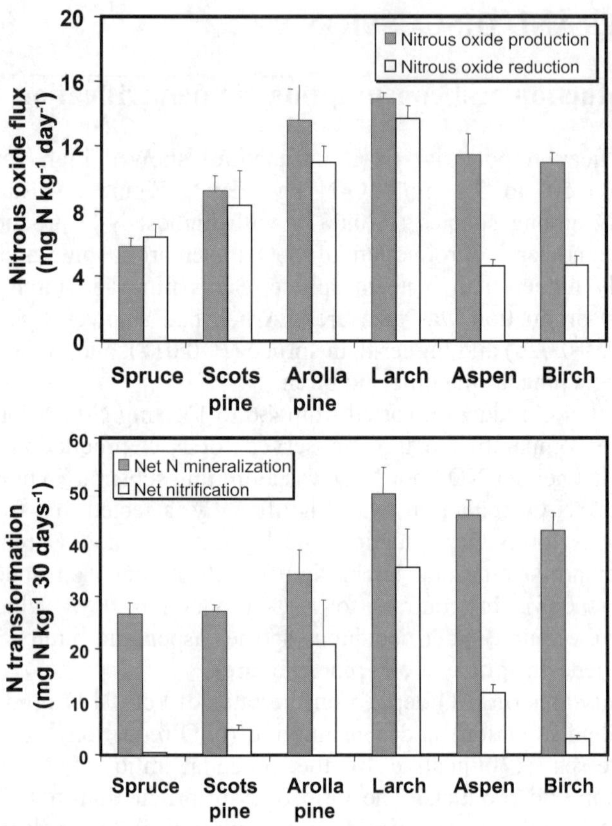

Figure 3. Potential rate of N_2O production (upper), measured in anaerobic conditions with addition of nitrate, glucose and acetylene and potential rate of N_2O reduction (anaerobic conditions, glucose and N_2O additions). Net N mineralization (lower) in 30 days aerobic incubation at constant soil moisture (60% WHC) and temperature regime (28 °C) and net nitrification.

Net N mineralization and nitrification

Inorganic N is important factor regulating the fluxes of greenhouse gases in many ways. We expected that the differences in net N mineralization and net nitrification might be helpful to explain the variation in the processes of greenhouse gas transformations. Mean values for net N mineralization ranged from 27 to 50 mg (NO_3^- + NH_4^+)-N kg^{-1} 30 $days^{-1}$ (Figure 3). The highest rate of net N mineralization was found under larch. Larch and arolla pine increased N mineralization significantly compared to spruce (P<0.001) and Scots pine (P=0.004), where the lowest rate was found. N mineralization under larch was also significantly higher than that under birch (P<0.001) and aspen (P=0.003).

Among all of the measured activities, net nitrification was most strongly affected by tree species (P<0.001), with means ranging over 2 orders of magnitude from 0.35 to 35.6 mg NO_3^--N kg^{-1} 30 $days^{-1}$ (Figure 3). Net nitrification tended to vary among species in the same order as net N mineralization: low activities of both processes occurred under spruce und Scots pine, intermediate values were found under aspen and birch, and the highest rates occurred under larch and arolla pine.

The contribution of net nitrification to net N mineralization depended strongly on tree species. Net nitrification constituted the smallest part of net N mineralization under spruce, just 1-3%. By contrast, net nitrification contributed to 90-100% of net N mineralization under larch and arolla pine, indicating that almost all of the net increase in inorganic N was due to NO_3^- accumulation.

CO_2 production from soil heterotrophic microorganisms

Soil respiration is the major source of ecosystem CO_2 efflux, and much of the efflux from the soil comes from heterotrophic microbes. Basal respiration (without glucose addition) varied within a very narrow range: 65 to 87 mg CO_2-C kg^{-1} day^{-1} (Figure 4). Tree species significantly (P=0.009) influenced basal respiration rate. arolla pine, larch and aspen increased basal respiration rates compared to Scots pine (P<0.020) and birch (P=0.010). Basal respiration under birch was significantly lower than under spruce (P=0.041). Even though the species effects were significant, the differences among species were not as large as for N transformations.

As expected, substrate-induced respiration (SIR) through glucose addition caused a 3-fold increase in respiration (Figure 4). SIR was also influenced by tree species (P<0.001). Arolla pine and larch increased SIR compared to other species (P<0.050). Intermediate values of SIR were found beneath spruce, aspen and birch. The lowest value was found under Scots pine, where SIR was significantly lower than under all other species (P<0.050). Overall, the effects of species on SIR were similar to effects on basal respiration rate.

The Siberian afforestation experiment showed that while effects of tree species on C respirations were significant, the differences among species were much smaller than the effects on N transformations. These results are in accord with other laboratory incubation studies (Scott 1998), and field measurements of CO_2 evolution (Paré, cited by Côté et al., 2000) which indicated no difference between coniferous and deciduous stands. Mikola (1985) studied 30- to 50 year-old spruce and birch forests growing on originally approximately similar sites and found no difference in soil respiration. Also no differences in CO_2 flux were found between an agricultural and forest soil, while much larger N_2O was produced in the agricultural field (van Bachove et al., 2000). These results support our

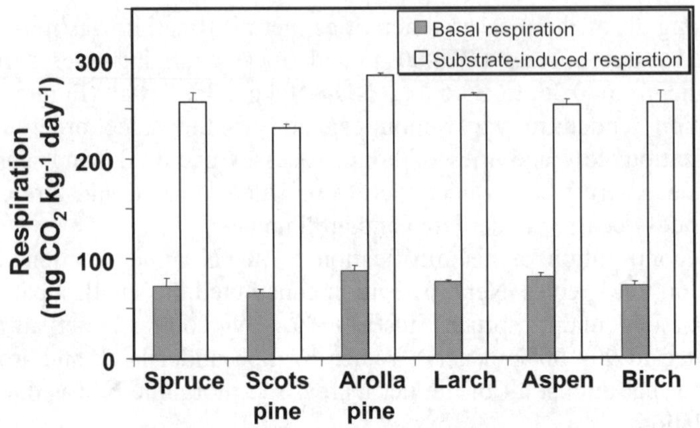

Figure 4. Activity of basal respiration in soil samples beneath six tree species (N=3) measured as CO_2 production during 24 h incubation at 60% of WHC and 28 °C and substrate-induced respiration (SIR) in the same soil samples measured as described above but with addition of glucose.

conclusion about stronger effects of tree species on N than on C transformation processes. However, this contrasts to other study where tree seedlings of three species (pine, spruce and birch) were growing under greenhouse conditions for four months and no differences between species were found on net N mineralization, nitrification and denitrification, while species did affect C mineralization (Priha et al., 1999b). Clearly, seedling studies cannot be used reliably to predict patterns in forests.

CH_4 oxidation

Temperate and boreal forest upland soils are a significant sink of atmospheric CH_4. Methane consumption in our soils varied from approximately 1 to 5 nmol CH_4 g^{-1} h^{-1} (Figure 5). Tree species strongly affected CH_4 consumption (P=0.004). Overall, birch had higher values than coniferous species (P<0.05). Aspen had higher values than Scots pine (P=0.033) and arolla pine (P=0.043). While the main effect of soil moisture had no significant effect on CH_4 oxidation (P>0.05), the interaction between species and moisture was significant (P=0.002). Increased soil moisture enhanced CH_4 consumption in soils under spruce (P=0.009) but reduced CH_4 consumption under Scots pine (P=0.04) and larch (P=0.002). Under other species soil moisture did not affect CH_4 consumption (P>0.05). Previous studies of the effects of soil moisture on CH_4 consumption also showed varying responses. For example, Adamsen and King (1993) showed an inverse relationship between gravimetric soil water content and CH_4-oxidation in coniferous soil, and Yavitt et al. (1995) also found that the lowest rate of CH_4 consumption was associated with the highest soil water content in

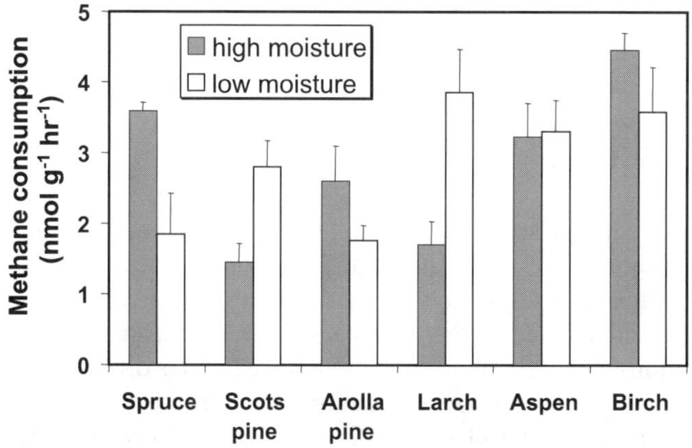

Figure 5. Rate of CH_4 consumption measured under the six tree species at two moisture contents (60 and 90% WHC). The initial concentration of CH_4 was 10 ml l^{-1} (n=6). The significant effects of soil moisture were observed beneath spruce and larch ($P<0.010$) and Scots pine ($P<0.05$).

hardwood forest soil. In contrast, Nesbit and Breitenbeck (1992) found that CH_4 oxidation was relatively insensitive to soil moisture (25 and 75% water-filled pore space) for swamp and forest soils. Wahlen and Reeburgh (1996) explained such differences in response to moisture by the physiological characteristics of the extant microbial communities and by differences in initial CH_4 concentrations. In our incubation experiment, soils were exposed to equal CH_4 concentration; thus, differences in microbial communities under different tree species is one plausible reason for different responses to soil moisture. Both methanotrophs and nitrifying bacteria are capable of CH_4 oxidation (Conrad, 1995), and the proportion of these groups is likely to vary among different soils (Gulledge et al., 1997). Thus, the different responses to soil moisture may be explained by plant species effects on the relative abundances of methanotrophs and nitrifiers and by differing sensitivities of these groups to soil moisture. For example, we have shown that net nitrification is highest under larch and lowest under spruce and Scots pine (Menyailo et al., 2002b, and this work). If nitrifying bacteria were mostly responsible for CH_4 consumption under larch, the increase in moisture resulted in inhibition of nitrification activity and decline in CH_4 consumption. It is more difficult to explain variation in response to increased moisture due to greater activity of methanotrophs under spruce and Scots pine, as the response was different (Figure 5). This may be due to a) the lack of a relationship between net nitrification rate and actual CH_4 oxidation by nitrifying bacteria or b) different response to increased moisture by different groups of methanotrophs.

These results provide evidence that the future predictions of CH_4 uptake in boreal and temperate forests should consider changes in tree species composition together with changes in soil moisture regimes. However, if global warming causes birch to replace coniferous species, potential CH_4 uptake will be higher regardless of soil moisture changes.

CONCLUSIONS

We have shown strong effects of tree species on soil microbiological processes responsible for the fluxes of greenhouse gases. The species effects are much larger than were previously reported (Priha et al., 1999a,b). Both species and depth had larger effects on N transformations than on C transformations. Denitrification potential varied under different species in the same way as did net N mineralization and net nitrification, while N_2O reduction rate did not. This caused a large difference between N_2O production and reduction rates under deciduous species. We predict that deciduous species will produce more N_2O in the field than conifers and that Siberian forests will produce more N_2O if global climate change results in changes in species composition. The rate of CH_4 oxidation is affected by tree species, but this effect interacts strongly with soil moisture. These results provide evidence that the future predictions of CH_4 uptake in boreal and temperate forests should consider changes in tree species composition together with changes in soil moisture regimes. Nevertheless, if, as predicted, global warming causes deciduous species to replace coniferous species, the uptake of CH_4 will be higher.

ACKNOWLEDGMENTS

This work was supported by the US Civilian Research and Development Foundation (grant RG1-2537-KY-03) and by the John D. and Catherine T. MacArthur Foundation (04-81300-000-GSS).

REFERENCES

Adamsen A P S, King GM 1993 Methane consumption in temperate and subarctic forest soils: rates, vertical zonation, and responses to water and nitrogen. Appl. Environ. Microbiol. 56, 485-490.

Binkley D 1994 The influence of tree species on forest soils: processes and patterns. *In* Proceedings of the Trees and Soil Workshop, Lincoln University, 28 February - 2 March 1994. Eds. D.J.Mead and I.S.Cornforth. pp 1-33. Arg. Soc. New Zealand Spec. Pub. No.10. Lincoln University Press, Canterbury.

Butterbach-Bahl K, Gasche R, Breuer L and Papen H 1997 Fluxes of NO and N_2O from temperate forest soil: impact of forest type, N deposition and of liming on the NO and N_2O emissions. Nutr. Cycl. Agroecosys. 48, 79-90.

Conrad R 1995 Soil microbial processes involved in production and consumption of atmospheric trace gases. *In* Advances in microbial ecology. Ed. J Gwynfryn Jones. pp.207-250. Plenum Press, New York, N.Y.

Côté L, Brown S, Paré D, Fyles J and Bauhus J 2000 Dynamics of carbon and nitrogen mineralization in relation to stand type, stand age and soil texture in the boreal mixedwood. Soil Biol. Biochem. 32, 1079-1090.

Gulledge J, Doyle A P, Schimel J P 1997 Different NH_4^+ inhibition patterns of soil CH_4 consumption: a result of distinct CH_4 oxidizer populations across sites? Soil Biol. Biochem. 29, 13-21.

FAO 1990 Soil map of the world, revised legend. FAO, Rome, Italy.

Matson P A, Vitousek P M, Livingstone G P, Swanberg N A 1990 Sources of variation in nitrous oxide fluxes in Amazonian ecosystems. J. Geophys. Res. 95, 16789-16798

Menyailo O and Huwe B 1999 Activity of denitrification and dynamics of N_2O release in soils under six tree species and grassland in central Siberia. J. Plant Nutr. Soil Sci. 162, 533-538.

Menyailo O V, Hungate B A, Zech W 2002a Tree species mediated soil chemical changes in a Siberian artificial afforestation experiment, Plant Soil 242, 171-182.

Menyailo O V, Hungate B A, Zech W 2002b The effect of single tree species on soil microbial activities related to C and N cycling in the Siberian artificial afforestation experiment. Plant Soil 242, 183-196.

Menyailo O V and Hungate B A 2003 Interactive effects of tree species and soil moisture on methane consumption. Soil Biol. Biochem. 35, 625-628.

Mikola M 1985 The effect of tree species on the biological properties of forest soil. Nat. Swed. Env. Protect. Board 3017, 1-29.

Nesbit S P and Breitenbeck G A 1992 A laboratory study of factors influencing methane uptake by soils. Agriculture, Ecosystems and Environment 41, 39-54.

Pastor J and Post W M 1988 Response of northern forests to CO_2-induced climate change. Nature 334, 55-58.

Priha O, Grayston S J, Pennanen T and Smolander A 1999a Microbial activities related to C and N cycling and microbial community structure in the rhizospheres of *Pinus sylvestris*, *Picea abies* and *Betula pendula* seedlings in an organic and mineral soil. FEMS Microbiol. Ecology 30, 187-199.

Priha O, Lechto T and Smolander A 1999b Mycorrhizas and C and N transformations in the rhizospheres of *Pinus sylvestris*, *Picea abies* and *Betula pendula* seedlings. Plant Soil 206, 191-204.

Priha O and Smolander A 1997 Microbial biomass and activity in soil and litter under *Pinus sylvestris*, *Picea abies* and *Betula pendula* at originally similar field afforestation sites. Biol. Fertil. Soils 24, 45-51.

Scott, N.A. 1998. Soil aggregation and organic matter mineralization in forests and grasslands: plant species effects. Soil Science Society of America Journal 62:1081-1089.

van Bachove E, Jones H G, Bertrand N and Prévost D 2000 Winter fluxes of greenhouse gases from snow-covered agricultural soil: Intra-annual and interannual variations. Global Biogeochem. Cycles 1, 113-126.

Wahlen S C, Reeburgh W S 2000 Effect of nitrogen fertilization on atmospheric methane oxidation in boreal forest soils. Chemosphere – Global Change Science 2, 151-155.

Wedin D A and Tilman D 1990 Species effects on nitrogen cycling: a test with perennial grasses. Oecologia 84, 433-441.

Yavitt J B, Fahey T J, Simmons J A 1995 Methane and carbon dioxide dynamics in a northern hardwood ecosystem. Soil Sci. Soc. of Am. J. 59, 796-804.

Chapter 19

THE FORMATION OF SOIL INVERTEBRATE COMMUNITIES IN THE SIBERIAN AFFORESTATION EXPERIMENT

I.N. Bezkorovaynaya
Sukachev Institute of Forest, Siberian Division, Russian Academy of Sciences, Akademgorodok, Krasoyarsk, 660036, Russia
e-mail: biosoil@forest.akadem.ru

ABSTRACT

An estimation of formation of soil invertebrate complexes was carried out on a common-garden provenance test with the major Siberian woody species: arolla pine *(Pinus sibirica)*, Scots pine *(Pinus silvestris)*, Siberian larch *(Larix sibirica)*, Norway spruce *(Picea abies, =Picea obovata)*, birch *(Betula fruticosa)*, and aspen *(Populus tremula)*. The experiment was established on a former agricultural field, so the formation of soil communities over the 30 year period combined the effects of both the reestablishment of forest conditions for soil formation and the influence of each tree species.

INTRODUCTION

The development of forest ecosystems entails reciprocal interactions between the trees and the soil communities. The trees provide the substrates processed by the soil biota, and the rates of nutrient release by the soil biota strongly influence the growth and success of trees. The structural and functional characteristics of the invertebrate communities are of key importance in forest ecosystem development (Szujecki, 1983). The success of soil animals depends strongly on the composition of the vegetation, in relation to the quantity and quality of litter produced by the vegetation, and also from the microenvironmental influences of the plants.

D. Binkley and O. Menyailo (eds.),
Tree Species Effects on Soils: Implications for Global Change, 307–316.

Soil invertebrates are a key heterotrophic link of biological cycling of organic matter and nutrients (Coleman and Crossley, 2003). The invertebrates also strongly influence plants by promoting interaction between many structural ecosystem components.

The Siberian Afforestation Experiment provided an opportunity to examine the effects of tree species and time on the soil animal communities. These experimental plots developed for 30 years under the influence of the six major Siberian tree species: arolla pine *(Pinus sibirica)*, Scots pine *(Pinus silvestris)*, Siberian larch *(Larix sibirica)*, Norway spruce *(Picea abies, = Picea obovata)* birch *(Betula fruticosa)*, and aspen *(Populus tremula)*. The planting of these species at a common location (with soils mixed and redistributed across the site before planting; see Shugalei, this volume) allowed us to look at the development and functioning of different forest ecosystem types (Shugaley et al., 1984).

METHODS

The soil communities were examined at three points in time after plantation establishment: 8-10 years, 20-25 years, and 30 years. Microarthropod communities were sampled with a 5-cm corer, with ten samples taken in each plot. Large invertebrates were sampled from five 25 x 25 cm quadrats on each plot, separated by horizon (O horizon, 0-5 cm, 5-10 cm, 10-20 cm, and 20-30 cm). These samples were evaluated for microarthropods using Tullgren funnels (Dunger, 1987). Ten subsamples were also taken at each sampling time to determine the numbers of large invertebrates by manual sorting (Gilyanov, 1987). An adjacent plowed field was used as a control to represent the likely starting conditions for the afforested plots.

RESULTS AND DISCUSSION

The soil fauna in the plowed field was characterized by low densities and low diversities of both small and large invertebrates (Table 1). *Collembola* averaged just 102 individuals/m^2 and were represented by only four species (and almost all of these belonged to the genus *Onychiurus*). These insects have a very narrow ecomorphological range, typically specific to plowed fields. The absence of an O horizon prevents the development of a more diverse community of *Collembola*. The density of large invertebrates was also low with only 38 individuals/m^2 (Figure 1), again with low diversity (Table 2).

The community of large invertebrates in the plow land was dominated by *Lasius niger* ants (50% of total numbers) and *Carabidae* (14% of total number). The trophic structure of the plow land was dominated by zoophages

Table 1. Dominant *Collembola* in the plow land and different-aged forest plantation (% of all *Collembola* found within each plot at each age).

Species	Plow land	8-10 yr old						30 yr old					
		Arolla pine	Larch	Scots pine	Spruce	Birch	Aspen	Arolla pine	Larch	Scots pine	Spruce	Birch	Aspen
Onychiurus sp.	9.4	10.8	51	31	20.3	38.5	41.3	4.3	6	14.3	10.4	5.4	10.8
O. sp.$_1$	78.1	38.6	19.3	48.1	47.7	33.5	49.5	9	2	44.5	13.6	1.8	2.6
O.taymiricus	0	3.8	0	0	0	0	0	6.7	10.4	4.2	2.6	9	2.6
O.subarcticus	9.4	0	0	3.3	2.6	2.6	3.3	0	6	6	3.8	9	2.1
O.subnemoratus	0	0	0	0	0	0	0	5.5	2	0	0	6.8	16
Anurida sp.$_2$	3.1	0	0	0	0	0	0	0	0	0	0	0	0
Folsomia quadrioculata	0	0	0	8.5	7.8	10	0	13.7	7.5	7.8	16.5	13.6	19.3
F. inoculata	0	2.1	0	0	0	0	0	2.6	0	0	2	1.1	0
F.taimyrica	0	0	0	0	0	0	0	0	0	0	0	4.3	2.1
Isotoma notabilis	0	19.3	10.4	0	7.8	0	0	28.4	26.8	2.6	19	28	29.3
I. viridis	0	16	0	0	0	0	0	9.6	10.4	0	16.5	4.3	3.8
I. maritime	0	0	5.4	0	0	0	0	0	5.5	3.3	0	0	2.6
I.bipunctata	0	0	0	0	0	7.3	0	0	2	5.4	3.8	9	2.1
Sminthurinus aureus	0	0	2	0	0	0	0	4.3	0	0	0	0	0
Lepidocyrtus cyaneus	0	2.6	0	0	0	0	3.3	6.7	3.3	2.6	2.6	0	3.8
Tullbergia sp.	0	0	5	0	0	0	0	2.6	2.6	3.3	0	0	0

Table 2. Group structure of large invertebrates in the plow land and different-aged forest plantation (% of all large invertebrates found in each plot by time).

		8-10 yr old						20-25 yr old						30 yr old					
Taxa	Plow land	Arolla pine	Larch	Scots pine	Spruce	Birch	Aspen	Arolla pine	Larch	Scots pine	Spruce	Birch	Aspen	Arolla pine	Larch	Scots pine	Spruce	Birch	Aspen
Lumbricidae	4.8	8.9	7.6	2.9	4.1	12.1	8.7	4.1	2.2	17.9	0.9	15.6	10.8	16.7	9.6	12.1	0	15.7	3.9
Enchytraeidae	0	0.8	0	0	0	0	0	12.1	51.3	8	38.7	40.4	44.5	8.3	53.9	18.2	3.6	39.8	66
Mollusca	0	1.8	2.9	29.4	0	2	10.7	0	0	0.5	0.5	4.4	0.9	0	0	3	0	4.8	2.9
Diplopoda	4.8	3.6	0	1.5	0	1	2.7	1.9	0	0	0	1.5	0	0	0	0	0	2.7	4.4
Chilopoda	0	0	0	0	0	0	0	2.4	1.9	5.4	1.9	2.2	2.7	8.3	9.7	21.2	7.4	4.8	0.5
Aranei	4.8	4.5	5.7	7.4	0	12.1	13.4	1.9	1.9	0.9	1.9	1.4	5.4	0	1.9	0	3.6	1.9	1.9
Opiliones	0	1.8	5.7	4.4	1.3	2	1.3	0	0	0	0	0	2.5	0	0	0	1.9	0	0
Orthoptera	2.3	0	0	0	1.3	0	0	0	0	0	0	0	0	0	0	0	0	0	0
Hemiptera	4.7	8	3.8	2.9	4.1	5.1	4	4.7	1.9	0	0	0	0.5	0	0	0	3.8	4.4	1.9
Aphidinea	0	4.5	2.9	1.5	1.4	1	0.7	0	0	0	0	0	0	0	0	0	0	0	0
Cicadinea	0	0.8	4.8	0	0	0	4	0	0	0	0	0	0	0	0	0	0	0	0
Coleoptera:																			
Carabidae	14.3	10.7	17.1	7.4	6.8	8.1	6.7	1.9	1.9	1.9	5.4	1	2.2	0	1.9	3	3.6	6.7	0
Scarabaeidae	0	0	0	0	0	1	0	0	0	2.2	0	0	0	0	0	0	0	1.9	0
Silphidae	0	0.9	0	0	0	0	0.7	0	0	0.5	0	0	0	0	0	0	0	0	0
Staphylinidae	2.4	10.7	6.7	11.8	2.7	11.1	18.1	5.9	4	11.4	6	3	8.8	16.7	1.9	21.2	21.4	1.9	1.9
Elateridae	2.4	4.5	1.9	4.4	2.7	4.1	0	5.4	0.5	23.9	2.7	3.3	3.9	41.7	13.5	6	10.7	4.4	2.9
Curculionidae	0	6.4	0	4.4	5.4	6.1	1.3	2.5	5.1	7	7.7	13.1	2.4	0	3.8	4.4	17.7	3.8	4.4
Chrysomelidae	0	0.8	5.7	2.9	0	0	3.4	0	2.2	1.3	0	0.9	0.7	0	0	0	0	1.9	0
Cantharidae	0	0.9	1.9	0	1.3	6.1	1.3	0	0	0	0.9	0	0	0	0	0	0	0	0
other	7.1	7.7	6.7	7.4	1.3	12.2	8.2	7.7	3.2	5.9	8.6	1.1	2.3	0	0	4.1	1.3	1	0
Formicidae	52.4	8	17.2	8.8	54.1	1	4.7	0	1.9	7	7	2.4	2.2	0	1.9	0	0	1.9	1.9
Diptera (l):																			
Muscidae	0	0	0.9	0	0	0	0	1.9	0.5	0	0	0.9	0	0	0	0	0	0	0
Tipulidae	0	5.4	1.9	2.9	13.5	12.1	10.1	4.4	2.9	4	10	5.4	0	0	0	3	8.8	1.9	0
Bibionidae	0	0.9	0	0	0	0	0	7.7	6.7	2.2	0	0	5	0	1.9	1.9	0	0	4.7
Cecidomyiidae	0	0.8	0.9	0	0	0	0	13.5	1.9	0	0	0	0	0	0	0	0	0	0
Tendipedidae	0	0	0	0	0	0	0	12.1	5.4	0	5.8	0	0	8.3	0	0	0	0	0
Rhagionidae	0	0	0	0	0	0	0	0	0	0	0	3.4	2.2	0	0	0	3.6	0	2.7
Sciaridae	0	0	0	0	0	0	0	6.7	0	0	0	0	0	0	0	1.9	10.7	0	0
other	0	7.6	5.7	0	0	2.9	0	3.2	3.5	0	2	0	3	0	0	0	1.9	0.5	0

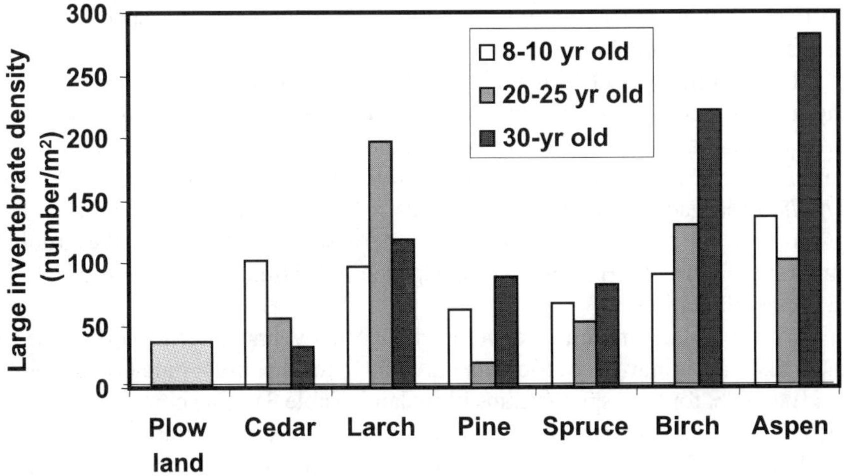

Figure 1. Large invertebrate density in the plow land and different-aged forest plantation.

(79%), with lower numbers of phytophages (17%) and saprophages (4%). After 8 to 10 years of tree influence, the *Collembola* species diversity was highest under spruce (27 species) and birch (25 species), while it was the lowest (7 species) under aspen (V.K. Dmitrienko, p. 87 in Shugaley et al. 1984). The differences in diversity among the tree species at this point in time were concentrated mostly in rare species and subdominants. The dominant species under the trees remained the same as in the plow land (*Onychiurus sp.* and *Onychiurus sp.1*), but the densities differed (Table 1).

The number of large invertebrates ranged from 1.5 to 3.0 times greater under the trees than in the plow land (Figure 1). The proportion of *Collembola* forms increased under trees with the development of the O

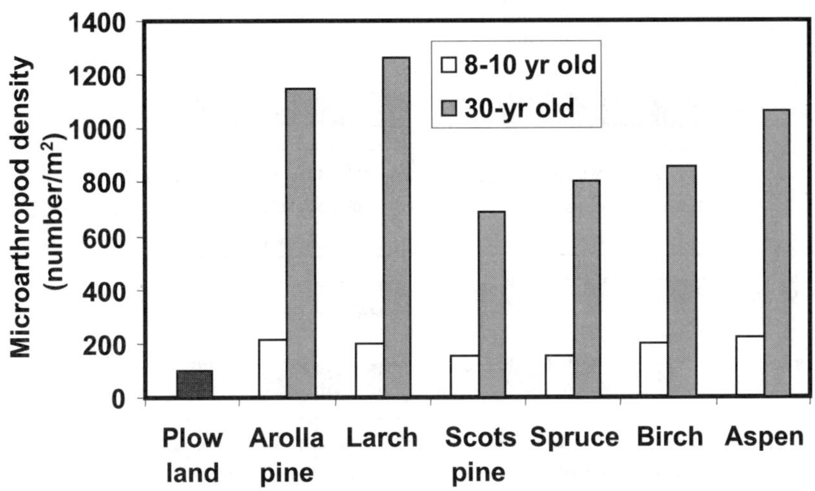

Figure 2. Microarthropod density in the plow land and different-aged forest plantation (only Collembola are for the plow land and 8-10 years old forest plantation).

horizon, though the community was still dominated by the species found primarily in mineral soils (Figure 3).

The number of soil animals in the dominant groups varied widely among the tree species. *Lumbricidae, Carabidae, Staphylinidae, Curculionidae, Lasius ants*, and *Tipulidae* larvae dominated under arolla pine; *Mollusca, Arachnidae, Carabidae*, and *Staphylinidae* under Scots pine, and ants and *Tipulidae* larvae under spruce (Table 2). As the O horizon developed, the proportion of saprophytic consumers increased 3-4 times in the trophic structure (Figure 4). However, zoophages still continued to prevail (51-67%) in all the plantations.

As the soil communities developed for 8-10 years under the influence of the trees, the similarity coefficients ranged 9% for the arolla pine (versus plow land) to 38% for the aspen (versus plow land, Table 3). The communities also differed among the tree species plots, with the greatest similarity between arolla pine and spruce, and least similarity between the two hardwood species (birch and aspen).

The soil communities continued to change over time, along with the increasing mass of the O horizon. At 20-25 years, the birch and aspen plots

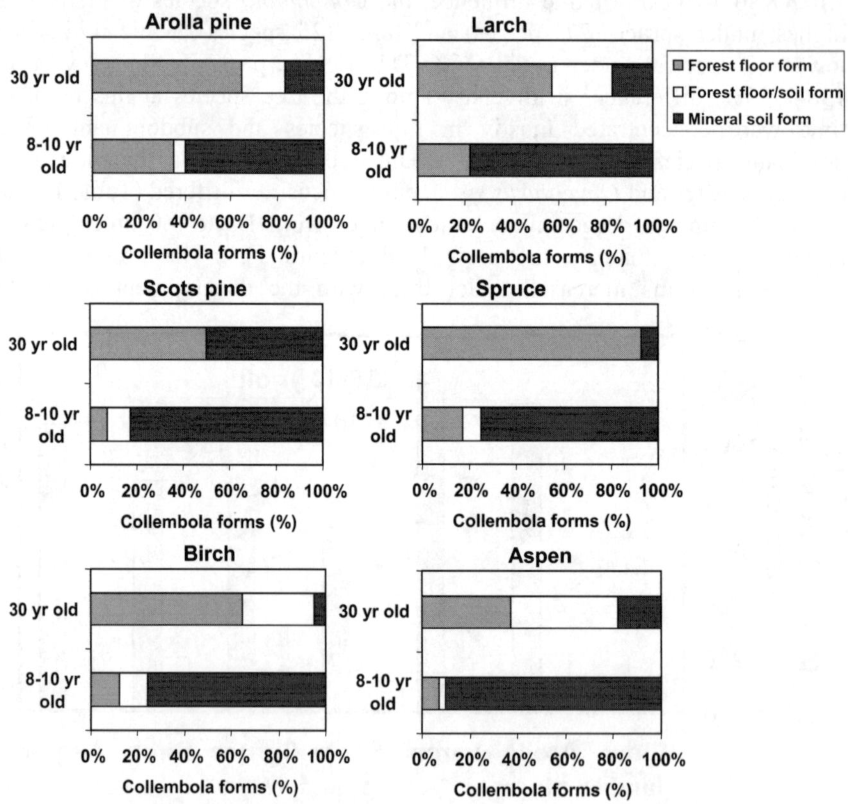

Figure 3. Collembola forms in forest plantation.

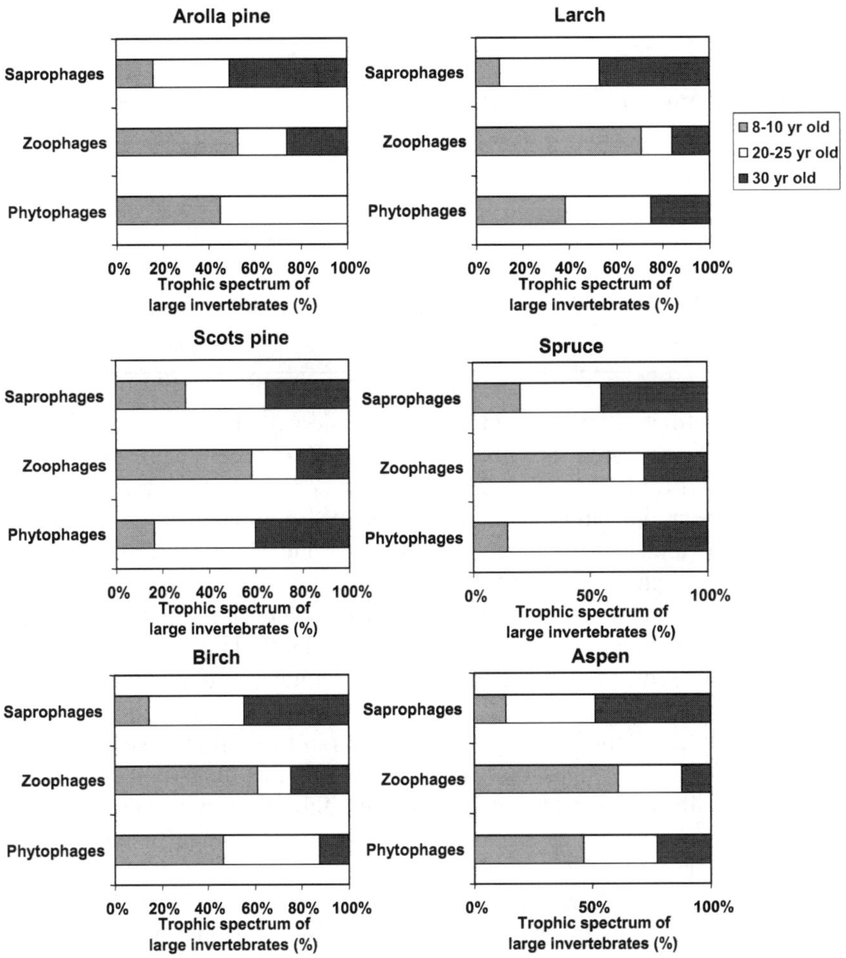

Figure 4. Trophic spectrum large invertebrates in forest plantation.

had the greatest number of invertebrates (102-196 indivduals/m^2). The conifer plots had only 20-56 individuals/m^2, not much higher than in the plow land. Most invertebrates (95%) occurred in the upper soil horizons, with the communities dominated by mineral soil taxa (*Lumbricidae, Enchytraeidae*, and *Diptera* and *Coleoptera* larvae; Table 2). By this age, the saprophage proportion increased up to 58-64% in the hardwood and larch plantations, and to 35-49% for other species (Figure 4). Phytophages comprised a substantial proportion (34-48%) in the Scots pine and spruce plots.

The density of microarthropods increased by 6-fold after 30-years compared to the plow land (Figure 2). *Collembola* accounted for 54 to 76% of the total numbers, and the major species differed among all the tree species (Table 1). The Scots pine stand had changed the least, with the *Collembola*

Table 3. The similarity coefficients for invertebrate community (% similarity between communities), 8-10 years after planting.

Tree species	Arolla pine	Scots pine	Larch	Spruce	Birch	Aspen	Plow land
Arolla pine	100	40	34.6	46.9	40.6	22.7	9.1
Scots pine		100	36.4	35.5	42.9	37.5	18.8
Larch			100	35.5	29	36.4	18.7
Spruce				100	48.6	25.9	14.8
Birch					100	23.1	16
Aspen						100	37.5

complex remaining dominant in the forest floor. The density of large invertebrates was greatest under birch and aspen, with over 200 individuals/m^2 (Table 2). The density under spruce, Scots pine and larch ranged between 82 and 118 individuals/m^2, compared with only 32 individuals/m^2 under arolla pine. *Lumbricidae* and *Enchytraeidae* began to prevail under larch, aspen, and birch, whereas *Chilopoda* and *Elateridae* larvae were dominant under arolla pine, Scots pine and spruce (Table 2). Saprophages continued to dominate under aspen, birch, larch, and arolla pine, but they comprised just 33% of the population under spruce and Scots pine (Figure 4).

The newly developed ecological conditions in the soil involved chemical and morphological changes that influenced both the quantity and species diversity of soil invertebrates. After 30 years, the soil communities differed substantially in each plot as shown by the low similarity coefficients for the invertebrate communities (Table 4).

The mass of the O horizon (forest floor) did not correlate with the number of invertebrates ($r = 0.22$; Table 5), probably because the quality of substrates matters much more than the quantity (Striganova, 1987). It is difficult to
matters much more than the quantity (Striganova, 1987). It is difficult to
gauge the substrate quality of organic materials for supporting soil biota (Bezkorovaynaya, 2002). Indeed, the chemical characteristics of the O horizon materials varied less than the invertebrate populations across the tree species.

Table 4. Quantitative similarity coefficients for invertebrate community (% similarity between communities after 30 years).

	Scots pine	Larch	Spruce	Birch	Aspen
Arolla pine	21.6	8.5	14.3	13.1	5.5
Scots pine		26.9	27.1	26.1	9.7
Larch			9.6	46.5	28.1
Spruce				12.1	3.2
Birch					31.9

The populations of invertebrates related weakly to the organic matter content of the mineral soil, accounting for 25% of the variation under birch, larch, and aspen, but less than 17% for the other species. Birch and aspen litter was humified rapidly and completely. Micromorphological analysis of upper soil horizon samples from the hardwood plantations showed a high content of excrement from large invertebrates, mainly humified coprolites from worms. This high coprolite content enhanced the microstructure of the upper mineral soil, particularly under birch. Coprolites were also important in the microstructure under larch, but in the other conifer plantations the organic material was poorly digested, leading to fewer biogenic microaggregates, less developed humification, and lighter-colored humus.

The majority of the soil animals were found in the 0-10 cm mineral soil, reflecting the overall microenvironmental effects (Table 5). Distribution of trophic groups of large invertebrates also showed patterns with relation to soil depth. Most zoophages (*Staphylinidae, Aranei, Carabidae*) occurred in the forest floor and its interface with mineral soil layers under all tree species. The density of saprophages followed a similar pattern, especially beneath the hardwood species, though individuals were occasionally found below 50 cm.

Table 5. Vertical allocation of soil large invertebrates, humus content and hydrothermic characteristics in 20-25 year old forest plantation.

Tree species	Soil layer (cm)	Density (individuals/m^2)	Organic matter (g/kg)	Temperature ^oC	Water content (mm)
Arolla pine	O horizon	4	438	12.3	-
	0-10	47.9	57	10.7	29.9
	10-20	2.2	51	7.8	30.7
	20-30	1	33	7.5	30.1
Larch	O horizon	13.6	379	12.4	-
	0-10	138.7	64	9.9	30.3
	10-20	35.6	55	9.5	31.3
	20-30	7	45	8.2	30.5
Scots pine	O horizon	5.1	420	11.2	-
	0-10	10	39	10.6	28.9
	10-20	2.4	37	9.5	29.3
	20-30	2.7	35	7.4	30
Spruce	O horizon	1.3	463	12.4	-
	0-10	42	45	12.2	30.2
	10-20	3.6	34	10.6	30
	20-30	5.3	31	8.3	28.9
Birch	O horizon	5.3	377	14.1	-
	0-10	104.9	37	13.4	31.3
	10-20	11	34	11.9	32
	20-30	9.3	29	10.2	31.6
Aspen	O horizon	11	329	12.3	-
	0-10	84.4	52	12.2	30.5
	10-20	4.4	42	11.1	31.9
	20-30	1.7	32	9.9	31.3

The majority of phytophages (*Curculionidae, Elateridae*) occurred in 5 cm to 20 cm mineral soil. The vertical distribution and migration of phytophagous beetles differ with the trophic preference of each group, and changes in preferences among life cycle phases.

Low coefficients of the correlation between invertebrate population density and soil water amount (r = 0.16-0.20) and temperature (r = 0.20-0.32) suggest differences in hydrothermal regime of soil among the plantations do not account for major differences in soil communities among the tree species. The seasonal changes in microenvironmental conditions can be very large, having more influence on invertebrate dynamics throughout the growing searson. For example, soil saprophages and some zoophages become active during periods with favorable combinations of heat and moisture in soil, which usually occur in spring and fall (Bezkorovaynaya et al. 2003). The activity of phytophages increases as vegetation develops through the season, with maximum populations in mid- to late-July. Relatively rapid heating of the forest floor and the upper soil horizons promotes early biological activity in soil under aspen, birch, and larch compared with the shadier conditions under the evergreen conifers. Large invertebrates become active in these plantations as early as the beginning of May, while the onset of their activity occurs in late May under arolla pine, spruce, and Scots pine.

REFERENCES

Bezkorovainaya I N 1999 Effect of some environmental factors on soil mesofauna formation in forest cultures. Eurasian Soil Sci+ 5, 593-600.

Bezkorovainaya I N and Yashihin G I 2003 Effect of hydrothermic conditions on invertebrate complexes in coniferous and deciduous cultures. Ecology (Russia) 1, 56-62.

Coleman D C and Crossley D A Jr 2003 Fundamentals of Soil Ecology. Academic Press, Burlington, Massachusetts, USA. 204 pp.

Dunger W 1987 Estimate of microarthropod (microfauna). *In* Quantitative methods in soil zoology. pp 26-510. Nauka, Moscow.

Gilyarov M S Estimate of large invertebrates (mesofauna). *In* Quantitative methods in soil zoology. pp 9-26. Nauka, Moscow.

Shugaley L S, Dmitrienko V K and Yashikhin G I 1984 Modeling the development of artificial forest biogeocenosises. Nauka, Novosibirsk. 152 pp.

Striganova B R 1980 Nutrition of soil saprophages. Nauka, Moscow. 244 pp.

Szujecki A and Szyszko J 1983 The process of forest soil macrofauna formation after afforestation of farmland. Warsaw Agricultural University Press, Warsaw. 196 pp.

Chapter 20

THE TRANSFORMATION OF PLANT RESIDUES UNDER DIFFERENT TREE SPECIES IN THE SIBERIAN AFFORESTATION EXPERIMENT

L.V.Mukhortova

Institute of Forest, Russian Academy of Sciences, Siberian Branch
660036, Krasnoyarsk, 36, Akademgorodok, Russia

ABSTRACT

The forest floor (O horizon) is a major pool of organic matter in many forests, comprised of a wide range of materials that vary in chemistry and morphology. The processing of fresh detritus progresses with gradual degradation under the influence of biotic and abiotic factors, leading to decreasing particle size and carbon:nitrogen ratios. I used multiple regression models of the dynamics of litter input and processing through the litter layer (Ol or Oi) into the fermented (Of or Oe) layer using experimental data from beneath six tree species. The carbon contained in annual litterfall inputs equaled 20- 26% of the total forest floor C content in conifer plots, compared with 48 to 55% in the aspen and birch plots. This material remains in the Ol layer for 2 to 12 months. The Of layer comprises 47 to 85% of the total C content of the forest floor, with residence times of 0.5 to 5 years.

INTRODUCTION

Soil is a locus of many processes that transform organic matter and cycle elements. The complexities of these transformations and flows have interested generations of researchers, and many challenges remain to segregate and study processes under natural conditions. The accumulation of organic matter depends on the input of fresh materials, the partial processing of these materials into carbon dioxide (CO_2) and residual molecules, as well

D. Binkley and O. Menyailo (eds.),
Tree Species Effects on Soils: Implications for Global Change, 317–335.

as the formation of new materials as byproducts of microbial processes. All of these depend on litterfall amount, its fractional composition, and decomposition intensity; and these depend in turn on forest type (including structure and density), bonitet (=soil productivity class), character of understory vegetation, and on environmental site conditions (Kulagina, 1997; Aleksandrova, 1980). Plant residues are decomposed at any time when temperature and moisture are favorable for microbial activity (Sokolov and Karpova, 1965; Kulagina, 1977; Karpachevsky, 1981, 1988; Grishina, 1986; Vedrova, 1995; Bezkorovainaya and Vishnyakova, 1966; Vedrova and Mindeyeva, 1998; Quals et al., 1991; Vitousek et al., 1994).

The importance of all these interacting factors are difficult to unravel by experimentation with isolated plant material (such as in litterbags), because methods may distort the rates of processes by modifying physical, chemical and biotic factors (Ganzhara and Orlov, 1993). In this chapter, I describe the dynamics of organic matter transformation in litter using data on its stock (pool size) and fractional composition through a year.

The progression of litter processing results in decreasing particle sizes, allowing increasing compression of the litter, which promotes moisture retention and favors biochemical processing (Sokolov and Karpova, 1965). The biochemical processing of the materials (sometimes called fermentation, although this does not imply anaerobiosis) reflects the intensity of microbiological activity, and has secondary importance (Bogatyrev, 1990).

The rate of decomposition of plant litter differs strongly among forests, depending in part on the chemical composition of the litter (including C:N), the physical structure of the liter, and environmental conditions (such as water, temperature, oxygen, and soil texture and chemistry). The actual impact of these factors may be mediated through effects on the soil animal community. Grishina (1986) rated the relative contribution of such factors in controlling the overall processing of litter: 10% abiotic factors, 10% microorganisms; 10% microfauna, and 70% mesofauna.

OBJECTS AND METHODS OF STUDY

Investigations of litter decomposition and transformation were carried out in the Siberia Afforestation Experiment (Shugalei, this volume), when the plantations were 25 years old. This common-garden plantation used all 6 tree species that commonly dominate forest landscapes in Siberia: arolla pine (*Pinus sibirica*), Scots pine (*Pinus sylvestris*), Siberian larch (*Larix sibirica*), Norway spruce (*Picea abies (=Picea obovata*), aspen (*Populus tremula*) and birch (*Betula fruticosa*). Seedlings of each species were planted in single plots at a common site where the top soil had been mixed (to increase homogeneity) and respread across the site. The soil is a dark-grey forest weakly podzolized gleyish soil developed on chestnut- brown clay. At age 25, the O horizon (forest floor) was developed well in all plots. The aspen and birch stands have well developed understories (ground layer) dominated by herbs, and a moss layer was well developed under spruce. The understory was very sparse beneath arolla pine, Scots pine and larch.

The field sampling included collecting the samples of the litter layer (L and F layers, or Ol and Of, or Oi and Oe) of the forest floor, 2- 3 times each month from May through November. Each sampling period included the collection of 10 samples (0.2 x 0.2 m) from each plot. Samples were divided into fractions in a soil sieve: leaves, branches >10 mm in diameter, branches < 10 mm in diameter, cones, bark, and "other". Each fraction was ground to pass through a 1 mm sieve, weighed and moisture content determined.

To determine microbial biomass, samples of soil and litter were separated into 5 fractions: 0l (Oi), 0f (Oe), as well as mineral soil from depths of 5 cm, 10 cm, and 20 cm. Microbial biomass concentration was determined in fresh taken samples, and results reported based on dry mass.

The concentrations of total and labile carbon (C) were determined on each sample using Anstet's method in modification of Ponomareva-Nikolaeva (Ponomareva and Plotnikova, 1975). Nitrogen concentration was determined in the same extraction by the Kjeldahl method (Arinushkina, 1970). Carbon concentration of labile organic matter was defined by successive extractions using water and 0.1 M NaOH. Microbial biomass was assessed in soil and litter by the rehydration method (Anonymous 1991).

Decomposition process of forest litter is shown as a unity of two simultaneously proceeding processes: mineralization up to simple chemical compounds (CO_2, CH_4, NO etc.) and synthesis of humus substances. The pattern in mass of the subhorizons of the forest floor can be expressed as:

$$C(t) = C_0 + b_1t + b_2t^2 + b_3t^3, \qquad (1)$$

where C – C stock in litter, g m^{-2}; t – period of time (days) beginning from spring litter; C_0 – carbon stock in litter at the initial moment.

The decomposition of litter entails the processing of material already within a pool (such as the Ol layer) that decays into another pool (such as Of), and also the replenishment of the first pool by the input of fresh material from litterfall. Assessments of decomposition within the pool need to account for both losses and gains. As fresh material from litterfall immediately begins to decompose, the previous portion (OC_1) equals:

$$OC_1 = C_1 - C_1V_d\,t_1,$$

where C_1 is organic matter of the previous litterfall portion in g/m^2, V_d is the specific rate of litterfall decomposition in mgC gC^{-1} day^{-1}, and t is the number of days between litterfall inputs. Thus, the total litterfall accumulation (ΣOC) during a year is:

$$\Sigma OC = (C_1 - C_1V_dt_1) + (C_2 - C_2V_dt_2) + ... +(C_n - C_nV_d\,t_n), \qquad (2)$$

or $\Sigma OC = C_1 + C_2 + ... + C_n - V_d (C_1t_1 + C_2t_2 + ... + C_nt_n)$ (3)

Assuming that decomposition follows first-order kinetics ($C_1 - C_1 V_d t_1 = C(t) = C_0 e^{-kt}$), then litterfall accumulation at any time would be:

$$C(t) = C_n + C_{n-1} e^{-kt} \quad (4)$$

where k is the decay constant (day^{-1}), C_n is the litterfall C mass which came at the evaluation moment in g/m^2, C_{n-1} is the same for the previous point in time, and t is the time period (days) between evaluations C_n and C_{n-1}.

The complex chemical composition of fresh litterfall leads to rapid decomposition of simple molecules, with overall rates of decomposition decreasing as the more recalcitrant molecules (such as cellulose and lignin) are broken down. Our field studies developed decomposition constants (Vedrova and Mindeyeva, 1998) described by the logarithmic relationship:

$$k(t) = b_0 + b_1 \lg(t) \quad (5).$$

Appropriate coefficients (b_0 and b_1) are given in the Table 1. If we put (5) into the equation (4) and make some transformations, then the fresh litterfall accumulation during a year is described by the model:

$$\Sigma OC(t) = OC_n + OC_{n-1}(t) e^{-k(t)\,t} \quad (6)$$

This polynomial model has the form of second and third order for Scots pine and spruce plots; a logistic curve for aspen and birch; exponential for larch; and linear for arolla pine.

Given these calculated changes in litter accumulation ($OC(t)$) and organic matter pools in plant detritus in the L ($LC(t)$) subhorizon over time, we can calculate their decomposition as $DL(t) = LC(t) - OC(t)$. This approach yielded polynomial patterns for the litter of all species except aspen, which fit a logistic pattern.

In order to follow the decomposition of organic matter in the fermentation layer (Of, or Oe), we should subtract the pool ($FC(t)$) the incoming material from the L layer (Table 1) which was transformed by fermentation processes ($LFT(t)$), giving:

$$LFT(t) = C_0L - DL(t),$$

where C_0L is the initial mass in the L layer; and DL (t) is the change in mass of the decomposing L layer. Therefore, the transformation of material from the L layer to the F layer will equal:

$$LFT(t) = (C_0L - DL(t)) - C_{CO2},$$

where C_{CO2} is the mineralized carbon (Figure 6). As a result of these equations, the decomposition of material in the F layer has the form:

$$DF(t) = FC(t) - LFT\ (t),$$

where $DF(t)$ is the loss of mass from the F layer through decomposition in gC/m^2, $FC(t)$ is the mass of the F layer in g C/m^2, and $LFT(t)$ is the rate of input of material from the L layer. These patterns were described by third- or fifth-order polynomials.

The models of dynamics of labile forms of organic matter in decomposing litter (DLLOM(t)) were calculated the same way as described above. Calculations were performed separately for the water-soluble and alkali (0.1 *M* NaOH) soluble forms.

RESULTS AND DISCUSSION

The O horizon was well developed after 25 years under all species. The majority of the material originated as aboveground litterfall from the trees (Table 2). Leaf materials accounted for the majority of material (78 to 88% of the totals). Branches comprised 4 to 6% of the material under a pine, but 9 to 16% under Scots pine, spruce, aspen, and birch. The total rate of aboveground litterfall differed by more than 3-fold between species, ranging from 94 g m^{-2} yr^{-1} under spruce to 346 g m^{-2} yr^{-1} under aspen.

Arolla pine, Scot's pine and larch plots had the greatest mass accumulated in the litter layer (Ol or Oi), with 1130 to 1640 g/m^2, compared with 400 to 740 g/m^2 under birch and aspen (Table 2). The annual inputs from litterfall under birch and aspen were similar to the rates under the conifers, so the difference in litter accumulation related more to decomposition rates than to input rates. Combining all layers of the forest floor, the conifer plots accumulated 380 to 710 g C/m^2, compared with just 160 to 230 g C/m^2 under the hardwoods (Table 3). All species the peak period of litterfall, declining over the following winter (Figure 1).

Table 1. Regression models for dynamics of litter decomposition processes

Tree species	Characteristics	Model	Correlation coefficient R^2	F
Arolla pine	Decomposition constant (k)	$k=0.0034-0.0028*lg(t)+0.0034$	0.87	F(1,168)=1159
	Litterfall accumulation (OC)	$OC=2.408+0.085*t$	0.98	F(1,9)=548
	Stocks of OL litter subhorizon (LC)	$LC=17.152-0.263*t+0.00259*t^2-0.000007*t^3$	0.57	F(3,7)=3.85
	OM transition from OL to OF (LFTM)	$LFTM=2.319+0.225*t-0.00078*t^2$	0.99	F(2,8)=40221
	OF subhorizon stock (FC)	$FC=81.490-0.311*t+0.007*t^2-0.000029*t^3$	0.54	F(3,7)=7.36
	Microbial biomass in OL	$LCMB=0.0442+0.000014*t^2-0.00000005*t^3$	0.66	F(2,8)=7.26
	Microbial biomass in OF	$FCMB=2.125+0.0003*t^2-0.000002*t^3$	0.62	F(2,8)=6.42
Scots pine	Decomposition constant (k)	$k=0.0026-0.0018*lg(t)+0.0026$	0.78	F(1,168)=612
	Litterfall accumulation (OC)	$OC=1.266+0.0404*t+0.000003*t^3$	0.99	F(2,8)=278
	Stocks of OL litter subhorizon (LC)	$LC=25.643-0.195*t+0.00090*t^2$	0.78	F(2,8)=13.9
	OM transition from OL to OF (LFTM)	$LFTM=0.598+0.234*t-0.0009*t^2$	0.99	F(2,8)=7145
	OF subhorizon stock (FC)	$FC=68.167+0.0027*t^2-0.000021*t^3$	0.56	F(2,8)=4.79
	Microbial biomass in OL	$LCMB=0.195+0.012*t-0.0004*t^2+0.000004*t^3-0.0000002*t^4$	0.97	F(4,6)=48.0
	Microbial biomass in OF	$FCMB=1.883+0.000003*t^3-0.0000001*t^4$	0.66	F(2,8)=7.89

Table 1. Regression models for dynamics of litter decomposition processes (continued).

Tree species	Characteristics	Model	Correlation coefficient R^2	F
Larch	Decomposition constant(k)	$k=0.0040-0.0034*lg(t)+0.0040$	0.86	F(1,168)=1012
	Litterfall accumulation (OC)	$OC=16.722+e^{(-8.39+0.046*t)}$	0.98	F(1,9)=638
	Stocks of OL litter subhorizon (LC)	$LC=13.596-0.306*t+0.0017*t^2$	0.67	F(2,8)=7.80
	OM transition from OL to OF (LFTM)	$LFTM=8.791+0.335*t-0.004*t^2+0.000017*t^3$	0.97	F(3,7)=817
	OF subhorizon stock (FC)	$FC=89.361-0.110*t+0.0017*t^2-0.0000022*t^3$	0.60	F(3,7)=12.0
	Microbial biomass in OL	$LCMB=2.757-0.093*t+0.0009*t^2-0.000003*t^3$	0.92	F(3,7)=26.5
	Microbial biomass in F	$FCMB=0.228+0.120*t-0.0015*t^2+0.000006*t^3$	0.88	F(3,7)=17.6
Spruce	Decomposition constant (k)	$k=0.0096-0.0087*lg(t)+0.0096$	0.79	F(1,168)=904
	Litterfall accumulation (OC)	$OC=6.434+0.191*t-0.00045*t^2$	0.99	F(2.8)=384
	Stocks of OL litter subhorizon (LC)	$LC=61.619-0.448*t+0.00001*t^3$	0.70	F(2,8)=9.24
	OM transition from OL to OF (LFTM)	$LFTM=2.029+0.898*t-0.0039*t^2$	0.98	F(2,8)=183
	OF subhorizon stock (FC)	$FC=40.861+1.165*t-0.309*t^2+0.0003*t^3-0.000002*t^4$	0.72	F(4,6)=3.86
	Microbial biomass in OL	$LCMB=0.509-0.048*t+0.0022*t^2-0.000035*t^3+0.0000002*t^4-0.0000000003*t^5$	0.96	F(5,5)=21.6
	Microbial biomass in OF	$FCMB=2.503+0.733*cos(-1.120+0.104*t)$	0.40	F(1,9)=6.07

Table 1. Regression models for dynamics of litter decomposition processes (continued).

Tree species	Characteristics	Model	Correlation coefficient R^2	F
Aspen	Decomposition constant (k)	$k=0.00009*t\ (t<=25)+0.0084-0.0029*lg(t)\ (t>=25)$	0.84	F(1,168)=987
	Litterfall accumulation (OC)	$OC=57.260/(1+10^{(-56.639*t)})$	0.99	F(1,9)=16084
	Stocks of OL litter subhorizon LC)	$LC=42.660+0.092*t-0.0019*t^2$	0.44	F(2,8)=6.47
	OM transition from OL to OF (LFTM)	$LFTM=(-31.026/(1+10^{(-6.8+0.05*t)}))+30.133$	0.99	F(3,7)=369
	OF subhorizon stock (FC)	$FC=51.516-0.835*t+0.009*t^2-0.000027*t^3$	0.55	F(3,7)=2.87
	Microbial biomass in OL	$LCMB=-0.395+0.000018*t^3-0.000002*t^4+0.000000007*t^5$	0.98	F(3,7)=216
	Microbial biomass in OF	$FCMB=3.044-0.00074*t^2+0.000005*t^3$	0.94	F(2,8)=64.7
Birch	Decomposition constant (k)	$k=0.0133-0.0116*lg(t)+0.0133$	0.83	F(1,168)=966
	Litterffall accumulation (OC)	$OC=100/(0.369+e^{(0.996*t)})$	0.99	F(1,9)=4032
	Stocks of OL litter subhorizon (LC)	$LC=48.869-1.359*t+0.014*t^2-0.000036*t^3$	0.87	F(3,7)=15.7
	OM transition from OL to OF (LFTM)	$LFTM=-1.219+1.994*t-0.039*t^2+0.00031*t^3-0.000002*t^4$	0.95	F(4,6)=26.0
	OF subhorizon stock (FC)	$FC=50.467-0.814*t+0.0052*t^2$	0.73	F(2,8)=11.1
	Microbial biomass in OL	$LCMB=1.261-0.0615*t+0.00130*t^2-0.000011*t^3+0.0000001*t^4$	0.99	F(4,6)=123
	Micrrobial biomass in OF	$FCMB=1.434-0.065*t+0.0014*t^2-0.000012*t^3+0.0000004*t^4$	0.98	F(4,6)=80.3

Table 2. Litterfall mass by components (g m^{-2} yr^{-1}), and mass of the litter layer of the forest floor (L, or Oi) (g m^{-2})

Plot	Leaves from within plot	Leaves of other tree species	Bark	Bran-ches	Cones	Other	Litter-fall total	Litter layer mass
Arolla pine	249	0.2	0.4	10.3	1.1	22.1	283	1643
Scots pine	246	3.9	13.5	16.8	6.8	9.6	296	1818
Larch	204	1.8	1.2	36.8	2.2	14.6	261	1803
Spruce	73	2.3	0.5	4.9	9.9	3.0	94	1130
Aspen	3059	0.4	0.04	30.8	-	10.5	347	739
Birch	190	0.03	1.0	37.9	-	7.8	237	405

The patterns of decomposition of each component of the forest floor are illustrated in Figures 2 through 7, and the parameters for the best-fit polynomials are in Table 1. The graphs in Figures 2 through 7 have the same format. The Y-axis represents some showed an increase in total forest floor mass in the autumn after maximum value for the C content of a given pool (relative to the total C stock in forest litter at the initial point in time, which is 100% on each graph), and the X-axis shows the progression through the year. The top left graph (A) is the carbon input to the Ol layer in litterfall (open circles, dashed line), and to the Of layer from the Ol layer (solid circles, solid line). However, the lines chart the net input for each period as the sum of litterfall recently and all previous times from the beginning of observation. The upper middle graphs (B) in Figures 2 through 7 show the dynamics of carbon stock in the Ol and Of layers, relative to the initial sum of Ol+Of. The top right graph (C) in Figures 2 through 7 is similar to the "B" graphs, except that the values are reduced by input coming in each time increment (DL(t)=LC(t)-OCE(t) and DF(t)=FC(t)-LFTM(t)). The lower graphs in Figures 2 through 7 show the changes in the pools of C that are soluble in water , soluble in 0.1M NaOH, and total labile, in the Ol and Of layers during decomposition.

The F layer under arolla pine, Scots pine, spruce and aspen decomposed more slowly than the L layer under the same species (Table 4). Despite the slower rate, a substantial percentage of the F layer is decomposed during each growing season: 30% for arolla pine, 19% for Scots pine, 26% for spruce, and 38% for aspen. In the birch plot, decomposition is so rapid that all material is completely decomposed in just 30 days. The patterns for individual species are highlighted in the following sections.

These models allow us calculate specific rates and intensities of the main transformation processes for litter (Table 4). The decomposition of Ol layer entails losses of C as CO_2, and input of material to the Of layer. The rate of C decomposition in the Ol layer under arolla pine, larch and birch matched the rate of input from litterfall. The rate of litterfall input was higher than decomposition under Scots pine and aspen, but lower than the decomposition rate in the Ol layer under spruce. The C content of the

Table 3. Patterns in forest floor (O horizon) mass of C (g m^{-2})

Tree species	O horizon layer	Sampling time										
		May 14	June 6	July 20	Aug. 3	Aug. 16	Aug. 31	Sep. 13	Sep. 27	Nov. 9	May 5	May 30
	L	130.6	97.8	90.9	88.1	49.7	60.0	58.0	119.3	113.0	85.8	127.4
Arolla	F	643.5	546.0	609.0	749.0	277.6	358.6	356.1	568.4	496.3	451.1	493.7
pine	H	-	-	-	-	238.4	252.3	270.1	158.2	207.9	-	-
	Sum	774.1	643.8	699.9	837.1	565.7	670.9	684.2	845.9	817.2	536.9	621.1
	L	163.5	127.9	120.4	133.6	99.1	86.1	103.6	107.6	90.9	104.8	136.7
Scots	F	464.3	368.0	578.2	563.7	359.9	354.1	372.5	565.4	439.3	553.3	542.9
Pine	H	-	-	-	-	239.1	200.8	220.8	212.9	182.7	-	-
	Sum	627.8	495.9	698.6	697.3	698.1	641.0	696.9	885.9	712.9	658.1	679.6
	L	-	-	-	-	-	-	-	91.2	84.4	81.1	73.2
Larch	F	641.7	642.4	666.6	651.7	515.8	538.9	417.7	720.2	878.3	525.2	617.4
	Sum	641.7	642.4	666.6	651.7	515.8	538.9	417.7	811.4	962.7	606.3	690.6
	L	184.1	107.2	105.0	112.3	73.8	74.5	79.9	172.4	181.9	190.2	173.9
Spruce	F	174.9	232.4	192.5	131.2	201.2	265.1	312.4	-	281.2	141.6	163.8
	Sum	359.0	339.6	297.5	243.5	275.0	339.6	392.3	410.5	463.1	331.8	337.7
	L	108.1	138.7	141.4	151.9	105.9	98.4	112.1	188.3	189.9	127.6	118.1
Aspen	F	112.7	100.9	130.9	132.2	108.7	106.3	109.0	255.9	175.3	157.9	154.2
	Sum	220.8	139.6	272.3	284.1	214.6	204.7	221.1	444.2	365.2	285.5	272.3
	L	76.6	69.4	73.3	80.0	112.1	115.8	57.2	134.4	316.1	256.8	177.0
Birch	F	-	-	-	-	-	-	64.2	181.0	-	-	-
	Sum	76.6	69.4	73.3	80.0	112.1	115.8	121.4	315.4	316.1	256.8	177.0

Table 4. Specific rate (mg C g^{-1} day^{-1}) and flux (g C m^{-2} day^{-1}) of organic matter input and decomposition in forest litters.

Processes	Arolla pine	Scots pine	Larch	Spruce	Aspen	Birch
Litterfall input:						
Specific rate	0.81	0.96	0.98	1.03	1.92	1.88
Flux	0.735	0.753	0.728	0.464	0.857	0.745
Decomposition of L layer:						
Specific rate	2.73	2.66	3.31	3.06	1.99	2.38
Flux	0.730	0.543	0.720	0.951	0.574	0.653
Input to the F layer:						
Specific rate	0.86	0.84	0.64	2.35	1.33	2.63
Flux	0.658	0.464	0.379	0.595	0.266	0.634
Decomposition of F layer:						
Specific rate	1.85	0.42	0.09	2.79	2.99	4.82
Flux	1.416	0.232	0.057	0.705	0.600	1.161
Decomposition of L+F layers:						
Specific rate	1.65	0.69	0.53	2.71	2.62	3.45
Flux	1.172	0.462	0.329	1.029	0.602	0.552

Of layer under Scots pine was increasing, as litterfall inputs were double the decomposition rate.

Arolla pine

Litterfall continues throughout the year under arolla pine. About 20% of the C content of the forest floor is comprised of litterfall from the previous year (Figure 2A). The model calculations indicate that fresh litterfall is completely processed into the Of horizon within 3 months. The Ol horizon does not disappear owing to continual renewal from fresh litter inputs. Water-soluble organic matter comprises about 0.6% of the total C in Ol layer (Figure 1 D). The rapid decline early in the season in the water-soluble C (C_{H2O}) results from rapid washout and utilization by microbes. By late August, water-soluble C reaches a minimum, at the same time that microbial biomass reaches a maximum (Figure 8). Compounds soluble in 0.5 M NaOH comprise about 2% of total C in the forest floor (Figure 2E). The seasonal decline in alkali-soluble C is slower than the decline in water-soluble C. The Of layer contains 80 to 85% of the total C in the forest floor, and about 21% of the Of C comes from the Ol layer each year (Figure 2A), and the Of loses about 24% of its C content through the growing season (Figure 2C). The water-soluble portion of the Of increases from 0.4 to 1.6% of total carbon as the growing season progresses, while the alkali-soluble pool increases from 12 to 18% (Figure 2D and E).

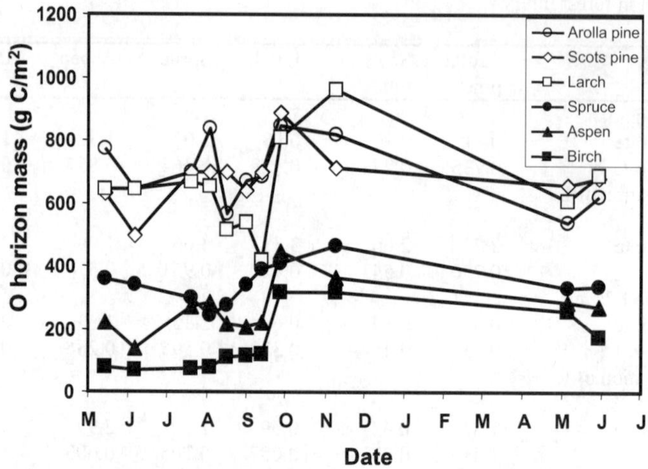

Figure 1. Time trends in forest floor (O horizon) mass.

Scots pine

Unlike arolla pine, the annual litterfall under Scots pine was concentrated in the late autumn (Figure 3A). The Ol layer comprises about a quarter of the total C of the forest floor (Figure 3B). Complete processing of litter from the OL into the Of takes about 12 months (Figure 3C), four-times longer than the processing of arolla pine. Slow decomposition appears to result from the hard pine needles of Scots pine being more resistant to decomposition because of great amount of complex "skeleton" compounds in cell walls. A thick wax cuticle is characteristic of Scots pine needles, comprised of up to 20% lignin, along with antibiotic substances and polyphenols that form stable complexes with proteins. Epidermal cell walls of pine needle are broken in fermentation layer of litter (Aristovskaya, 1980). As decomposition progresses, the C content of the Ol declines by 11%, as the C content of the Of increases by 20% (Figure 3C). A portion of the C is lost as CO_2, so the net increase in C in the Of may represent the accumulation of fungal mycelia. The pools of soluble C increase from 0.2 to 1.8% (water soluble) and 16.5 to 19% (alkali soluble, Figure 3D and E). Microbial mass in the Of layer contains an order of magnitude more C than in the Ol layer (Figure 8).

Larch

The annual litterfall beneath deciduous larch comes mostly in late autumn (Figure 4A). The pulse of litterfall in autumn gives a maximum C content of the Ol from autumn to winter, and a rapid decline in summer as it is processed into the Of (Figure 4B). A thin network of white mycelia covers the surfaces of the larch needle soon after abscission. The C content of the Of shows two peaks: one in spring from the processing of needles from the previous autumn, and another in late autumn as moist conditions lead to the leaching of C from the Ol into the Of (Figure 4A). This pattern is mirrored by the increasing pool of water-soluble C in the Of layer, and the increase in microbial biomass (Figures 3D, 7). The C content of the microbial biomass increases at this point from 3 to 8% of total carbon in litter.

Spruce

The rate of litterfall under spruce was uniform through the year, with current litterfall accounting for 26% of the C content of the forest floor (Figure 5A). The half-life for needle residence in the Ol was just 44 days, and the C content of the Ol layer dropped by half as the growing season progresses (Figure 4B). The Ol layer beneath spruce had higher concentrations of soluble C than the Ol under other species (Figure 4D and E). The processing of material in the Ol layer led to an increase in the C content of the Of by late August and early September (Figure 4B). Decomposition is rapid in the spring and early summery for material in both the Ol and Of layers (Figure 4C). After this period, inputs in litterfall exceed decomposition losses, and the C contents increase. The soluble fractions of the C pool matched the dynamics of microbial biomass (Figure 8), which is a common pattern (Khudyakov Y.P., 1972; Parinkina O.M.,1989).

Aspen and birch

The input of C to the Ol layer from litterfall under aspen and birch had low rates through the summer, high rates in late autumn, followed by low rates (Figure 6A, 7A). About half of the total carbon stock in the Ol layer fell during a 4-8 week period. The C content of the forest floor was divided almost evenly between the Ol and Of layers (Figures 6B, 7B).

Birch litter decomposed very quickly (Figure 7C), with 83% of fresh litter processed into either CO_2 or into the Of layer within 2 months. Aspen litter showed slower initial decomposition (Figure 6C), perhaps reflecting less faunal processing as a result of differences in the structure of cell walls, a thicker cuticle layer, and higher concentrations of tannins. Solubilization of compounds in water increase decomposition by saprophages (Figure 6D). Their active participation at decomposing was reflected by higher biomass in comparison with others tree species (924 and 553 mg m^{-2}, respectively under birch and aspen), as well as by great amounts of insect frass (Bezkorovainaya and Vishnyakova, 1996; Vedrova, 1997).

The Ol horizon fully disappeared within the first few months after litterfall, and the C content of the F layer under birch decreased by 74% in 60 days during the spring and summer, followed by a level period as inputs from the Ol layer increase (Figure 7B, C).

The forest floors under aspen and birch had much higher concentrations of water-soluble C throughout the summer (Figure 6D, 7D). The water-soluble C showed a decrease early in the growing season, increasing in August; this increase corresponded with intensive decomposition of litterfall from the previous year under aspen, and the breakdown of more complex compounds under birch. The August increase in water-soluble C was matched by a burst of microbial biomass (Figure 8).

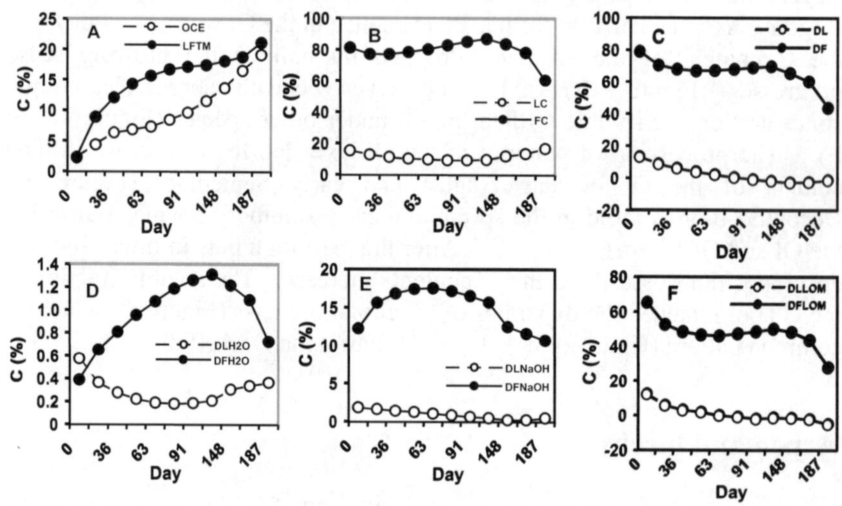

Figure 2. Arolla pine. Carbon dynamics of organic matter in litter subhorizons during the growing season (see the second paragraph of Results and Discussion for explanation). A: incoming; B: pool or stock; C: decomposition; D: water soluble organic matter; E: 0.1 *M* NaOH soluble organic matter; E: labile organic matter.

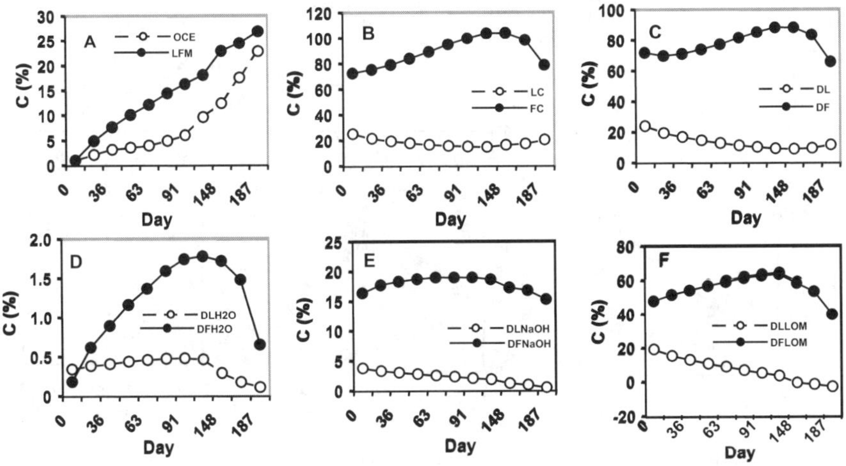

Figure 3. Scots pine. Carbon dynamics of organic matter in litter subhorizons during the growing season (see the second paragraph of Results and Discussion for explanation). A: incoming; B: pool or stock; C: decomposition; D: water soluble organic matter; E: 0.1 *M* NaOH soluble organic matter; E: labile organic matter.

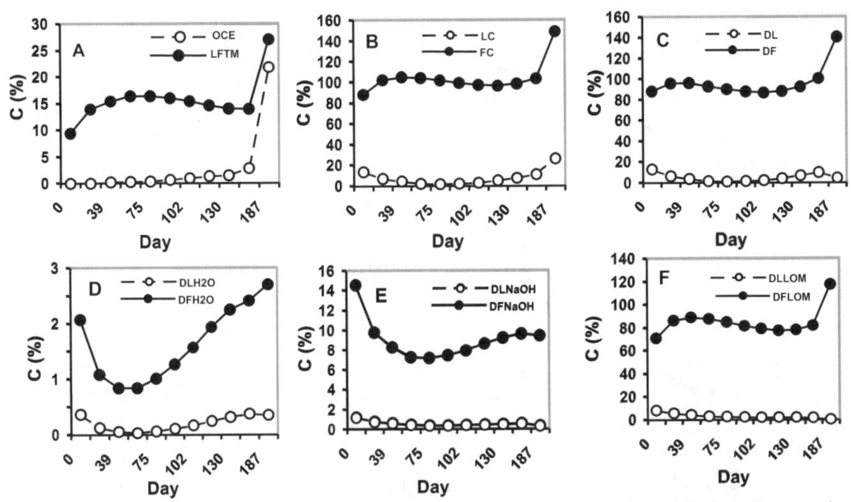

Figure 4. Larch. Carbon dynamics of organic matter in litter subhorizons during the growing season (see the second paragraph of Results and Discussion for explanation). A: incoming; B: pool or stock; C: decomposition; D: water soluble organic matter; E: 0.1 *M* NaOH soluble organic matter; E: labile organic matter.

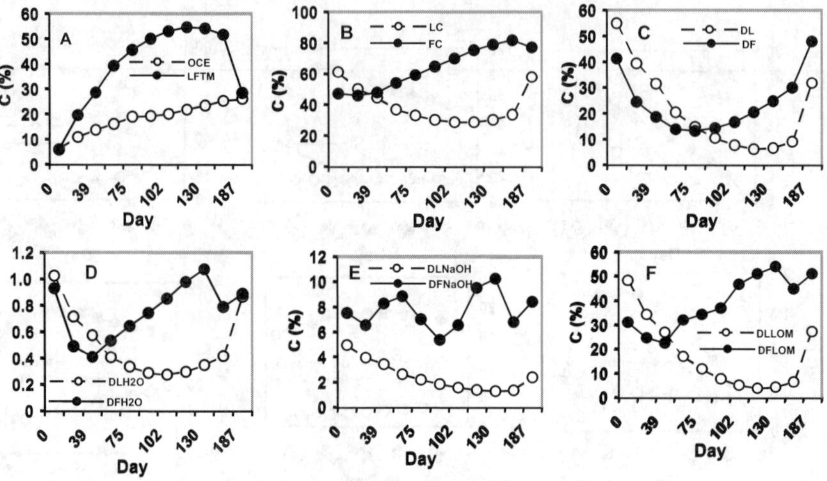

Figure 5. Spruce. Carbon dynamics of organic matter in litter subhorizons during the growing season (see the second paragraph of Results and Discussion for explanation). A: incoming; B: pool or stock; C: decomposition; D: water soluble organic matter; E: 0.1 *M* NaOH soluble organic matter; E: labile organic matter.

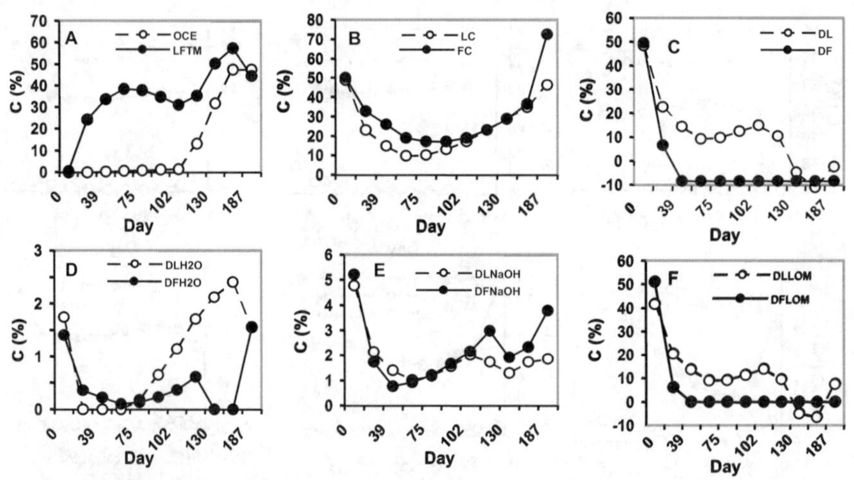

Figure 6. Aspen. Carbon dynamics of organic matter in litter subhorizons during the growing season (see the second paragraph of Results and Discussion for explanation). A: incoming; B: pool or stock; C: decomposition; D: water soluble organic matter; E: 0.1 *M* NaOH soluble organic matter; E: labile organic matter.

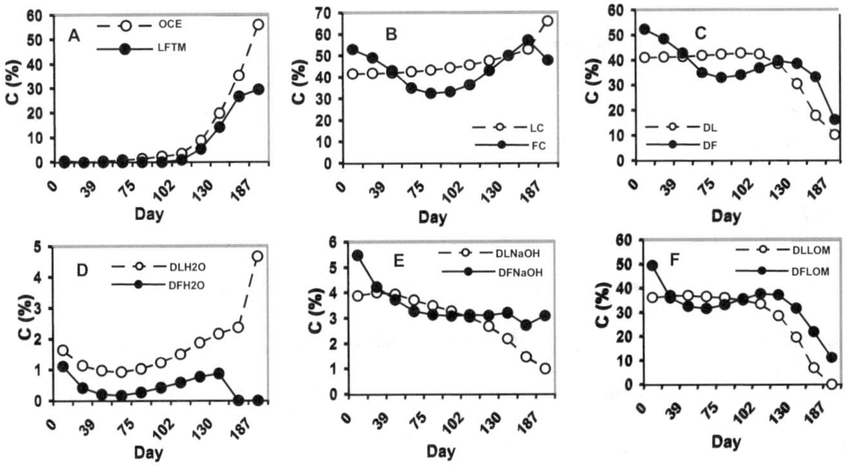

Figure 7. Birch. Carbon dynamics of organic matter in litter subhorizons during the growing season (see the second paragraph of Results and Discussion for explanation). A: incoming; B: pool or stock; C: decomposition; D: water soluble organic matter; E: 0.1 *M* NaOH soluble organic matter; E: labile organic matter.

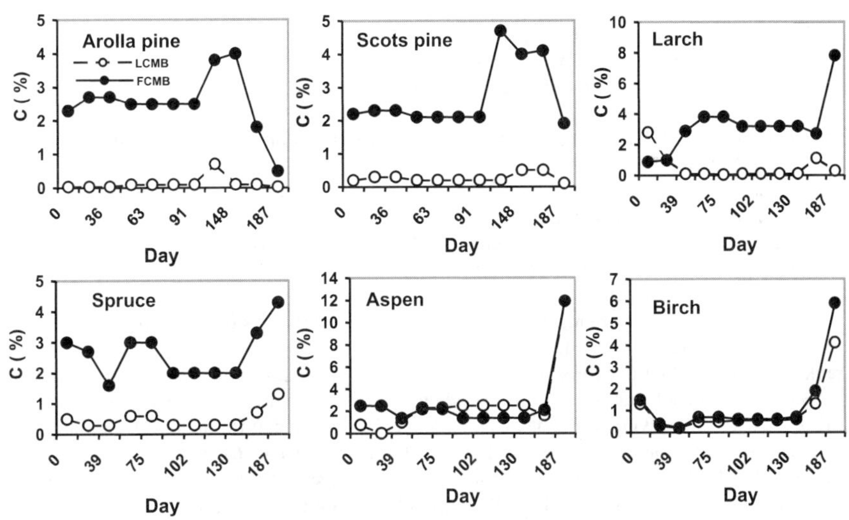

Figure 8. Dynamics of microbial biomass share in total organic matter stock of litter in subhorizons during the growing season.

ACKNOWLEDGEMENTS

This research was supported by Russian Scientific Foundation, Projects No 02-04-49459 and 03-04-20018 BNTS.

REFERENCES

Anonymous 1991 Methods of soil microbiology and biochemistry. Publ. House of Moscow State University, pp.24-27.

Alexandrova L N 1980 Soil organic matter and its transformation processes. L Nauka (Science) 288p.

Arinushkina E V 1970 Manual in soil chemical analysis. M. Publ. House of Moscow State University. 487p.

Berg B and Ekbohn G 1991 Long- term decomposition in a Scots pine forest. VII. Litter mass-loss rates and decomposition patterns in some needle and leaf litter types. Can. J. Bot. 69, 1449-1456.

Bezkorovainaya I N and Vishnyakova Z V 1996 Soil biota participation in litter destruction in forest stands. Lesovedenie (Forest Science) N2, 53-61.

Bogatyrev L G 1990 On classification of forest litters. Pochvovedenie (Soil Science) N3, 118-127.

Das P K, Nath S, Mykhopadhyay N and Banerjee S K 1993 Decomposition of litters and their effect on physical- chemical and microbial properties of soil. Proc.Indian. Nat. Sci. Acad. B. 59, N5, 517-524.

Ganzhara N F 1983 Factors determining the levels of relative stabilization of humus amount, stock and composition in soils. *In* Organic matter and soil fertility. M. pp.17-24.

Grishina L A 1986 Humus formation and humus soil state. Publ. House of Moscow State University. 244p.

Karpachevsky L O 1981 Forest and forest soils. M.: Lesnaya promyshlennost' (Forest Industry) 264p.

Karpachevsky L O 1973 Some peculiarities of forest leaffall decomposition //Problems of Forest Soil Science. Nauka (Science) M. 51-65.

Karpachevsky L O 1988 Litter – a particular biological horizon of forest biogeocenosis. *In* Litter role in forest biogeocenoses. Abstracts of papers at the All Union Meeting. Pp.88-89. M.: Nauka (Science).

Kononova M M 1966 Soil organic matter. Its nature, properties and study methods. M. Publ. House of Academy of Sciences of the USSR, 314p.

Kravkov S P 1978 Materials for studying decomposition processes of plant remains insoil. *In* Bio- and agrochemistry of soil processes. Pp.67-102. Leningrad: Nauka (Science).

Kravkov S P 1978 Studying of the dead vegetation cover role in soil formation. *In* Bio- and agrochemistry of soil processes. Pp.103-127. Leningrad: Nauka (Science).

Kulagina M A 1977 Litterffall and litter role in biological cycle of substances. *In* Forests of the Middle Priangarie region. Pp.164-191. Novosibirsk: Nauka (Science).

Lyul'kovich I N 2000 Cycle of chemical elements in 25- aged pine and aspen stands. *In* Soil organic matter and yield: Collection of sci. papers of young researchers. Pp.68-77. Krasnoyarsk State Agrarian University. Krasnoyarsk.

Orlov A Ya 1966 Importance of dying off tree roots in substance turnover in the forest. General Biology N1, 40-48.

Orlov D S 1998 Soil organic matter of Russia. Pochvovedenie (Soil Science), №9, 1049-1057.

Panikov N S, Sadovnikova L K and Fridland E V Non- specific compounds of soil humus. Moscow: Publ. House of Moscow State University. 144 p.

Persson K 1985 Dynamics of thin roots of forest trees. Ecology. N1, 33-39.

Ponomareva V V and Nikolaeva T A 1965 Methods for studying organic matter in peat and bog soils. Pochvjvedenie (Soil Science) 5, 88-95.

Sokolov D F and Karpova I V 1965 Litter decomposition rate and impact of its decomposition products on soil humus amount and composition in complex pine forest. *In* Forests of the Podmoskovie region . Moscow: Nauka (Science) 63-102.

Qualls R G, Haines B L and Swank W T 1991 Fluxes of dissolved organic nutrients and chemical substances in a deciduous forest. Ecology 72(1), 254-266.

Titlyanova A A, Bulavko G I, Mironycheva-Tokareva N P and Khvoshchevskaya M F 1994 Organic carbon stock in soils of West Siberia. Pochvovedenie (Soil science) N10, 49-53.

Titlyanova A A and Tesarzhova M 1991 Regimes of biological turnover. Novosibirsk, Nauka (Science) 150p.

Tuev N A 1989 Microbiological processes of humus formation. Moscow: Agropromizdat. 239p.

Vedrova E F 1995 Transformation of plant remains in 25-aged stands of the main forest forming tree species of Siberia. Lesovedenie (Forest Science) N4, 13-21.

Vedrova E F Carbon balance in pine forests of Krasnoyarsk forest- steppe// Lesovedenie (Forest Science) N5, 51-59.

Vedrova E F and Mindeyeva T N 1998 Intensity of carbonic acid gas producing at forest litter decomposition. Lesovedenie (Forest Science) N1, 30-41.

Vitousek P M, Turner D R, Parton W J and Sanford R L 1994 Litter decomposition on the Mauna Loa Environmental matrix, Hawai'i: Patterns mechanisms and models. Ecology 75(2), 418-429.

Zvyagintsev D G, Bab'eva I P, Dobrovol'skaya T G, Zenova G.M, Lysak L V and Mirchink T G 1993 Vertical- layer organization of microbial communities in forest biogeocenoses. Microbiologia (Microbiology) 62, N1, 5-36.

Chapter 21

TREE DIVERSITY AND SOIL BIOLOGY: A NEW RESEARCH PROGRAM IN FRENCH GUYANA

Jacques Roy[1], Stephan Hättenschwiler[1], and Anne-Marie Domenach[2] and the other participants to the *Diprotroflux guyanensis* program[3]
[1]*Centre d'Ecologie Fonctionnelle et Evolutive, Centre National de la Recherche Scientifique, F-34293 Montpellier cedex 5, jacques.roy@cefe.cnrs.fr*
[2] *UMR Ecofog, BP 709, 97387 Kourou Cedex, Guyane Française*
[3] *see Acknowledgements*

INTRODUCTION

The composition of plant communities is highly affected by human activities, either directly through land management practices in agriculture and forestry, or indirectly through increasing concentrations of CO_2, changing temperatures, and high rates of N deposition. The influences of these changes on biodiversity and the functioning of ecosystems has been intensively studied the past ten years, mainly in herbaceous ecosystems. Experiments using randomly assembled communities found that plant species and functional-group richness have positive effects on primary production and nutrient retention (see synthesis in Loreau et al. 2001, 2002; Kinzig et al. 2002). Recent interpretations of these experiments showed that sampling effect as well as complementarity and facilitation among species were the mechanisms involved (Loreau and Hector 2001, Tilman et al. 2001, van Ruijven and Berendse 2003). More understanding of the role of biodiversity is still to be gained from such experiments, through the manipulation of the diversity of trees instead of herbs, trophic levels other than primary producers, and through the simultaneous manipulations of diversity at several trophic levels. Longer term experiments that will look also at the role of biodiversity for the maintenance of ecosystem functioning under changing or disturbed environmental conditions are also needed.

D. Binkley and O. Menyailo (eds.),
Tree Species Effects on Soils: Implications for Global Change, 337–348.

In most of these biodiversity experiments, the composition of the vegetation explained a larger percentage of the variability than did the richness *per se* (Tilman et al. 2001, Spehn et al. 2005) . The identity (and specific biological characteristics) of the species present in a community has a stronger impact on the functioning of the ecosystem than the number of species. As described in other chapters in this volume, the capacity of plants to influence soil processes has been known for a long time (Dokuchaev 1879), and some generalizations between groups of species (such as conifers and deciduous species) have been supported and some have not (Zinke 1962; Binkley and Giardina 1998, Augusto et al. 2002). Some general hypotheses were developed in the late 20th Century about vegetation characteristics and soil properties (e.g. Pastor and Post 1986, Chapin 1991, 1993, Hobbie 1992, van Breemen 1993). They state that for most ecosystems, there is a feedback between soil fertility and the characteristics of the primary producers. Plants from infertile habitats are slow growing and bear low quality leaves (high C/N and phenol content) with a slow turn-over and decomposition rates. All these characteristics are hypothesized to reinforce the low fertility of soils. The opposite occurs in fertile habitats with plant of high quality leaves reinforcing soil fertility. However, in such comparisons made along fertility gradients, the impacts of initial soil characteristics and species traits are confounded. In the context of biodiversity changes in unmanaged ecosystems as well as of forestry and agriculture practices, we need to identify, for a given initial substrate, the impact of species identity (as well as diversity) on soil biology and ecosystem physiology. An experimental approach is best suited for such an objective (Van Cleve et al. 1991; Hobbie 1992) because many factors are confounded when the relationship between productivity and biodiversity is analysed *in natura* (Roy 2001, Vila et al. 2005). Such an experimental approach has not yet been largely developed (but see Wedin and Tilman, 1990; Craine et al. 2002; Lovett et al. 2004, and other chapters in this volume).

Using a collection of 16 species of local trees established as monocultures 20 years ago in the humid tropical forest of French Guyana, we are starting a program to analyse the relationship between the DIversity of primary PROducers, TROphic webs and soil FLUXes (*Diprotroflux guyanensis*). Our specific aims are to analyse:

i) the ecological and physiological determinants of litter quality;
ii) the impact of tree species identity and diversity on decomposition and soil functioning; and
iii) the responses in the structure and activity of the faunal and microbial communities.

THE EXPERIMENTAL SET-UP IN FRENCH GUYANA

The study area is located in the Northern part of French Guyana, in the experimental forest zone of Paracou (CIRAD-Forêt) located between Kourou and Sinnamary, about 80 km west of Cayenne (5°20'N, 52°50'W, 40 m elevation). The climate is characterized by an average annual precipitation of 2200 mm with a long dry season from September to December (monthly precipitation below 100 mm) and a short dry season in February or March. The soil is a well drained acid (pH 4.5) Oxisol developed on migmatite and shales.

Twenty four local tree species were planted in monocultures by CIRAD-Forêt to determine their growth performances under silvicultural practices. One hectare of natural forest was cleared, stumps pulled out with bulldozer and the soil ploughed. The monoculture plots (20 x 20 m) were contiguous and arranged in two rows of 12 (Figure 1). Seeds were germinated in greenhouses and seedlings planted in 1983 and 1984 at a density of 49 trees per plot (3 x 3 m spacing). Understory vegetation was cut mechanically and manually up to 4 times a year during tree establishment, and then once or twice a year. Sixteen out of the 24 planted species had a very low mortality and present a well developed canopy with leaf area indexes of 3 to 5. Our study focuses on these 16 species: *Carapa guianensis , Caryocar glabrum, Dicorynia guianensis(L), Diplotropis purpurea(L), Eperua falcate(L), Goupia glabra, Hymenaea courbaril(L), Peltogyne venosa(L), Pletonia insignis, Protium insigne, Qualea coerulea, Simarouba amara, Sterculia excelsia, Symphonia globulifera, Virola melinonii, Vochysia densiflora.* Among the 5 legumes (*L*) 4 were previously analysed for root and leaves characteristics and only *D. purpurea* was found to have nodules (Guehl et al. 1998).

In order to separate the aboveground (mainly litter) from the underground (rhizodeposition) impacts of the tree species on soil biology, we manipulated litter and rhizodeposition. In the centre of each monoculture, 4 plots (3x3m each) have been identified with two of them trenched (20 cm wide, 60 cm deep) and two un-trenched. On half of each of the 4 plots, litter is regularly removed. Three litter traps have been installed in each monoculture stand.

Since the plots are contiguous, there are zones were 2, 3 or 4 species interact, allowing study of the influence of species diversity. In addition to the 16 monocultures, we identified 14 plots with litter being a mixture of 2 species, 4 plots with litter mixture of 3 species and 2 plots with litter mixture of 4 species (Figure 1). The roots of the different species extend a lot further than their canopies. We can nevertheless supposed that the dominance of the root of a single species decreases progressively, and the evenness increases progressively in the 1, 2, 3 and 4 species plots.

This research program takes advantage of having species planted 20 years ago. This is a long enough period for the feedbacks between species characteristics and soil to establish. Having a large number of species in

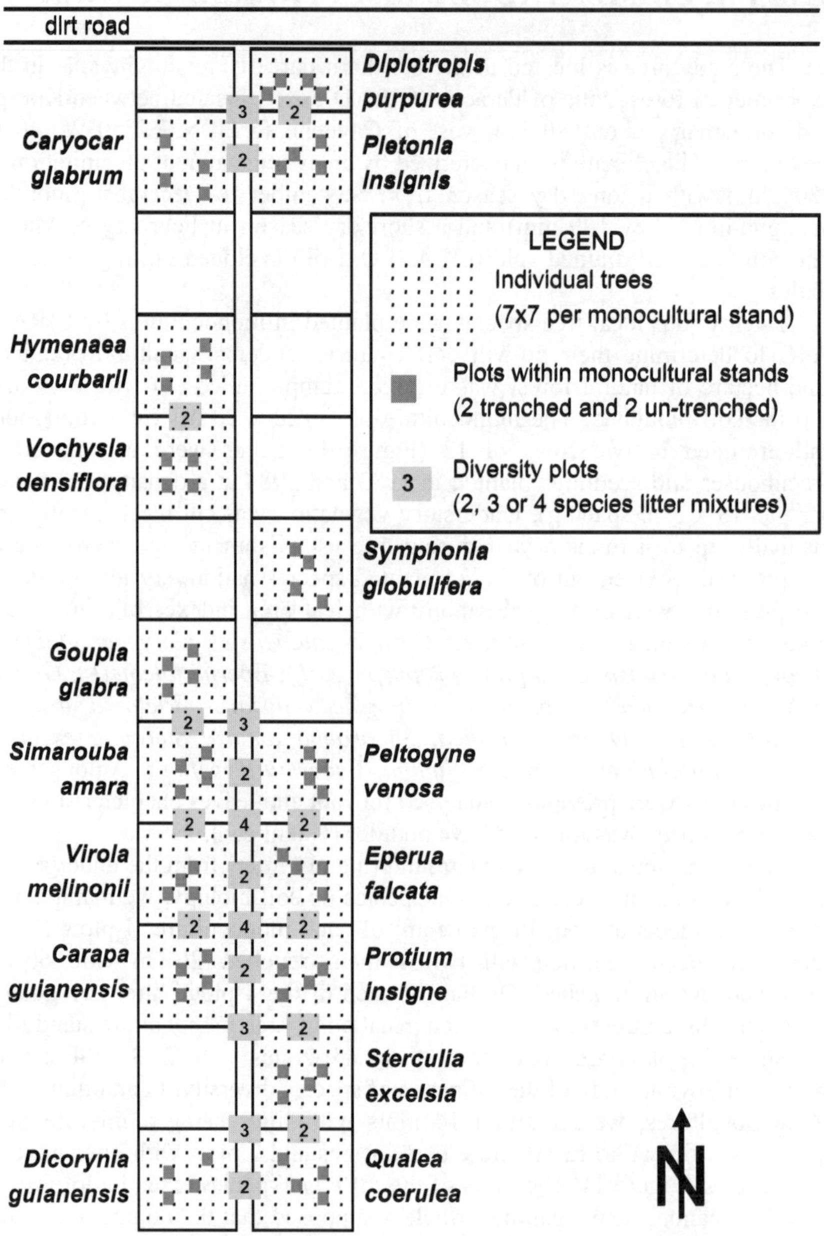

Figure 1. Lay-out of the experiment at Paracou.

monocultures (16) should allow to draw relationships between species characteristics (leaf properties for example) and soil parameters independently of species identity, which is not possible when only a few monocultures are compared. The confidence in these relationships will be tested by measuring some of the studied parameters for a number of species in other forest plantations of French Guyana. Direct comparisons between monocultures will be impaired by the lack of replicated monoculture stands. New experimental tree plantations manipulating diversity at various levels and with adequate replications have recently been installed (Sherer-Lorenzen et al. 2005).

RESEARCH QUESTIONS

Functional species traits

Testing the impact of every tree species on soil biology is not feasible. An alternative is to classify the species into functional groups, with species within a given group having a similar effect on ecosystem processes. Such an approach has been used by several groups (Tilman 2001, Hooper et al. 2002, Naeem and Wright 2003). Grouping of species into classes has often been based on species properties (ruderal, stress tolerant, competitive: Grime 1977), on the characteristics of tissues (specific leaf mass, N content, leaf longevity: Westoby et al. 2002; Wright et al. 2004; N nutrition: Roggy et al. 1999) and on the species responses to environmental conditions (disturbance: Lavorel et al. 1997). How well can these classes represent the responses of ecosystems? Some leaf traits do indeed correlate with leaf decomposition rate (Cornelissen et al. 1999; Pérez-Harguindeguy et al. 2000) and to nitrogen mineralization (Craine et al. 2002), but we need to relate species traits (in particular leaf and root characteristics) to a larger range of soil parameters to see if species traits relate to a full suite of soil parameters or if functional typologies are function specific.

Across the 16 species of our experiment, we will relate leaf and root chemistry as well as the form of N acquisition (nitrate, ammonium, symbiotic) to a range of physical, chemical, and biological soil characteristics. Correspondence analysis and multiple regressions will be used to determine the most relevant species traits and the specificity of the species trait – soil parameter relationships. The mechanisms behind these relationships will be studied, for example by looking at the relative role of leaf litter and rhizodeposition, at the impact of specific phenolic compounds on mineralization or at the potential inhibition of some microbial functions by the roots. The relationship between litter characteristics and soil processes will also been analysed by simulations with the CANTIS (Carbon And Nitrogen Transformations In Soils) mechanistic model (Garnier et al. 2003).

Leaf chemistry and decomposition

Locally, the chemical and physiological characteristics of leaves may relate to the ecological status of the species (pioneer vs. late successional species, sun vs. shade tolerant species, Strauss-Debenedetti and Bazzaz 1996). Litter chemistry is determined both by the chemistry of green leaves and by the nutrient resorption. The latter has not been given enough attention despite its major role in nutrient conservation both at the individual and ecosystem level (Aerts 1990, Killingbeck 1996, Eckstein et al. 1999).

Mature green leaves and recently fallen senescent leaves will be analyzed for a set of characteristics (specific leaf area, and concdentrations of C, N, P, soluble compounds, hemicelluloses, cellulose, lignin, total and specific phenols) and resorption will be calculated (Van Heerwaarden et al. 2003). These parameters will be related to the growth rate and ecology of the species derived from a large number of floristic analysis previously conducted in French Guyana. Preliminary data indicate a large range of litter traits among the 16 species studied (Table 1).

Leaf decomposition will be studied *in situ* using nets periodically deposited on the forest floor in order to determine the decomposition rate per strata and cohort of fallen leaves. Overall measurements will allow the calculation of the litter mean residence time. Litter bags with different types of litter and mesh size will also be used. Laboratory decompositions under standard conditions will be conducted in order to parameterize the CANTIS model and to compare these species with an existing data base (e.g. Trinsoutrot et al. 2000). The capacity of various decomposition models to simulate the actual decomposition rate will be tested and potential improvements will be investigated.

Soil processes and trophic interactions

Soil processes are largely driven by microbial activities, and understanding the influence of litter quality on microbial activities requires an analysis of the interactions between the C and N cycles (Hungate *et al.*, 1996; Hodge *et al.*, 2000). This can be done by measuring the respired CO_2 and the

Table 1: Extreme characteristics of fallen leaves harvested in December on the floor of the 16 monoculture stands.

	Nitrogen (%)	C/N	Phosphorus (‰)	Phenols (%)		Specific mass (g/m^2)
				soluble	insoluble	
Minimum	0.72	31	13	3	34	83
Maximum	2.03	72	74	109	141	162

main regulating steps of the N cycle (ammonification, nitrification and denitrification). We also have now the molecular tools for the analysis of the diversity of the bacterial communities fulfilling these functions (Tiedje et al., 1999). In addition to usual physical and chemical soil characterization, microbial activities (potential respiration, ammonification, nitrification and denitrification) will be measured on soil and litter samples from the monoculture stands. The structure of microbial communities will be analyzed by DNA fingerprinting applied on soil extracted DNA. The A-RISA technique will be used (Ranjard et al., 2003). The structure of the nitrifying and denitrifying communities will be analysed (Laverman et al., 2001; Philippot. 2002) according to the variability of these functions under the different monocultures.

More generally, the soil properties depend on the structure and functioning of the soil webs (de Ruiter at al. 1994, Beare et al. 1995), but the analysis of the factors regulating these webs remains a challenge (Scheu & Setälä 2002, Wardle 2002). We will the use two approaches to analyse soil food webs in a subset of 9 among the 16 monoculture stands. The first one is the measurement of the natural abundance of the stable isotopes of carbon and nitrogen ($^{13}C/^{12}C$ and $^{15}N/^{14}N$) in the major organisms at different trophic levels. This will allow the determination of their diet and their respective trophic position (Scheu & Falca 2000, Post 2002, Schmidt et al. 2004). The second approach, called ecological stoechiometry (Sterner and Elser 2002), is based on the principle that the ratio of elemental components (mainly C/N/P) of the consumers body is constrained by the ratio of these components in the primary producers. Recent studies have shown that the C/N/P ratios of species result from evolutionary strategies and vary according to environmental constraints (Sterner and Elser 2002, Klausmeier et al. 2004). Having large variations of these ratios among our 16 species (table 1), we can expect selective impacts on the communities of detritivores with subsequent consequences on the biogeochemical cycles. This will be studied by direct measurements of these ratios at different trophic levels as well as by theoretical modeling using adapted versions of previously developed models (Daufresne and Loreau 2001; Klausmeier et al. 2004).

Biodiversity

Biodiversity changes and their impacts are at the heart of this program (as described above). Biodiversity is studied with three research focuses. One is to analyse the extent and mechanisms behind the influence of tree species identity on the diversity of the main components of the biota (understory vegetation, soil macrofauna and micro-organisms). Biodiversity gradients across the 16 monoculture stands will be related to tree characteristics like leaf chemistry and in particular elemental stochiometry (see above) to suggest regulating mechanisms. The second focus is to attempt to analyse the consequences of these diversity changes on soil processes, especially for some specific categories of microorganisms. The third focus is to address the long term (20 years) role of the number of species on soil processes using the 1, 2, 3 and 4 litter species plots. A complementary experiment to this third focus is run by S. Hättenschwiller, using soil microcosms installed in the nearby natural forest prepared with litter from 4 of the 16 species. Available literature indicates that litter species diversity can have important implications for decomposition processes but that predictability is low (Hättenschwiler 2005). In this new experiment, all combinations of the litter of the 4 species are considered with adequate replication, together with allowing or not the access to macrofauna. Laboratory experiments will contribute to have a more comprehensive mechanistic approach to this diversity-soil processes relationship.

Recycling pathways and ecosystem productivity

Nutrients allocated to green leaves are recycled through 4 parallel pathways: herbivory (feces and dead bodies), throughfall, foliar resorption and litter decomposition. Research often focuses exclusively on decomposition, but the fraction of nutrients recycled trough each pathway may be of similar magnitude. This fraction varies with the nutrient considered, the species and the climatic conditions. For example, leaf nitrogen recycling is estimated to be 10% through herbivory, 5% through throughfall, 40% through resorption and 45% through litterfall and subsequent decomposition (Zimmermann et al 2002, McClaugherty 1983, Lowman 1992). Moreover, these pathways differ greatly in term of recycling rate (fast for resorption, slow for litter decomposition) and in term of probability of loss either for the individual or for the ecosystem (low for resorption, high for decomposition). The impact of litter decomposition rate on ecosystem production is generally considered obvious, but it has not been thoroughly studied. Some ecosystem models show that an increase in decomposition rate should have a positive impact only in the short term, but that long term fertility is related to the loss of nutrient from the system in mineral and organic forms (de Mazancourt et al. 1998, Loreau 1998, Daufresne and Hedin submitted).

A theoretical exploration of the consequence of having different fractions of the nutrients recycled through the 4 pathways will be undertaken for various conditions of nutrient loss, using an adapted version of the model developed by Daufresne and Hedin (submitted). Measurements will be taken in 9 of the 16 monocultural stands to estimate the interspecific variation in these recycling pathways. Preliminary results show a strong positive relationship across species between C/N ratio of the litter and N foliar resorption, suggesting that resorption is a key determinant of litter quality. If confirmed, the results suggest that a low litter quality is not to be always associated with poor recycling of nutrients. On the contrary, if it is the consequence of a high resorption, overall nutrient recycling may be improved since taking away of nutrients through competition in the soil and leaching is reduced.

ACKNOWLEDGEMENTS

The Diprotroflux guyanensis program is partly funded by the French national program ACI/ECCO/PNBC, and includes collaborators: Bonal D., Goret J.-Y., Josephzoon I., Ponton, S., Roggy J.-C.(Ecofog, INRA, CNRS, Kourou, Guyana) ; Brechet L., Coûteaux E., Jorgensen H., Lensi R., Pinay, G., Roy J., Schimann H., Sonié L. (CEFE, CNRS , Montpellier); Bertrand I., Garnier P., Recous S.(Unité d'Agronomie INRA, Laon-Reims-Mons) ; Maraun M., Scheu S.(Technische Universitat, Darmstadt) ; Henault C., Phillipot L., Ranjard L.(Microbiologie INRA, Dijon) ; Daufresne T.(Princeton University, Princeton) ; Bouillet J.-P.(Ecosystèmes Tropicaux, CIRAD, Montpellier) ; Brechet C., Guehl J;-M.(Ecologie Forestière, INRA, Nancy).

REFERENCES

Aerts R. 1990 Nutrient use efficiency in evergreen and deciduous species from heathlands. Oceologia 84, 391-397.

Augusto L, Ranger J, Binkley D and Rothe A 2002 Impact of several common tree species of European temperate forests on soil fertility. Ann. For. Sci. 59, 233-254.

Beare, MH, Coleman DC, Crossley DA, Hendrix PF, and Odum EP 1995 A hierarchical approach to evaluating the significance of soil siodiversity to siogeochemical cycling. Plant Soil 170, 5-22.

Binkley D and Giardina C 1998 Why do tree species affect soils? The Warp and Woof of tree-soil interactions. Biogeochem. 42, 89-106.

Chapin FS III 1991 Effect of multiple environmental stresses on nutrient availability and use. *In* Response of plants to multiple stresses. Eds. HA Mooney and EJ Pell., pp. 67-88. Academic Press, San Diego.

Chapin FS III 1993 The evolutionary basis of biogeochemical soil development. Geoderma 57, 223-227.

Cornelissen, JHC, Perez-Harguindeguy N, Diaz S, Grime JP, Marzano B, Cabido M, Vendramini F and Cerabolini B 1999 Leaf structure and defence control litter decomposition rate across species and life forms in regional floras on two continents. New Phyt. 143, 191-200.

Craine JM, Tilman D, Wedin D, Reich P, Tjoelker M and Knops J 2002 Functional traits, productivity and effects on nitrogen cycling of 33 grassland species. Func. Ecol. 16, 563-574.

Daufresne T and Hedin LO (in review). Plant coexistence depends on ecosystem nutrient cycles : extension of the resource-ratio theory.

Daufresne T and Loreau M 2001 Ecological stoichiometry, primary producer-decomposer interactions, and ecosystem persistence. Ecology 82, 3069-3082.

De Mazancourt C, Loreau M, and Abbadie L 1998 Grazing optimization and nutrient cycling: When do herbivores enhance plant production? Ecology 79, 2242-2252.

De ruiter PC, Neutel AM, and Moore JC 1994 Modeling food webs and nutrient cycling in agroecosystems. Trends Ecol. Evol. 9, 378-383.

Dokuchaev VV 1879 Abridged historical account and critical examination of the principal soil classifications existing. Trans Petersburg Soc. Naturalists 1, 64-67.

Eckstein RL, Karlsson PS, and Weih M 1999 Leaf life-span and nutrient resorption as determinants of plant nutrient conservation in temperate-artic regions. New Phytol. 143, 177-190.

Garnier P, Neel C, Aita C, Recous S, Lafolie F and Mary B 2003 Modelling carbon and nitrogen dynamics in a bare soil with and without straw incorporation. Eur. J. Soil Sci. 54, 555-568.

Guehl JM, Domenach AM, Bereau M, Barigah TS, Casabianca H, Ferhi A, and Garbaye J 1998. Functional diversity in an Amazonian rainforest of French Guyana: a dual isotope approach (delta N-15 and delta C-13). Oecologia 116, 316-330.

Grime JP 1977 Evidence for existence of 3 primary strategies in plants and its relevance to ecological and evolutionary theory. Am. Nat. 111, 1169-1194.

Hättenschwiler S 2005 Effects of tree species diversity on litter quality and decomposition. *In:* Forest diversity and function in temperate and boreal systems. Eds. M. Scherer-Lorenzen, C. Körner, and E.-D. Schulze. Pp. 149-164. Springer, Berlin.

Hobbie SE 1992 Effects of plant-species on nutrient cycling. Trends Ecol. Evol. 7, 336-339.

Hodge A, Robinson D and Fitter A 2000 Are microorganisms more effective than plants at competing for nitrogen? Trends Pl. Sci. 5, 304-308.

Hooper DU, Solan M, Symstad AJ, Diaz S, Gessner MO, Buchmann N, Degrange V, Grime JP, Hulot FD, Mermillot-Blondin F, Roy J, Spehn EM and van Peer L 2002 Species diversity, functional diversity, and ecosystem functioning. *In:* Biodiversiy and ecosystem functioning. Synthesis and perspectives. Eds. M. Loreau, S. Naeem and P. Inchausti. pp. 195-208. Oxford University Press, Oxford.

Hungate BA, Canadell J, and Chapin FS III 1996 Plant species mediate changes in soil microbial N in response to elevated CO_2. Ecology 77, 2505-2515.

Klausmeier CA, Litchman E, Daufresne T, and Levin SA 2004 Optimal nitrogen-to-phosphorus stoichiometry of phytoplankton. Nature 429, 171-174.

Killingbeck KT 1996 Nutrients in senesced leaves: keys to the search for potential resorption and resorption proficiency. Ecology 77, 1716-1727.

Kinzig AP, Pacala S, and Tilman D 2002 The functional consequences of biodiversity: empirical progress and theoretical extensions. Princeton University Press, Princeton.

Laverman AM, Speksnijder A, Braster M, Kowalchuk GA, Verhoef HA, and van Verseveld HW 2001 Spatiotemporal stability of an ammonia-oxidizing community in a nitrogen-saturated forest soil. Micr. Ecol. 42, 35-45.

Lavorel S, McIntyre S, Landsberg J, and Forbes TDA 1997 Plant functional classifications: from general groups to specific groups based on response to disturbance. Trends Ecol. Evol. 12, 474-478.

Loreau M 1998 Ecosystem development explained by competition within and between material cycles. Proc. Roy. Soc. London Series B-Biol. Sci. 265, 33-38

Loreau M, and Hector A 2001 Partitioning selection and complementarity in biodiversity experiments. Nature 412, 72–76.

Loreau M, Naeem S, and Inchausti P 2002 Biodiversity and ecosystem functioning: synthesis and perspectives. Oxford University Press, Oxford.

Loreau M, Naeem S, Inchausti P, Bengtsson J, Grime JP, Hector A, Hooper DU, Huston MA, Raffaelli D, Schmid B, Tilman D, and Wardle DA 2001 Biodiversity and ecosystem functioning: current knowledge and future challenges. Science 294, 804–808.

Lovett GM, Weathers KC, Arthur MA, and Schultz JC 2004 Nitrogen cycling in a northern hardwood forest: Do species matter? Biogeochemistry 67, 289-308.

Lowman MD 1992 Leaf growth dynamics and herbivory in 5 species of Australian rain-forest canopy trees. J. Ecol. 80, 433-447.

McClaugherty C 1983 Soluble polyphenols and carbohydrates in throughfall and leaf litter decomposition. Acta Oecol.-Oecol. Gener. 4, 375-385.

Naeem S, and Wright JP 2003 Disentangling biodiversity effects on ecosystem functioning: deriving solutions to a seemingly insurmountable problem. Ecol. Let. 6, 567-579.

Pastor J, and Post WM 1986 Influence of climate, soil moisture, and succession on forest carbon and nitrogen cycles. Biogeochem. 2, 3-27.

Perez-Harguindeguy N, Diaz S, Cornelissen JHC, Vendramini F, Cabido M, and Castellanos A 2000 Chemistry and toughness predict leaf litter decomposition rates over a wide spectrum of functional types and taxa in central Argentina. Plant Soil 218, 21-30.

Post DM 2002. Using stable isotopes to estimate trophic position: models, methods, and assumptions. Ecology 83, 703-718.

Philippot L 2002 Denitrifying genes in Bacterial and Archaeal genomes. Biochem. Biophys. Acta 1577, 355-376.

Ranjard L, Lejon D, Mougel C, Scherer L, Merdinoglu D and Chaussod R 2003 Sampling strategy in molecular microbial ecology: influence of soil sample size on DNA fingerprinting analysis of fungal and bacterial communities. Environ. Microbiol. 5, 1111-1120.

Roggy JC, Prevost MF, Gourbiere F, Casabianca H, Garbaye J, and Domenach AM 1999 Leaf natural N-15 abundance and total N concentration as potential indicators of plant N nutrition in legumes and pioneer species in a rain forest of French Guiana. Oecologia 120, 171-182.

Roy J 2001 How does biodiversity control primary productivity ? *In:* Global Terrestrial Productivity. Eds. Roy, J., Saugier, B., Mooney, H.A. pp. 169- 186. Academic Press, San Diego.

Sherer-Lorenzen M, Potvin C, Koricheva J, Schmid B, Hector A, Bornik Z, Reynolds G, and Schulze E.-D. 2005. The design of experimental tree plantations for functional biodiversity research. *In*: Forest diversity and function. Temperate and boreal systems. Eds. M. Scherer-Lorenzen, C. Körner, and E.-D. Schulze. pp. 347-376. Springer, Berlin.

Spehn EM, Hector A, Joshi J, Scherer-Lorenzen M, and Schmid B 2005 Ecosystem effects of biodiversity manipulations in European grasslands. Ecol. Mon., in press.

Scheu S, and Falca M 2000 The soil food web of two beech forests (*Fagus sylvatica*) of contrasting humus type: stable isotope analysis of a macro- and a mesofauna-dominated community. Oecologia 123, 285-296.

Scheu S, and Setala H 2002 Multitrophic interactions in decomposer communities. *In:* Multitrophic level interactions. Eds. T. Tscharntke and B. A. Hawkins. pp. 223-264. Cambridge University Press, Cambridge.

Schmidt O, Curry JP, Dyckmans J, Rota E, and Scrimgeour CM 2004 Dual stable isotope analysis (delta C-13 and delta N-15) of soil invertebrates and their food sources. Pedobiologia 48, 171-180.

Sterner RW and Elser J 2002 Ecological Stoichiometry. The biology of elements from molecules to the biosphere. Princeton University Press, Princeton.

Strauss-Debenedetti S and Bazzaz F 1996 Photosynthetic characteristics of tropical trees along successional gradients. *In:* Tropical forest plant ecophysiology. S. S. Mulkey, R. L. Chazdon and A. P. Smith. pp. 162-186. Chapman and Hall, New York.

Tiedje JM, Asuming-Brempong S, Nusslein K, Marsh TL, and Flynn SJ 1999 Opening the black box of soil microbial diversity. Appl. Soil. Ecol. 13, 109-122.

Tilman D. 2001 Functional diversity. *In:* Encyclopedia of biodiversity. Ed. S. A. Levin. pp. 109-120. Academic Press, New York.

Tilman D, Reich PB, Knops J, Wedin D, Mielke T and Lehman C 2001 Diversity and productivity in a long-term grassland experiment. Science 294, 843–845.

Trinsoutrot I, Recous S, Bentz B, Lineres M, Cheneby D, and Nicolardot B 2000 Biochemical quality of crop residues and carbon and nitrogen mineralization kinetics under nonlimiting nitrogen conditions. Soil Sci. Soc. Am. J. 64, 918-926.

Van Breemen N 1993 Soils as biotic constructs favouring net primary productivity. Geoderma 57, 183-211.

Van Cleve K, Chapin FS III, Dyrness CT and Viereck LA 1991 Element Cycling in Taiga Forests - State-Factor Control. Bioscience 41, 78-88.

Van Heerwaarden LM, Toet S, and Aerts R 2003 Current measures of nutrient resorption efficiency lead to a substantial underestimation of real resorption efficiency: facts and solutions. Oikos 101, 664-669.

van Ruijven J, and Berendse F 2003 Positive effects of plant species diversity on productivity in the absence of legumes. Ecol. Lett. 6, 170–175.

Vila M, Inchausti P, Vayreda J, Barrantes O, Gracia C, Ibanez JJ, and Mata T 2005 Confounding factors in the observational productivity-diversity relationship in forests. *In:* Forest diversity and function. Temperate and boreal systems. Eds. M. Scherer-Lorenzen, C. Körner and E.-D. Schulze. pp 65-86. Springer, Berlin.

Wardle DA 2002. Linking the aboveground and belowground components. Princeton University Press, Princeton.

Wedin DA and Tilman D 1990 Species effects on nitrogen cycling - a test with perennial grasses. Oecologia 84, 433-441.

Westoby M, Falster DS, Moles AT, Vesk PA and Wright IJ 2002 Plant ecological strategies: some leading dimensions of variation between species. Ann. Rev. Ecol. Syst. 33: 125-159.

Wright IJ, Reich PB, Westoby M, Ackerly DD, Baruch Z, Bongers F, Cavender-Bares J, Chapin FS III, Cornelissen JHC, Diemer M, Flexas J, Garnier E, Groom PK, Gulias J, Hikosaka K, Lamont BB, Lee T, Lee W, Lusk C, Midgley J, Navas ML, Niinemets U, Oleksyn J, Osada N, Poorter H, Poot P, Prior L, Pyankov VI, Roumet C, Thomas SC, Tjoelker MG, Veneklaas EJ and Villar R 2004 The worldwide leaf economics spectrum. Nature 428, 821-827.

Zimmermann S, Braun S, Conedera M and Blaser P 2002 Macronutrient inputs by litterfall as opposed to atmospheric deposition into two contrasting chestnut forest stands in southern Switzerland. For. Ecol. Manage. 161, 289-302.

Zinke PJ 1962 Pattern of influence of individual forest trees on soil properties. Ecology 43, 130-137.

Index